Multimodal Interaction with W3C Standards

Deborah A. Dahl
Editor

Multimodal Interaction with W3C Standards

Toward Natural User Interfaces to Everything

Springer

Editor
Deborah A. Dahl
Conversational Technologies
Plymouth Meeting, PA, USA

ISBN 978-3-319-42814-7 ISBN 978-3-319-42816-1 (eBook)
DOI 10.1007/978-3-319-42816-1

Library of Congress Control Number: 2016952224

Printed on acid-free paper

This Springer imprint is published by Springer Nature
The registered company is Springer International Publishing AG
The registered company address is: Gewerbestrasse 11, 6330 Cham, Switzerland

To the memory of my parents,
Robert and Marilyn Dahl

Foreword

The world of computing is changing rapidly, as we move from desktop computers to tablets, mobile phones, watches, rings, and other things such as devices containing embedded computers or sensors connected to the Internet. New modes of input to computers are evolving from keyboarding and clicking to speaking, gesturing, seeing, and sensing our actions and behaviors. Future applications will integrate information from one or more user inputs and determine the appropriate response.

Standards are one approach to taming the exponential growth of this complexity and opportunity. Standard APIs hide the complexity of how hardware and software platforms work. These APIs enable application developers to access the functions provided by platforms and to create new and exciting applications. Given a standard API, different platform developers will optimize internal processes for specialized hardware or software, enabling application developers to (1) choose the best platform available for their application; (2) implement software on multiple platforms, using the same API; and (3) switch from one platform to another as new platforms become available and old platforms become obsolete.

This book introduces existing and potential standards for multimodal technologies and provides examples of how these standards can integrate multiple modes of input for user interaction with existing applications. It also provides suggestions about the features, functions, and capabilities that new platforms might provide to application developers.

If you are a platform developer, this book will help you determine which new features you might add to existing platforms or include in future platforms. Standard APIs enable platform developers to enter the marketplace with new platforms that compete with established platforms.

If you are an application developer, this book will help you understand which new features will be available on future hardware and software platforms and which APIs will be available to access those features. With this insight, you can plan how to use these new features and functions when creating new applications.

If you are a visionary and plan road maps for exciting innovative applications, this book will help you conceptualize, design, and implement future applications.

V.P., Larson Technical Services, James A. Larson
Co-program Chair,
SpeechTEK Conference

Preface

From tiny fitness trackers to huge industrial robots, we are interacting today with devices in shapes, sizes, and capabilities that would have been hard to imagine when the traditional graphical user interface (GUI) first became popular in the 1980s. As we move further and further away from the classic desktop paradigm, with input limited to mouse and keyboard, and a large screen as the only output modality, it is becoming increasingly apparent that the decades-old GUI interface is a poor fit for today's computer-human interactions. While the growth of touch interfaces has been especially dramatic, we are now also starting to see applications that make use of many other forms of interaction, including voice, handwriting, emotion recognition, natural language understanding, and object recognition.

As these forms of interaction (modalities) are combined into systems, the importance of having standard ways for them to communicate with each other and with application logic is apparent. The sheer variety and complexity of multimodal technologies makes it impractical for most implementers to handle the full range of possible modalities (current and future) with proprietary APIs.

To address this need, the World Wide Web Consortium (W3C) has developed a comprehensive set of standards for multimodal interaction which are well-suited as the basis of interoperable multimodal applications. However, most of the information about these standards is currently available only in the formal standards documents, conference presentations, and a few academic journal papers. All of these can be hard to find and are not very accessible to most technologists. In addition, papers on applications that use the standards are similarly scattered among many different resources.

This book will address this gap with clearly presented overviews of the full suite of W3C multimodal standards, as well as some relevant standards from other standards bodies. In addition, to illustrate the standards in use, it also includes descriptions of implemented platforms and case studies of applications that use the standards. Finally, a future directions section discusses new ideas for other standards as well as for new applications.

Plymouth Meeting, PA, USA Deborah A. Dahl

Acknowledgments

The standards described in this book would not have been possible without the vision and support of the World Wide Web Consortium. From the first W3C Voice Workshop in 1998 in Cambridge, Massachusetts where the idea of a standard language for voice applications was initially discussed, to the present, the W3C has consistently supported the goal of providing powerful, comprehensive, and usable standard languages to support voice and multimodal application development. Despite the fact that it is easy to think of the web as just focused on graphical interaction, the W3C has supported work that takes a broader perspective on human-computer interaction than simply responding to graphical interaction events. This broader perspective includes voice, ink, and gesture-based interaction as well as the more abstract idea of interaction management. As the Chair of the Multimodal Interaction Working Group, I have had the privilege of working with many talented and dedicated individuals over the years. For many years of support, I would like to thank Philipp Hoschka, the W3C domain leader for the Ubiquitous Web throughout the history of both the Voice Browser and Multimodal Interaction Working Groups. In addition, we would also like to thank Judy Brewer, domain leader for the Web Accessibility Initiative, for discussions of how these standards can be used for enhancing accessibility. We are also of course very grateful to Sir Tim Berners-Lee, the director of the W3C, for inventing the web and founding the W3C in the first place.

Our W3C team contacts have helped steer the Voice Browser and Multimodal Working Group chairs through the W3C process. They've helped the groups prepare documents, understand the stages of a W3C standard, and helped us find relevant points of interaction with other W3C working groups and other standards bodies. We are very grateful to Dave Raggett, Max Froumentin, Kazuyuki Ashimura, and Matt Wormer for their work over the years as team contacts for the Voice Browser Working Group and the Multimodal Interaction Working Group.

The Voice Browser Working Group was the first W3C working group that worked on user interface standards beyond the graphical user interface, specifically focusing on voice interaction. I am extremely grateful for the inspiring leadership of the Voice Browser chairs, Jim Larson, Dan Burnett, and the late Scott McGlashan. Jim Larson is an amazing meeting facilitator who can always cut to the chase and push a group toward consensus. Jim is never afraid to ask the question that everyone's been thinking about. Dan has an incredible mastery of detail and a deep knowledge of the standards and is a careful and thorough leader who always makes sure that everyone has a chance to express their opinion.

The late Scott McGlashan served as cochair of the Voice Browser Working Group. He also led the VoiceXML 2.0 and speech grammar efforts. Scott was a rare person who combined brilliant technical skills with efficient and organized leadership. Scott managed the hundreds of VoiceXML change requests with efficiency, dispatch, and a keen eye for opportunities for consensus.

The editors of standards documents play a key role in making the standards possible. From the first steps after a working group decides to create a standard for a particular area, through the final publication of an official standard, the editors guide the participants through the process, even when the work is contentious or, worse, tedious. Inevitably, in the sometimes lengthy standards development process, there will be times when companies decide that they can no longer support work on the standard or when key working group members change companies and can no longer participate. It is the editors who step up to compensate for changes in participation, helping to bring new people on board and making sure that the work continues. In the Voice Browser Working Group, the editors in chief at the time of the final publication of each standard were:

1. VoiceXML: Scott McGlashan
2. VoiceXML 2.1: Matt Oshry
3. Speech Recognition Grammar Specification: Andrew Hunt and Scott McGlashan
4. Semantic Interpretation for Speech Recognition: Luc Van Tichelen and Dave Burke
5. Speech Synthesis Markup Language: Dan Burnett, Mark Walker, and Andrew Hunt
6. Pronunciation Lexicon Specification: Paolo Baggia
7. Call Control Markup Language: R. J. Auburn
8. State Chart XML: Jim Barnett

The Multimodal Working Group editors at the time of final publication of the standards include:

1. MMI Architecture: Jim Barnett
2. EMMA: Michael Johnston
3. InkML: Stephen Watt and Tom Underhill
4. EmotionML: Felix Burkhardt and Marc Schroeder
5. Discovery and Registration: Helena Rodriguez (in progress)

Developing an official W3C standard is a tremendous amount of work which requires participation from many team members. Since the beginning of the Voice

Browser Working Group in 1999, there have been dozens of participants in the Voice Browser and Multimodal Interaction Working Groups. Some have only attended a few calls, but many members of both working groups have spent countless hours helping to define the standards and write the documents. Others have worked on the important task of developing implementation tests and testing. The standards would not have been possible without their efforts.

I am also very grateful for the colleagues who took time from their many other commitments to review the chapters in this book and to provide helpful comments to the authors. The reviewers were Nuno Almeida, Masahiro Araki, Paolo Baggia, Jim Barnett, Harry Bunt, Felix Burkhardt, Dan Burnett, Carlos Duarte, Jesus Garcia Laborda, Michael Johnston, James Larson, Teresa Magal-Royo, Rob Marchand, Jean-Claude Moissinac, Miguel Oliveira e Silva, Catherine Pelachaud, Stefan Radomski, Vikram Ramanarayanan, B. Helena Rodriguez, Dirk Schnelle-Walka, Samuel Silva, David Suendermann-Oeft Antonio Teixera, and Raj Tumuluri. Their thoughtful comments improved the quality of the book immeasurably. I am especially grateful to several of the reviewers who reviewed extra papers when a paper needed an additional review, sometimes on very short notice. I would like to thank Mary James, senior editor at Springer, for first suggesting this book and helping me through the process. Our project coordinator at Springer, Brinda Megasyamalan, always had a quick and helpful response to every question.

Finally, I would like to thank my family: my husband, Richard Schranz, who has always been happy to join me at W3C meetings in interesting places; my daughter Sarah Schranz-Oliveira; my son-in-law Nuno Cancela Oliveira; and my son Peter Schranz, for all their love and support.

Contents

Part II Implementations

Part III Applications

Part IV Future Directions

Introduction

Multimodal Interaction Standards

We live in an era of increasingly accelerated technology advances. Factors like Moore's Law, smaller and faster devices and networks, improvements in the ability of technologists to cooperate on shared research interests, and the disaggregation of capabilities that has given us things like web services and cloud computing, all combined with an investment market that catalyzes creativity, innovation, and entrepreneurship, are combining to drive technology at an incredible rate. Within this world of technology advancement, the multimodal component technologies that are the subject of this book, including speech recognition, touch, typed and gesture interpretation, and natural language understanding, have also become dramatically more capable.

Another, crucially important, factor in the accelerating development of technology innovation infrastructure has been standards. Standards such as HTTP and HTML make it possible for small organizations and even individuals to contribute to larger projects. They reduce duplication of effort across projects, and they make it possible for components developed by different organizations to interoperate. In only a few years, we have come to take such things as voice interfaces and voice search on our mobile devices for granted. But even more exciting possibilities are ahead of us as new interaction modalities become available and as these natural interface capabilities become more interoperable. Only the largest organizations can possibly have enough expertise in all the multimodal interaction technologies to provide a complete proprietary platform for multimodal applications. But if systems are composed of interoperable components, smaller organizations can contribute components to larger systems.

Standards are the foundation of cross-vendor interoperability. The goal of this book is to provide an introduction to existing and potential standards for multimodal technologies, to showcase some examples of the standards in action, and to provide some suggestions for where multimodal standards can go in the future. It is hoped that this book will inspire developers, both in start-ups and in large organizations, to take advantage of standards in order to make existing applications more reliable and efficient, as well as to make possible completely new types of applications. In addition, we also hope to inspire organizations and individual developers to work on improving the standards and to think of new ways that multimodal technologies can work together to create innovative and disruptive applications for the twenty-first century. In this book, the focus is on the interfaces between components that enable multimodal components to work together in systems, rather than the design of multimodal applications. Design considerations for multimodal user interfaces are addressed in recent related books such as [1, 2].

Speech and Multimodal Interaction

The vision of natural spoken language interaction with computers was present from the very beginning of computers, starting with Turing's original 1950 paper [3] describing the idea that people could interact with computers using natural language. However, one fact about speech and natural language interaction with computers that has been proven repeatedly in the over 60 years since the publication of Turing's paper is that vision is cheap. The difficulty of the technical problem has been dramatically underestimated, and there have been countless predictions of how soon we would be able to talk to our computers "just like talking to a person." In fact, the user experience with even the most sophisticated of today's systems is far from the experience of talking to another person. The technology has certainly improved dramatically in the last few years, however, and we do now have real time, spoken interaction with systems.

Human interaction with computers was originally all about making the humans accommodate themselves to the computer's limitations. Only a few highly skilled professionals were able to use computers. As computers become easier to use, their benefits become available to more and more people.

Keyboards and graphical displays greatly improved human-computer interaction over earlier technologies such as punched cards. The computer-human interface took a tremendous leap forward with the introduction of the now very familiar graphical interface. It could be argued that, between the introduction of the graphical user interface by the Xerox Alto in 1973 and the introduction of Apple Siri 2011, a period of 38 years, the human-computer interface was completely based on the graphical desktop metaphor. The graphical user interface made it possible for nonprofessionals, or at least non-enthusiasts, to use computers, greatly extending their value.

Nevertheless, the current graphical web is not easy for everyone to use, yet nearly everyone can benefit from the kinds of information and functionality that the web and indeed computers, in general, can provide. Some people have trouble seeing a screen, typing, moving a mouse, or understanding complex text. For these users, alternative modalities of interaction are crucial. Spoken and haptic output can help users avoid the requirement of having to see a screen, speech input can help people with motor issues, and interaction paradigms like directed dialog can help those who are unable to understand complex language.

The touch interface introduced with smartphones and tablets further extended the power of computers to more kinds of users. Now even very young children can use tablets. Finally, with the addition of high-quality speech-based interfaces, the reach of computers is further broadened, and with ambient devices like the Amazon Echo, anyone who can speak can access the computer, without even having to find a tablet or phone. The computer is simply available, whenever it is needed.

Multimodal interaction is a big part of the opening up of computers to a broader population, because, through speech, it enables people to use smaller and cheaper devices with small screens and keyboards or even no screen and no keyboard.

Multimodal Interaction Standards

The idea of combining voice and graphical interaction has been explored in the research literature for many years. The 1980 paper "Put that there" [4] which discussed combining gesture and speech was an early example. Multimodal output combining graphical and text output was described in 1990 by Feiner and McKeown [5]. Research on multimodal interaction continued through the 1990s, for example, in the work of Sharon Oviatt, Philip Cohen, and their collaborators [6–8]. This initial work focused largely on research and did not concern itself with interoperability across different systems.

The goal of creating shared interface standards that could make it easier for teams to collaborate in the development of multimodal systems emerged as part of the DARPA Galaxy project [9, 10]. Going beyond a single project, the idea of formalizing multimodal interaction standards through a standards body was raised at the beginning of the World Wide Web Consortium (W3C) Voice Browser Working Group [11] in 1999. Multimodal interaction was originally a subgroup of the Voice Browser Group. The Voice Browser Working Group was working on standards for unimodal voice systems, including standards for voice dialogs, speech grammars, text to speech markup, and natural language understanding. Many of these standards provided the foundations for later multimodal standards.

Two industry proposals for integrating voice with HTML were developed in 2002. These were Speech Application Language Tags (SALT) and XHTML + Voice. In response to this interest, the W3C decided to create a separate working group focused specifically on standards for multimodal systems, the Multimodal Interaction Working Group [12]. Because the two industry proposals only covered the limited use case of integration of voice into HTML, the Multimodal Interaction Working Group decided to take a more abstract and generic approach that could be extended to many different interaction modalities. This group adopted the Voice Browser's efforts at developing a natural language standard, Natural Language Semantics Markup Language (NLSML) [13], which expanded to cover multimodal use cases and became Extensible Multimodal Annotation (EMMA) [14, 15]. The Multimodal Interaction Working Group went on the develop standards for multimodal architectures, emotion annotation, and electronic ink, in addition to EMMA. This book will describe these standards, talk about implemented platforms, discuss applications of the standards, and, finally, present ideas for future multimodal standards and their applications.

How the Standards Fit Together

This book discusses the standards for natural computer-human interaction which have been developed over the past 17 years, primarily within the World Wide Web Consortium Voice Browser Working Group and Multimodal Interaction Working Group. Together, the standards are designed to support multimodal interaction from the initial user input, through analysis and interpretation, to the final system output to the user.

We can divide the standards into three categories: languages for controlling processors, protocols for communications among components, and formats for representing input and output. To see the overall organization of how the standards fit together in systems, we will refer to Figs. 1 and 2. These figures show how multimodal systems for input and output are composed of components that accept and interpret input, prepare and present output, and conduct an interaction between the user and the system.

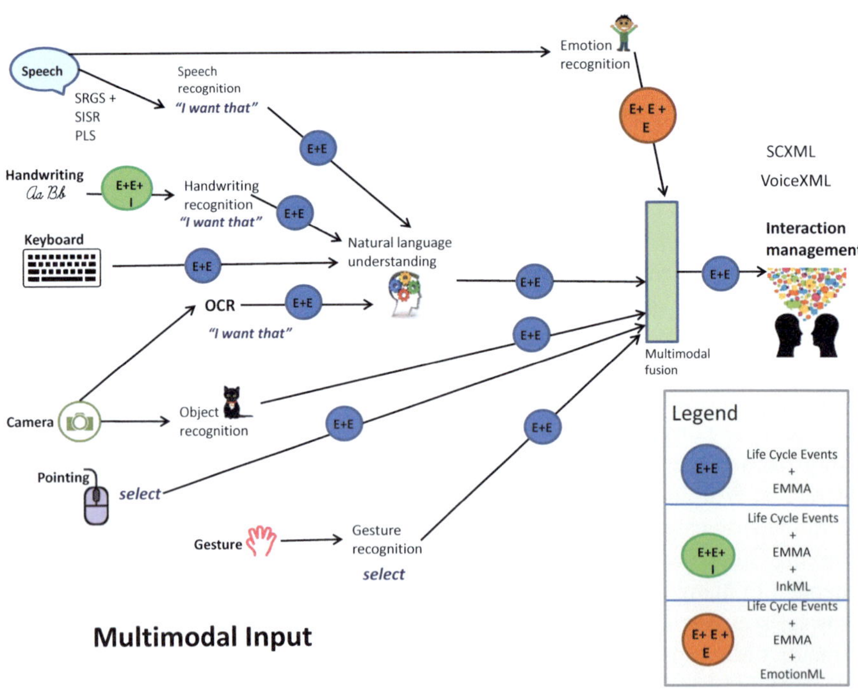

Fig. 1 Multimodal inputs and some associated standards

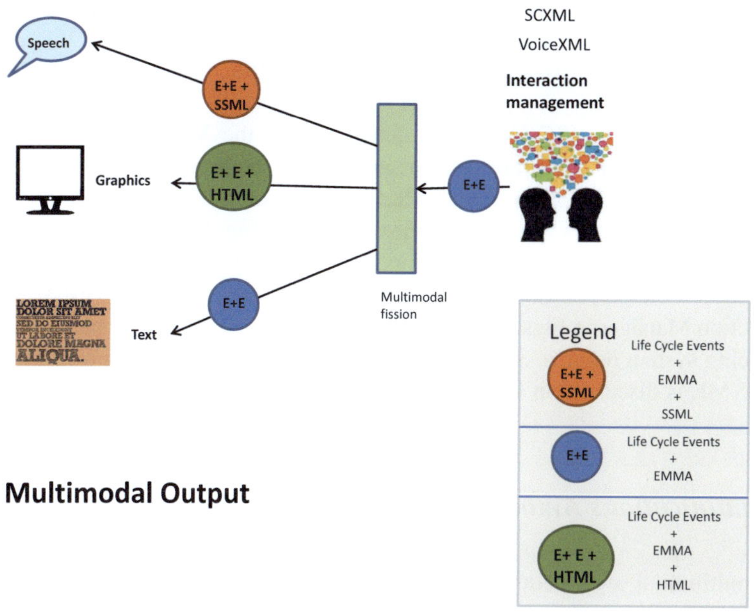

Fig. 2 Multimodal outputs and some associated standards

Formats for Controlling Processors

These are declarative formats that define the operations for a processor. With the exception of State Chart XML (SCXML), they are all related to voice processing. The voice standards include the following:

1. Pronunciation Lexicon Specification (PLS) defines an XML format for describing the pronunciations for words which can be used with both speech recognizers and speech synthesizers.
2. Speech Recognizer Grammar Specification (SRGS) defines two formats for defining context-free speech grammars: an XML format and an ABNF format.
3. Semantic Interpretation for Speech Recognition (SISR), used in conjunction with SRGS, is a format for defining the semantic interpretation of recognized speech.
4. Speech Synthesis Markup Language (SSML) is an XML format for representing instructions for a text to speech synthesizer.
5. VoiceXML is a declarative XML format for specifying spoken form-filling dialogs, including defining prompts (which may include SSML instructions for prompts rendered with TTS or instructions to play audio files), referring to speech grammars defined with SRGS, and specifying the information (fields or slots) to be filled in during the course of a dialog. VoiceXML documents are interpreted by VoiceXML platforms which conduct the actual dialogs with users.

In Fig. 1, we see the voice recognition standards next to the speech recognition component, since they are used by a speech recognizer.

The voice standards—VoiceXML, SRGS, SISR, PLS, and SSML—are discussed in Chap. 2.

State Chart XML (SCXML) was developed by the Voice Browser Working Group, but plays an important role in multimodal applications as an Interaction Manager. SCXML is a declarative XML formalization of Harel State Charts. It can be used to define state chart-based processes, including, but not limited to, voice-based and multimodal dialogs. One important use case of SCXML is its use as an Interaction Manager, or dialog controller, in multimodal systems. It is shown as the Interaction Manager in Figs. 1 and 2, reacting to events triggered by user inputs and producing system outputs. SCXML is discussed in Chap. 5, and an implementation of SCXML is discussed in Chap. 10.

Communications Among Components

The multimodal architecture, discussed in Chap. 1, defines the behavior of the components of a multimodal system (the Interaction Manager and the Modality Components) and specifies the format that they must use to communicate with each other (Life Cycle Events). The first "E" in the circles in Figs. 1 and 2 stands for "Events" and refers to a Life Cycle Event going between an Interaction Manager and a Modality Component. The Life Cycle Events consist of basic operations that an Interaction Manager might use to control components—operations like "Start," "Cancel," "Pause," "Resume," and so on. The events are defined as having standard fields, including, for example, the source and destination of the event and a context ID that associates the event with a particular context. Application-specific information for a particular operation is contained in the "Data" field. There is no specific syntax defined for Life Cycle Events, although there are informative examples in XML contained in the specification. Events could be represented in XML, JSON, or binary objects, depending on the requirements of the application. Similarly, the events are transport-agnostic. Although there are informative HTTP examples in the specification, other protocols could be used, for example, WebSockets, which would be particularly applicable because of its bidirectional nature.

Representation of Input and Output

Representing Uninterpreted Input

Uninterpreted input is the initial user input, which is often in a binary form as audio, images, or video. In most cases, existing formats are available for representing this kind of input, so there was no need to define new standards for these formats. The single exception in the multimodal standards suite is Ink Markup Language (InkML) [16, 17], designed to represent the initial electronic input from a user, using a stylus or finger. In addition to representing ink itself (in the <trace> element), InkML also provides a rich set of metadata that preserves the appearance of the original input. Some examples of InkML metadata include channels for color, width, pen orientation, and timing for each sample point within a trace. Use cases for InkML include remote collaboration for mathematicians, efficient storage of signatures and other handwriting, archiving of academic lectures, and representing handwriting recognition results along with the original handwriting. This last use case involves combining InkML with EMMA and is shown in Fig. 1 in the modality path that starts with "Handwriting."

Representing Interpreted User Input

Extensible Multimodal Annotation

EMMA is the primary format in the MMI standards suite for representing user inputs, as shown by the circles marked "E + E" (Events + EMMA), "E + E + I" (Events + EMMA + InkML), and "E + E + E" (Events + EMMA + EmotionML) in Fig. 1. EMMA is designed to represent user input, as it is processed in a sequence of stages, in a uniform format, starting with an initial result from a processor like a speech recognizer or handwriting recognizer. Additional levels of interpretation can be provided by processes such as natural language understanding or sentiment analysis. Each successive stage in processing augments and enriches the growing EMMA document with its results. The fact that EMMA is a uniform representation across modalities also facilitates the fusion of inputs from different modalities.

EMMA is available as a speech recognition output format in the MRCP v2 [18] protocol for controlling speech recognizers and synthesizers as well as the Web Speech API [19] proposal for using speech recognition and speech synthesis in browsers. EMMA 2.0 can also represent system output, as shown in Fig. 2, and as described below.

EMMA is discussed in detail in Chap. 3.

Emotion ML

Another format for representing a specific type of interpreted user input is Emotion Markup Language (EmotionML). EmotionML is an XML format for representing emotions expressed by a variety of means (face, voice, text, etc.). EmotionML can be embedded in EMMA, which simplifies the task of fusing emotion recognition results with utterance interpretation results. EmotionML can also be used for describing affective output to be rendered as output by a system (e.g., by TTS), as described below.

EmotionML is discussed in detail in Chap. 4.

Representing the Semantics of System Output

Figure 2 looks at the standards from the perspective of output. Starting from an intended system output produced by an Interaction Manager, represented in EMMA 2.0, the intended output can be separated into distinct output modalities, or fissioned, into more concrete representations appropriate for presentation to a user.

EMMA 2.0

EMMA 2.0 includes the ability to represent the semantics of system output in addition to user input. This means that it is now possible to represent both the human and system sides of the dialog in a single, standard, uniform format. It also supports multimodal fission, by allowing the Interaction Manager to create a high-level, abstract representation of the system response, which can be progressively refined and divided into modality-specific presentations through different stages of processing. In effect, this process is the inverse of the process of handling user inputs defined in EMMA 1.0. In addition to supporting more maintainable systems by dividing the tasks of determining system intent from rendering the intent in specific modalities, this architecture supports accessibility by making it easier to present content in different forms for different users.

Overview of Sections

Part I Standards

The book begins with a description of the standards themselves. The formal standards documents can be very technical and are often written in "Standardsese" that has the goal of defining the standards very precisely, making it possible for widely dispersed teams to create interoperable implementations. In some cases, this goal has the consequence that the resulting standards document is difficult to understand. Standards documents try to eliminate any ambiguity about what an implementation must do to be conformant to the standard, and they often refer to a chain of underlying standards and standards bodies that may not be easy to follow. These characteristics are important for implementers, but they can make it difficult for other readers, including students, developers, and business people, to understand just what the standard is for and how they can use it. Tutorials and primers for many standards exist, but they are scattered among various books, presentations, and journal articles.

In Part I of this book, we bring together overviews of seven W3C, ITSCJ and ISO standards in what we hope is a more introductory and human-friendly format than the official standards documents. Part I covers the multimodal architecture, the suite of speech standards, EMMA, EmotionML, and SCXML. ISO dialog standards, which go deeper into the semantics of actual utterances that does EMMA, are reviewed in Chap. 6. A multimodal standard published by ITSCJ (Information Technology Standards Commission of Japan) is described in Chap. 7. The standard described in Chap. 7 is similar to the W3C standard, but provides a more fine-grained set of events and processing layers. Part I also includes Web Real-Time Communications (WebRTC) [20], developed by the W3C Web Real-Time Communications Working Group, in Chap. 8.

Part II Platforms

Part II discusses implementations of the standards in platforms intended to support multiple applications. Three platforms are discussed. Chapter 9 describes what is probably the most comprehensive implementation of the multimodal architecture, developed by Openstream. The Openstream implementation supports interactive spoken multimodal dialog systems and multimodal annotation for human-human interaction. Chapter 10 provides insight into an implementation of SCXML for resource-constrained devices, which could prove to be very valuable for applications in the Internet of Things. Chapter 11 describes an implementation of a multimodal architecture conformant modality component for natural language understanding and explains how the EMMA and multimodal architecture standards can be used to wrap native processing components so that their functionalities can be accessed in a standard way. Chapter 12 describes HALEF, an open source, standards-based, dialog system platform that is currently primarily voice-based, but which is moving toward incorporating multimodal capabilities.

Part III Applications

Part III includes two chapters on applications that make use of the standards. Chapter 13 describes applications in ambient-assisted living. Chapter 14 describes using SRGS and EMMA for a simple approach to text and audio alignment in bilingual books.

Part IV Future Directions

Standards for these dynamic technologies evolve. This evolution is a balance between gaining the benefits of vendor-independent interoperability and at the same time without constraining innovation. Even in the familiar GUI world of web browsers, this tension is reflected in the many differences between the major browsers in their implementations of HTML. We often see "you must use X browser to view this page." Websites like http://www.caniuse.com are needed to provide guidance on the state of HTML features and Javascript APIs in various browsers and their plans for implementation. It is inevitable that, as standards are implemented and used, limitations become apparent and people start to notice where many vendors are doing similar things, thus showing where standardization is possible. Plug-ins and polyfill Javascript libraries start to emerge as developers realize that they need functionalities that are not included in current standards. One vendor implements a feature that everyone recognizes as valuable, and then other vendors add it, opening up the possibility of a new standard feature. Industry consortia start to coalesce to promote new ideas. New devices and new functionalities give rise to new use cases, which in turn stimulate new ideas about functionalities that can be standardized. Eventually, a formal standard can emerge, or the ideas can continue as de facto standards.

Part IV talks about directions in the development of future standards. Chapter 15 describes standards that will be needed in the Internet of Things as dynamically configured systems become more prevalent. As users move into and out of environments, modality services will be able to advertise their availability and status. They can become part of an application as needed and then can be detached from the application when they are no longer needed or when they are no longer available. Chapter 16 talks about current and future ideas for multimodal applications in foreign language testing. Chapter 17 extends the use cases to include multi-device applications. Chapter 18 describes an approach to interaction management based on semantic data models, referring to semantic web standards OWL [21] and RDF [22]. Finally, Chap. 19 extends the multimodal architecture to include more detail, specifically adding components for multimodal fission and fusion.

References

1. Oviatt, S., & Cohen, P. R. (2015). *The paradigm shift to multimodality in contemporary computer interfaces*. San Rafael CA: Morgan and Claypool.
2. Shaked, N., & Winter, U. (Eds.). (2016). *Design of multimodal mobile interfaces*. Berlin: Walter De Gruyter Inc.
3. Turing, A. (1950). Computing machinery and intelligence. *Mind, 59*, 433–460.
4. Bolt, R. (1980). Put-that-there: Voice and gesture at the graphics interface. *Computer Graphics, 14*, 262–270.
5. Feiner, S. K., & McKeown, K. R. (1990). Coordinating text and graphics in explanation generation. In: *AAAI-90: Proceedings of the 8th National Conference on Artificial Intelligence, vol I* (pp. 442–449). AAAI Press/The MIT Press.
6. Oviatt, S. L. (1999). Ten myths of multimodal interaction. *Communications of the ACM, 42*, 74–81.
7. Oviatt, S., DeAngeli, A., & Kuhn, K. (1997). Integration and synchronization of input modes during multimodal human-computer interaction. In: *Proceedings of Conference on Human Factors in Computing Systems (CHI '97)* (pp. 415–422). New York, NY: ACM Press.

8. Cohen, P. R., Oviatt, S. L. (1995). The role of voice input for human-machine communication. *Proceedings of the National Academy of Sciences, 92*, 9921–9927.
9. Bayer, S. (2005). Building a standards and research community with the galaxy communicator software infrastructure. In: D. A. Dahl (Ed.), *Practical Spoken Dialog Systems, vol 26. Text, speech and language technology* (pp. 166–196). Dordrecht: Kluwer Academic Publishers.
10. Seneff, S., Lau, R., & Polifroni, J. (1999). Organization, communication, and control in the Galaxy-II Conversational System. In: *Proceedings of the Eurospeech 1999*, Budapest, 1999.
11. W3C W3C Voice Browser Group Home Page. http://www.w3.org/Voice/.
12. W3C. (2004). Multimodal Interaction Working Group Home Page. http://www.w3.org/2002/mmi/.
13. Dahl, D. A. (2000). Natural language semantics markup language for the speech interface framework. W3C. http://www.w3.org/TR/nl-spec/.
14. Johnston, M., Baggia, P., Burnett, D., Carter, J., Dahl, D. A., McCobb, G., et al. (2009) EMMA: Extensible MultiModal Annotation markup language. W3C. http://www.w3.org/TR/emma/. Accessed November 9, 2012.
15. Johnston, M., Dahl, D. A., Denny, T., & Kharidi, N. (2015). EMMA: Extensible MultiModal Annotation markup language Version 2.0. World Wide Web Consortium. http://www.w3.org/TR/emma20/. Accessed December 16, 2015.
16. Watt, S. M., Underhill, T., Chee, Y. -M., Franke, K., Froumentin, M., Madhvanath, S., et al. (2011). Ink markup language (InkML). World wide web consortium. http://www.w3.org/TR/InkML. Accessed November 27, 2012.
17. Watt, S. M. (2007) New aspects of InkML for pen-based computing. In: *International Conference on Document Analysis and Recognition*, (ICDAR), Curitiba, Brazil, September 23–26, 2007. IEEE Computer Society, pp. 457–460.
18. Shanmugham, S., & Burnett, D. C. (2008). Media resource control protocol version 2 (MRCPv2). Internet Engineering Task Force. http://tools.ietf.org/html/draft-ietf-speechsc-mrcpv2-16.
19. Shires, G., & Wennborg, H. (2012). Web speech API specification. https://dvcs.w3.org/hg/speech-api/raw-file/tip/speechapi.html. Accessed May 31, 2016.
20. Bergkvist, A., Burnett, D. C., Jennings, C., & Narayanan, A. (2016). *WebRTC 1.0: Real-time communication between browsers*. World Wide Web Consortium. http://www.w3.org/TR/webrtc/. Accessed November 28, 2012.
21. W3C OWL Working Group. (2012). *OWL 2 web ontology language document overview* (2nd ed.). World Wide Web Consortium. http://www.w3.org/TR/owl2-overview/. Accessed April 9, 2014.
22. Antoniou, G., & van Harmelen, F. (2004). *A semantic web primer*. Cambridge, MA: MIT Press.

Part I
Standards

Chapter 1
Introduction to the Multimodal Architecture Specification

Jim Barnett

Abstract The W3C's Multimodal Architecture standard is a high-level design featuring loosely coupled components. Its goal is to encourage interoperability and re-use of components, without enforcing any particular approach to building multimodal applications. This paper offers an overview of the architecture, outlining its components and the events they use to communicate, as well as giving basic examples of how it can be applied in practice.

1.1 Overview

Many standards emerge in areas where the technology is stable and industry participants think that they understand the field well enough to be able to codify existing best practices. However the consensus within the Multimodal Interaction Working Group of the W3C was that best practices for multimodal application development had not yet emerged. The group therefore took it as its task to support exploration, rather than trying to codify any particular approach to multimodal applications. The goal of the Multimodal Architecture and Interfaces standard [1] is to encourage re-use and interoperability while being flexible enough to allow a wide variety of approaches to application development. The Working Group's hope is that this architecture will make it easier for application developers to assemble existing components to get a base multimodal system, thus freeing them up to concentrate on building their applications.

As part of the discussions that lead to the Multimodal Architecture, the group considered existing multimodal languages, in particular SALT [2] and HTML5 [3]. SALT was specifically designed as a multimodal language, and consisted of speech tags that could be inserted into HTML or similar languages. HTML5 in turn has multimodal capabilities, such as video, which were absent from earlier versions of

J. Barnett (✉)
Department of Architecture Team, Genesys, Daly City, CA, USA
e-mail: jim.barnett@genesys.com

D.A. Dahl (ed.), *Multimodal Interaction with W3C Standards*,
DOI 10.1007/978-3-319-42816-1_1

HTML. One problem with this approach is that it is both language- and modality-specific. For example, neither SALT nor HTML5 supports haptic sensors, nor do they provide an extension point that would allow them to be integrated in a straightforward manner. Furthermore, in both cases overall control and coordination of the modalities is provided by HTML, which was not designed as a control language. Multimodal application developers using HTML5 are thus locked into a specific graphical language with limited control capabilities and no easy way to add new modalities. As a result of these limitations, HTML5 is not a good framework for multimodal experimentation.

The Multimodal Working Group's conclusion is that it was too early to commit to any modality-specific language. For example, VoiceXML [4] has been highly successful as language for speech applications, particularly over the phone. However there is no guarantee that it will turn out to be the best language for speech-enabled multimodal applications. Therefore the Working Group decided to define a framework which would support a variety of languages, both for individual modalities and for overall coordination and control. The framework should rely on simple, high-level interfaces that would make it easy to incorporate existing languages such as VoiceXML as well as new languages that haven't been defined yet. The Working Group's goal was to make as few technology commitments as possible, while still allowing the development of sophisticated applications from a wide variety of re-usable components. Of necessity the result of the Group's work is a high-level framework rather than the description of a specific system, but the goal of the abstraction is to let application developers decide how the details should be filled in.

We will first look at the components of the high-level architecture and then at the events that pass between them.

1.2 The Architecture

The basic design principles of the architecture are as follows:

1. The architecture should make no assumptions about the internal structure of components.
2. The architecture should allow components to be distributed or co-located.
3. Overall control flow and modality coordination should be separated from user interaction.
4. The various modalities should be independent of each other. In particular, adding a new modality should not require changes to any existing ones.
5. The architecture should make no assumptions about how and when modalities will be combined.

The third and fourth principles motivate the most basic features of the design. In particular item 3 requires that there be a separate control module that is responsible for coordination among the modalities. The individual modalities will of course need their own internal control flow. For example, a VoiceXML-based speech recognition component has its own internal logic to coordinate prompt playing, speech recognition, barge-in, and the collection of results. However the speech recognition component should not be attempting to control what is happening in the graphics component. Similarly the graphics component should be responsible for visual input and output, without concern for what is happening in the voice modality. The fourth point re-enforces this separation of responsibilities. If the speech component is controlling speech input and output only, while the graphics component is concerned with the GUI only, then it should be possible to add a haptic component without modifying either of the existing components.

The core idea of the architecture is thus to factor the system into an Interaction Manager (IM) and multiple Modality Components (MCs).

The Interaction Manager is responsible for control flow and coordination among the Modality Components. It does not interact with the user directly or handle media streams, but controls the user interaction by controlling the various MCs. If the user is using speech to fill in a graphical form, the IM would be responsible for starting the speech Modality Component and then taking the speech results from the speech MC and passing them to the graphics component. The IM is thus responsible for tracking the overall progress of the application, knowing what information has been gathered, and deciding what to do next, but it leaves the details of the interactions in the various modalities up to the MCs. A wide variety of languages can be used to implement Interaction Managers, but SCXML [5] is well suited to this task and was defined with this architecture in mind.

The Multimodal Architecture also defines an application-level Data Component which is logically separate from the Interaction Manager. The Data Component is intended to store application-level data, and the Interaction Manager is able to access it and update it. However the architecture does not define the interface between the Data Component and the IM, so in practice the IM will provide its own built-in Data Component.

Modality Components are responsible for interacting with the user. There are few requirements placed on Modality Components beyond this. In particular, the specification does not define what a "modality" is. A Modality Component may handle input or output or both. In general, it is possible to have "coarse-grained" Modality Components that combine multiple media that could be treated as separate modalities. For example, a VoiceXML-based Modality Component would offer both ASR and TTS capabilities, but it would also be possible to have one MC for ASR and another for TTS. Many Modality Components will have a scripting interface to allow developers to customize their behavior. VoiceXML is again a good example of this. However it is also possible to have hard-coded Modality Components whose behavior cannot be customized.

Note that to the extent that HTML5 is a multimodal language, it acts both as an Interaction Manager and as a Modality Component. The W3C Multimodal Architecture tries to provide more flexibility so that an application can use HTML5 as a graphical MC without being restricted to its limited control flow capabilities.

It is also possible to nest Modality Components. That is, multiple MCs plus an Interaction Manager can look like single MC to a higher-level Interaction Manager.

This design can be useful if some MCs need to be tightly coupled with each other, while they are more loosely coordinated with others. For example, ASR and TTS modalities are usually tightly coupled to coordinate prompt playing with recognition and barge-in. If a system was working directly with individual ASR and TTS Modality Components, it might want to couple them closely using a separate Interaction Manager. The resulting complex Modality Component would offer prompt and recognize capabilities to the larger application, similar to a native VoiceXML Modality Component.

In addition to the Interaction Manager and Modality Components, the architecture contains a Runtime Framework, which provides the infrastructure necessary to start and stop components, as well as enabling communication. The Runtime Framework contains a Transport Layer which must provide reliable, in-order delivery of events between the IM and the MCs. The Transport Layer might be HTTP for loosely coupled and distributed components or something proprietary for co-located and tightly coupled components. Overall, the Multimodal Architecture and Interfaces specification provides little detail on the Runtime Framework. However the separate Discovery and Registration specification [6] is filling in part of this gap.

Security is important for multimodal applications since they will often be dealing with sensitive information such as credit card numbers. However security is outside the scope of the Multimodal Architecture and Interfaces specification. The W3C Multimodal Working Group assumes that developers will consult the relevant security specifications when building their systems.

A diagram of the W3C Multimodal Architecture, taken from the specification [1], is given below (Fig. 1.1).

Overall, the W3C's Multimodal Architecture should look fairly familiar (lack of originality is considered a *good* thing in standards group work). One model for this design is the DARPA Hub Architecture, also known as the Galaxy Communicator [7]. The architecture can also be taken as an instance of the Model/View/Controller paradigm (especially when the data model is separate). Specifically, the Modality Components represent the view, while the Interaction Manager is the Controller.

The goal of this design is to chop the system up into pieces that are likely to make good re-usable components. For example, there are a number of open source SCXML interpreters that can be used as Interaction Managers. Similarly, an open source VoiceXML interpreter can be used as a Modality Component. On the other hand, the looseness of definition of Modality Components including the lack of a clear definition of "modality" is designed to allow room for experimentation while still providing enough structure so that the results of innovation can be re-used.

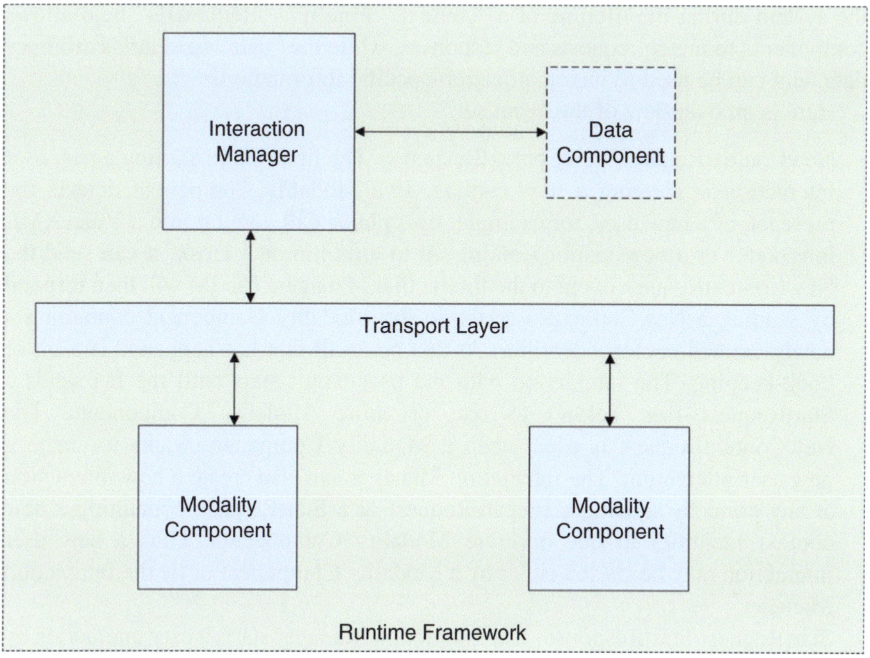

Fig. 1.1 Components of the multimodal architecture

1.3 The Interfaces

In addition to specifying the overall architecture, the Multimodal Architecture and Interfaces specification defines a set of events that are exchanged by the components. Since the specification does not commit to what the Modality Components are or to how they are implemented, the event set is high-level and generic, but still sufficient to build real-world applications. In keeping with the high-level nature of the event set, the events are defined as an abstract set of "fields," which any actual implementation would have to map onto a concrete syntax such as XML or JSON. The transport for the events is also not defined.

The majority of events are defined in request/response pairs. Certain fields are common to all events. The "target" field contains the address of the intended recipient, and allows the underlying messaging infrastructure to deliver the event. The "source" field gives the address of the sender, and the recipient of an event should be able to send a response to this address. The "context" field identifies a particular interaction with a user. For example, most VoiceXML and SCXML interpreters can handle multiple simultaneous sessions, so the "context" field allows the interpreters to determine which user session that event belongs to. The specification does not define the duration of a "context," but the Interaction Manager and all the Modality Components that are interacting with the user share the same "context," and it is possible for individual Modality Components to join and leave the system during the lifetime of a "context." Finally a "RequestID" field allows components to match requests and responses, while the "data" field holds arbitrary data, and can be used to pass application-specific information.

Here is an overview of the event set:

- NewContextRequest/NewContextResponse. The first step in starting a new user interaction is creating a new context. If a Modality Component detects the presence of a new user, for example, by a phone call coming into a VoiceXML interpreter or a new visitor walking up to a multimodal kiosk, it can send the NewContextRequest event to the Interaction Manager. The IM will then respond by sending a NewContextResponse to the Modality Component containing a newly created context identifier. At this point all that has happened is a bit of book-keeping. The interaction with the user won't start until the IM sends a StartRequest (see below) to one or more Modality Components. The NewContextRequest is used when a Modality Component wants to create a new user interaction. The Interaction Manager can also create a new interaction at any point by sending a PrepareRequest or a StartRequest containing a new context identifier to one or more Modality Components. Thus a new user interaction may be started either by a Modality Component or by the Interaction Manager.
- StartRequest/StartResponse. The Interaction Manager starts a user interaction by sending a StartRequest to one or more Modality Components. The Modality Components return a StartResponse to acknowledge that they have begun running. The StartRequest contains two mutually exclusive fields, Content and ContentURL, that are used to instruct the Modality Component how it should interact with the user. It is most natural to think of these fields as specifying the markup that the Modality Component should execute. For a VoiceXML interpreter, for example, the Content field would contain an in-line specification of VoiceXML markup, while the ContentURL field would specify the URL to download the markup from. However Modality Components need not be controlled by markup. For ones that are not, the Content or ContentURL fields could contain platform-specific parameters or commands that would modify or control the behavior of the Modality Component. It is also possible to have a hard-coded Modality Component that runs the same way no matter what is specified in these fields.
- PrepareRequest/PrepareResponse. The PrepareRequest event is an optional event that the Interaction Manager can send before the StartRequest. It contains the same Content or ContentURL fields as the StartRequest. The purpose of the PrepareRequest is to allow a Modality Component to get ready to run by, e.g., downloading markup, compiling grammars, or any other operations that will allow it to start immediately upon receipt of the StartRequest. The PrepareRequest is useful for Modality Components that need to be tightly synchronized, for example, a text-to-speech engine that reads out a message while a graphical display highlights the words as they are spoken. If we simply send StartRequests to both components, it might take one longer to get going than the other, so coordination

will be smoother if the PrepareRequest allows both to prepare to start with minimal delay. The PrepareResponse is sent by the Modality Component back to the Interaction Manager to acknowledge the PrepareRequest.

- DoneNotification. This event is not part of a request/response pair, though it can be considered to be a delayed response to the Start Request. It is sent by a Modality Component to the Interaction Manager to indicate that it has finished its processing. For example, a text-to-speech system would send it when it had finished playing out the text specified in the StartRequest, or a VoiceXML interpreter would send a DoneNotification when it had finished executing its markup. (In this case, the VoiceXML interpreter might include an EMMA [8] representation of the recognition results in the event.) Not all Modality Components have a built-in concept of termination, so the DoneNotification is optional. For example, a simple graphical component might keep displaying the information that it had been told to display indefinitely until it received a new Start Request telling it to display different information. Such a component would never send a DoneNotification.
- PauseRequest/PauseResponse. The Interaction Manager may send a PauseRequest to a Modality Component, asking it to suspend its interactions with the user. The Modality Component then responds with a PauseResponse once it has paused. If a Modality Component is unable to pause, it will send a PauseResponse containing an error code.
- ResumeRequest/ResumeResponse. The Interaction Manager may send a ResumeRequest to any Modality Component that it has previously paused. The Modality Component will return a ResumeResponse once it has resumed processing. It is an error for the Interaction Manager to send a ResumeResponse to a Modality Component that has not previously been paused.
- CancelRequest/CancelResponse. The Interaction Manager may send a CancelRequest to any Modality Component telling it to stop running. The Modality Component will stop collecting user input and return a CancelResponse.
- ExtensionNotification. This event is intended to carry application- or platform-specific logic. Either the Interaction Manager or the Modality Components may send it, and no response is required (though the recipient could reply with another ExtensionNotification). This event includes a “name” field, which holds the name of the application- or platform-specific event, as well as an optional “data” field, which can hold an application- or platform-specific payload. A re-usable component, whether an Interaction Manager or a Modality Component, should document the set of ExtensionNotifications that it expects to send and receive, as well as their semantics.
- ClearContextRequest/ClearContextResponse. The Interaction Manager can send a ClearContextRequest to a Modality Component to indicate that the particular context/interaction is finished. The Modality Component is not required to take any specific action in response to this event, but normally it would free up any resources it has allocated to the interaction (cached grammars, adapted voice models, etc.) The Modality Component then responds with a ClearContextResponse.

- StatusRequest/StatusResponse. This event may be sent by either the Interaction Manager or a Modality Component, and is intended to provide keep-alive functionality. The recipient will reply with the StatusResponse event with a "status" field containing "alive" or "dead." (If the recipient doesn't respond at all, it is obviously dead.) The "context" field in the StatusRequest event is optional. If it is present, the recipient will reply with the status of that specific context/interaction. (A status of "alive" means that the context is still active and can receive further events.) If the "context" field is absent, the recipient replies with the status of the underlying server. In this case, a status of "alive" means that the server is able to process new contexts/interactions.

1.4 Some Examples

As an example of how this event set could be used in practice, consider a simple application running on a hand-held device consisting of a form that can be filled out either by speech or by typing in the values of fields in the GUI. This application would consist of a GUI Modality Component, a Speech Modality Component, and the Interaction Manager. The Modality Components gather the values of the fields and return them to the Interaction Manager, which will process then and submit the form when it is complete.

The event flow for starting the application and filling out a single field by speech would be as follows:

1. The IM sends a StartRequest event to the GUI MC.
2. The GUI MC displays the form and returns a StartResponse event.
3. The user selects a field by tapping on it. The GUI MC sends an ExtensionNotification with the name of the field to IM.
4. The IM sends a StartRequest to the Speech MC. The selected field will be specified in-line in the "Content" field or via a URL in the "ContentURL" field.
5. The Speech MC starts listening for speech and sends a Start Response.
6. The user speaks the value of the field. The Speech MC returns a DoneNotification containing the recognition result.
7. The IM sends an ExtensionNotification to the GUI MC specifying the value of the field (taken from the DoneNotification). The GUI MC updates its display with this value (Fig. 1.2).

If it takes the Speech Modality Component an appreciable amount of time to load and compile its grammars, and response time is a concern, the application could be modified so that the Interaction Manager would send multiple PrepareRequests to the Speech MC at start-up time, allowing it to prepare its grammars before the GUI MC displayed the form. In this case, the Speech MC would send a separate PrepareResponse for each request, and the IM would wait for all the responses before sending the StartRequest to the GUI MC. Events 1–7 would then occur in the same order as shown above.

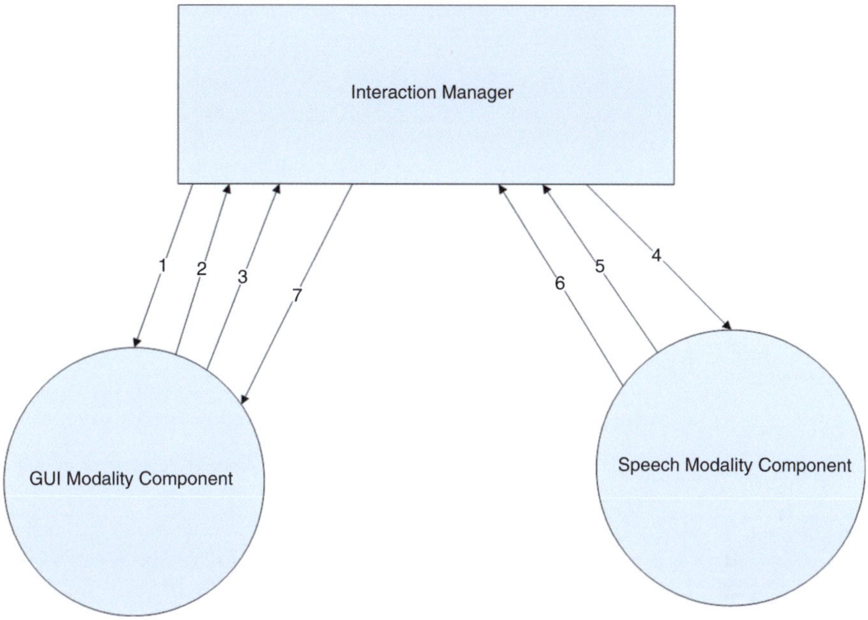

Fig. 1.2 Event sequence for filling a single field by voice

Now suppose that the user types in the value rather than speaking it. Events 1–5 remain the same, but this time it is the GUI MC that returns the value to the IM. The resulting event flow is as follows:

1. The IM sends a StartRequest event to the GUI MC.
2. The GUI MC displays the form and returns a StartResponse event.
3. The user selects a field by tapping on it. The GUI MC sends an ExtensionNotification with the name of the field to IM.
4. The IM sends a StartRequest to the Speech MC. The grammar for the selected field will be specified in-line in the "Content" field or via URL in the "ContentURL" field.
5. The Speech MC starts listening for speech and sends a Start Response.
6. The user types the value of the field. The GUI MC returns an ExtensionNotifcation containing the value. (Unlike the Speech MC, the GUI MC does not return values in a DoneNotification because it will keep running—that is, displaying the form—after it returns the result.)
7. The IM sends a CancelRequest to the Speech MC.
8. The Speech MC stops listening for speech and returns a CancelResponse (Fig. 1.3).

Since the user is using two modalities, there is a possibility of conflict, for example, if the user types one value and speaks another for a given field. It is the Interaction Manager's job to resolve such problems. It should keep its own Data

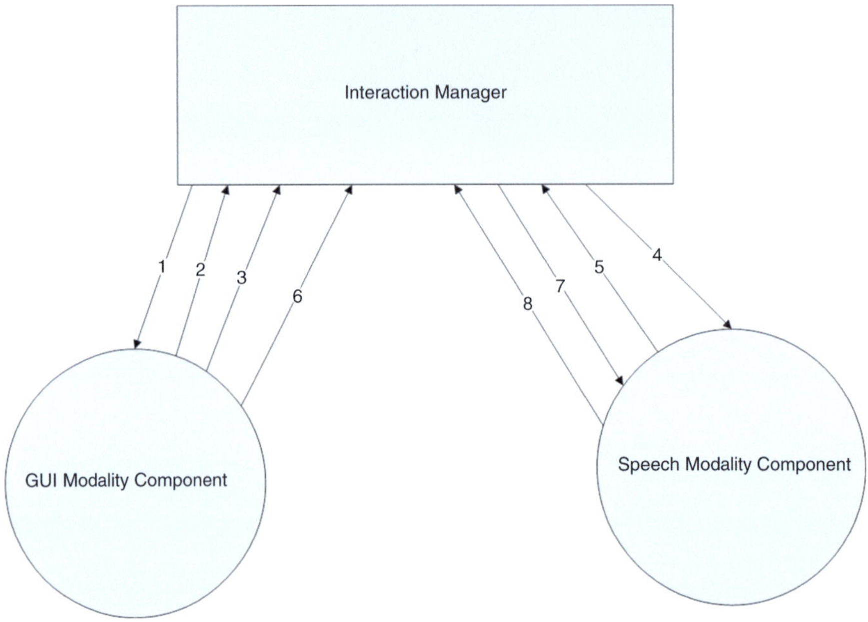

Fig. 1.3 Event sequence for filling a single field via the GUI

Component updated with the current state of the form. If a Modality Component sends the IM a value for a field that already has a value in the IM's Data Component, the IM knows a conflict has arisen. It is up to the application developer to decide what heuristic to use to resolve such conflicts since the Multimodal Architecture and Interfaces specification does not attempt to incorporate or enforce any particular approach to user interface development.

It is possible to modify the application so that the speech recognition doesn't follow the GUI field by field. Given a Modality Component that supports VoiceXML, the Interaction Manager can send it a StartRequest with a VoiceXML script that can capture the entire form. The VoiceXML Modality Component will now prompt the user for the various fields and gather input independent of what the GUI is doing. In fact, the user can speak the value for one field while simultaneously typing in the value of another. A sample event flow for such an application is given below:

1. The IM sends a StartRequest to the GUI MC.
2. The IM sends a StartRequest to the VoiceXML MC, either specifying the VoiceXML script in-line via the "Content" field, or by URL via the "ContentURL" field. (This event could also have been sent before the StartRequest to the GUI MC.)
3. The GUI MC displays the form and returns a StartResponse. (Depending on the timing of the application, this event could arrive at the IM before it sends the StartRequest to the VoiceXML MC.)

4. The VoiceXML MC loads the VoiceXML script and starts prompting the user for input. It then sends a StartResponse to the IM.
5. The VoiceXML MC obtains the value of one field and returns it to the IM in an ExtensionNotification Event.
6. The IM notifies the GUI MC of the field value with an ExtensionNotification event. The GUI MC updates its display accordingly.
7. The GUI MC obtains the value for a field and notifies the IM of it with an ExtensionNotification event.
8. The IM sends the field value to the VoiceXML MC in an ExtensionNotification event. If the VoiceXML MC updates its internal data model with this value, the Form Interpretation Algorithm [9] will ensure that it does not prompt the user for the value of this field.

   . . ... the user continues filling out the form using both modalities. . ...

9. The VoiceXML MC obtains the value for the final field and returns it to the IM in a DoneNotification.
10. The IM sends an ExtensionRequest to the GUI MC with the final value. The GUI MC updates its display (Fig. 1.4).

At the end of the event sequence shown above (i.e., after event ten reaches the GUI MC), the GUI MC is displaying the completed form, and the VoiceXML MC has stopped running. It is up to the application developer what happens next. The Interaction Manager will presumably submit or save the form. If more information

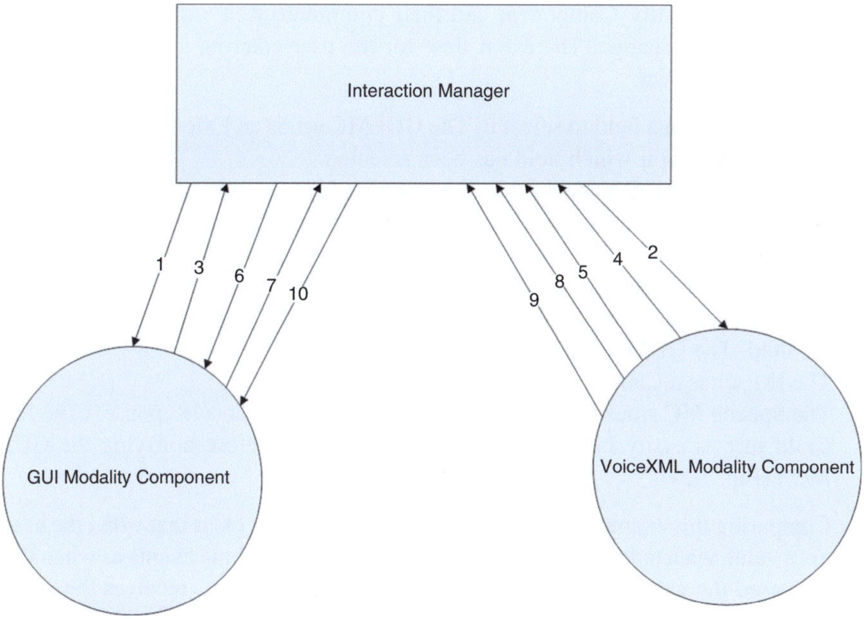

Fig. 1.4 Event sequence for filling multiple fields with voice and GUI

needs to be gathered, the Interaction Manager could send new StartRequests to both the GUI MC and the VoiceXML MC to continue the interaction with the user. One subtlety to note is the importance of the ContextID. If the new StartRequests contain the same ContextID as those in events 1–10, the Modality Components will view these requests as a continuation of the earlier interaction. Thus both Modality Components will keep any user adaptation they have performed. For example, the VoiceXML Modality Component will keep any speaker adaptation it has done to its voice models, while the GUI Modality Component will keep any display adjustments or special fonts that the user has selected. On the other hand, if the Interaction Manager sends ClearContextRequests before the StartRequests or simply uses a new ContextID in the StartRequests, the Modality Components will treat the requests as the start of a new interaction, possibly with a new user.

1.5 Adding a New Modality Component

As is clear from these examples, Modality Components do not communicate directly with each other, but only with the Interaction Manager. The value of this loose coupling becomes clear when a new Modality Component is introduced. Suppose the application is extended with a tablet capable of performing handwriting recognition. This Handwriting Modality Component will also communicate only with the Interaction Manager, sending it results via ExtensionNotifications. Neither the GUI nor the Speech Components need to be modified to work with the Handwriting Modality Component and their communication with the Interaction Manager will not change. The event flow for the user entering a field value with handwriting is shown:

1. The user taps on a field to select it. The GUI MC sends an ExtensionNotification to the IM telling it which field has been selected.
2. The IM sends the StartRequest to the Speech MC.
3. The Speech MC starts recognizing and returns a StartResponse.
4. The user writes out the value of the field using a stylus. The Handwriting MC sends this result back to the IM in an ExtensionNotification.
5. The IM sends an ExtensionNotification to the GUI MC containing the value for the field. The GUI updates its display.
6. The IM sends a CancelNotification to the Speech MC.
7. The Speech MC stops listening for speech and sends a CancelReponse. (The IM could just as easily have cancelled the Speech MC before notifying the GUI MC.) (Fig. 1.5)

Comparing this example to the first and second ones, it is clear that when the user enters a value via handwriting, the Speech MC receives the same events as when the user entered the value via the GUI MC. Similarly, the GUI MC receives the same event as when the user entered the value with speech. In fact, each Modality Component knows only that some other component has provided a value for the

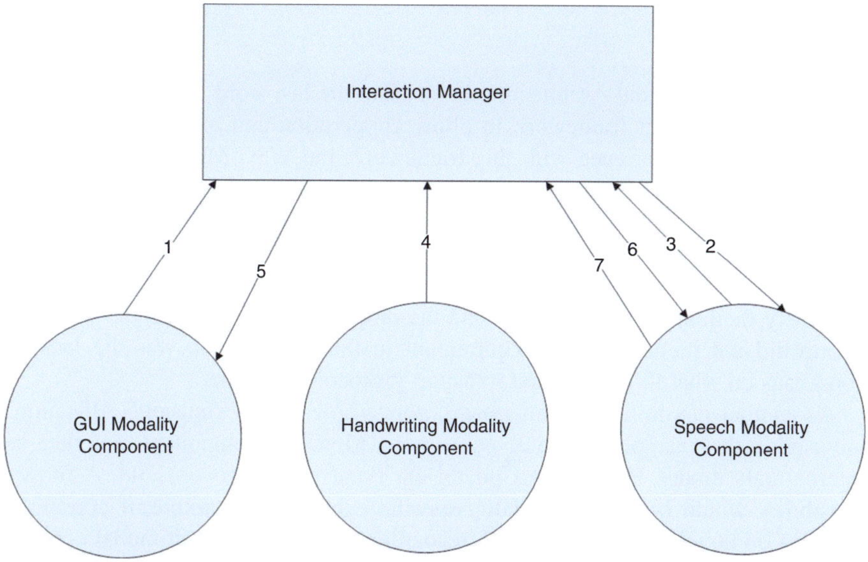

Fig. 1.5 Event sequence with handwriting component added

field in question. Only the Interaction Manager is aware of the new Modality Component. It is clear from this that the key to a successful implementation of the Multimodal Architecture and Interfaces specification is a powerful and flexible Interaction Manager, particularly one with good event handling capabilities. Given such an Interaction Manager, the design of the individual Modality Components is significantly simplified.

One important feature of the examples is the prevalence of ExtensionNotification events. The Multimodal Working Group felt that it was too early to define specific interfaces to modalities, with the result that ExtensionNotification carries a lot of the modality-specific logic. A Modality Component that supports this architecture will likely specify a lot of its interface in terms of ExtensionNotifications. For example, the GUI Modality Component's API specification might say that it will update the value of a field upon receipt of an ExtensionNotification event with name="fieldValue" and data="fieldname=value." One reason SCXML is a good candidate for an Interaction Manager language is that it has the ability to send and receive events with a variety of payloads, so that an SCXML interpreter can work with different Modality Components without requiring additional coding or integration work, particularly if the Modality Components support HTTP as an event transport.

1.6 Conclusion

The W3C's Multimodal Architecture is far from the last word on the subject. It is intended as an initial framework to allow cooperation and experimentation. As developers gain experience with this framework, the W3C Multimodal Working can modify it, extend it, or replace it altogether if a superior alternative emerges. The event set is quite high-level and will undoubtedly need to be refined if it becomes widely used. One obvious step would be to require support for a specific event syntax and transport protocol (for example, XML over HTTP). This would obviously facilitate interoperability and the only reason the Multimodal Working Group did not include such a requirement in the specification was the lack of consensus on what the syntax and transport protocol should be.

As another possibility for refinement, notice how often ExtensionNotifications are used in the examples given above to tell a Modality Component to update its internal data model. Perhaps an UpdateData event would prove useful. A further possibility would be to add modality-specific events. For example, if consensus emerges on how to manage a speech recognition system in a multimodal context, then a speech-specific event set could be defined.

Similarly, the multimodal architecture is quite high-level and will need to be articulated further. One possibility would be to add an Input Fusion component to the Interaction Manager. Consider the case of a user who says "I want to go here" and clicks on map. The utterance "I want to go here" will be returned by the speech Modality Component while the click will be captured by a graphical Modality Component. To understand the utterance, the system must combine the input from the two modalities and resolve "here" to the location on the map that the user clicked on. Right now this sort of combination is one of the many responsibilities of the Interaction Manager, but it might make sense to have a component that specialized in this task. Such a component would also be responsible for resolving conflicts between the voice and graphics Modality Components that were noted in the examples above, namely when the user types and speaks a value for the same field at the same time. See [10] in this volume for a more detailed discussion of how such a component might work. The Multimodal Working Group considered adding an Input Fusion component to the architecture, but decided that it didn't make sense to try to standardize the interface to such a component when there was not good enough agreement at the time about how it should work.

Finally, it is clear that many existing languages aren't easy to use as Modality Components because they don't allow fine-grained control. Both HTML and VoiceXML are designed to be complete stand-alone interfaces and it is not easy for an external component like an Interaction Manager to instruct a web browser what part of a page to display, or to tell a running VoiceXML interpreter to pause jump to another part of a form. Modality Component languages will fit into the W3C multimodal architecture much more easily if they are designed to accept asynchronous updates to both their data models and their flow of control.

References

1. Barnett, J., Bodell, M., Dahl, D., Kliche, I., Larson, J., Porter, B., et al. (2012). Multimodal architecture and interfaces. W3C Recommendation. http://www.w3.org/TR/mmi-arch/.
2. SALT Forum (2002). Speech Application Language Tags. http://xml.coverpages.org/SALT-FinalSpecificationV10.zip.
3. Hickson, I., Berjon, R., Faulkner, S., Leithead, T., Navarra, E., O'Connor, E., et al. (2014). HTML5. W3C Recommendation. https://www.w3.org/TR/html5/.
4. Oshry, M., Auburn, R., Baggia, P., Bodell, M., Burke, D., Burnett, D., et al. (2007). Voice Extensible Markup Language (VoiceXML) 2.1. W3C Recommendation. https://www.w3.org/TR/2007/REC-voicexml21-20070619/.
5. Barnett, J., Akolkar, R., Auburn, R., Bodell, M., Burnett, D., Carter, J., et al. (2015). State Chart XML (SCXML) State Machine Notation for Control Abstraction. W3C Recommendation. https://www.w3.org/TR/scxml/.
6. Rodriguez, B. H., Barnett, J., Dahl, D., Tumuluri, R., Kharidi, N., & Ashimura, K. (2015). Discovery and registration of multimodal modality components: State handling. W3C Working Draft. https://www.w3.org/TR/mmi-mc-discovery/.
7. Galaxy Communicator (2003). http://communicator.sourceforge.net/sites/MITRE/distributions/GalaxyCommunicator/docs/manual/.
8. Johnston, M., Baggia, P., Burnett, D., Carter, J., Dahl, D., McCobb, G., et al. (2009). EMMA: Extensible MultiModal Annotation markup language. W3C Recommendation. http://www.w3.org/TR/2009/REC-emma-20090210/.
9. McGlashan, S., Burnett, D., Carter, J., Danielsen, P., Ferrans, J., Hunt, A., et al. (2004). Voice Extensible Markup Language (VXML) Version 2.0. Appendix C. W3C Recommendation. https://www.w3.org/TR/voicexml20/#dmlAFIA.
10. Schnelle-Walka, D., Duarte, C., & Radomski, S. (2016). Multimodal fusion and fission within the MMI architectural pattern. In D. Dahl (Ed.), *Multimodal Interaction with W3C Standards: Toward Natural User Interfaces to Everything*. New York, NY: Springer.

Chapter 2
The Role and Importance of Speech Standards

Paolo Baggia, Daniel C. Burnett, Rob Marchand, and Val Matula

Abstract Within only a few years the landscape of speech and DTMF applications changed from being based on proprietary languages to being completely based on speech standards. In that, a role of primary importance was played by W3C Voice Browser Working Group (VBWG). This chapter describes this change, the implications, and highlights the standards created by the W3C VBWG, as well as the benefits that these standards can induce in many other application fields, including multi-modal interfaces.

2.1 Introduction

A strong wind of change was sweeping the stuffy world of Interactive Voice Response (IVR) and speech applications in general. This call for change developed in the very last years of the last century, resulting in a key event—the workshop on "Voice Browsers" held in Cambridge, MA on 13 October 1998 [1]. The workshop was sponsored by the W3C, and it raised huge interest in the standardization of voice application technologies. The direct result was the birth of a W3C Working Group—the Voice Browser Working Group (W3C VBWG [2]), formed to create an interconnected family of standards. This chapter offers a short introduction to most of the W3C VBWG standards and also describes their close relationship with the W3C Multimodal Interaction Working Group (W3C MMI [3]).

P. Baggia (✉)
Department of Enterprise, Nuance Communications, Inc., Torino, Italy
e-mail: paolo.baggia@nuance.com

D.C. Burnett
StandardsPlay, Lilburn, GA, USA

R. Marchand
Genesys, Markham, ON, Canada

V. Matula
Avaya Inc., Santa Clara, CA, USA

D.A. Dahl (ed.), *Multimodal Interaction with W3C Standards*,
DOI 10.1007/978-3-319-42816-1_2

Several factors combined to drive this change; the most relevant ones are

- The development of an IVR application was cumbersome and required the use of proprietary IDEs that were bound to individual vendors. At the time of the formation of the VBWG, IVR technology was proprietary and there was virtually no chance to exchange expertise or application assets between them.
- Speech technologies were very limited in their use; only simple commands, menu options, and sequences of digits were allowed. However, the core speech technologies were rapidly evolving to be more powerful and flexible and to allow a new generation of speech applications.
- Voice interactions were limited to simple menu navigation, with no flexibility to allow more advanced dialog capabilities. Their implementation was clumsy.
- But the most powerful factor was the advent of the Internet era: the HTTP protocol, the HTML language, and the flourishing of web sites. All these advances were based on public standards, while the world of voice applications was missing the opportunity to follow these new trends.

It was this combination of factors that drove the creation of the W3C VBWG, changing forever the world of voice applications. With the Working Group, a large number of companies now had a place to work together on this standardization effort, strongly motivated by a common interest. These companies included: speech technology makers (at that time L&H, Philips, Nuance, SpeechWorks, Loquendo, and Entropic), research laboratories (MIT, Rutgers, AT&T Bell Labs, and CSELT), large telephone companies (Lucent, AT&T, BT, Deutsche Telekom, France Telecom, Telecom Italia, Motorola, and Nokia), large software and hardware companies (Microsoft, HP, Intel, IBM, and Unisys), newly formed Voice Platform companies (PipeBeach, Voxpilot, Vocalocity, VoiceGenie, and Voxeo), hosting and developer studios (HeyAnita, BeVocal, and Tellme), IVR vendors (Avaya, Genesys, Comverse, and CISCO), and many more.

In the meantime, an industry organization named the VoiceXML Forum [4], created by AT&T, Lucent, Motorola, and IBM, proposed a new language called VoiceXML 1.0 [5] and started to evangelize its adoption. The newly created W3C VBWG, led at various times by Jim Larson (Intel), Scott McGlashan (PipeBeach and HP), and Dan Burnett (Voxeo and StandardsPlay), selected VoiceXML 1.0 as the starting candidate to inspire the creation of the standard to come.

With the standardization of VoiceXML and related technologies now firmly in the hands of the W3C, the VoiceXML Forum took on a complementary role in the evolution of VoiceXML. The Forum took on responsibilities including:

- Education: The Forum developed tutorials, sample applications, and produced a monthly e-zine for the developer community [6].
- Evangelism: The Forum marketed VoiceXML as an emerging standard to the IVR community, both on-line and at industry conferences.
- Conformance: Perhaps the most critical role of the Forum was the development of conformance test suites (based on the W3C specification Implementation

Reports[1]) for VoiceXML, SSML, and SRGS. The Forum also provided independent third-party conformance test certifications.
- Developer certification: The Forum created a VoiceXML Developer Certification program, including several test suites and access to third party certification testing. This helped to build a developer community.
- Technology and tools evaluation: The Forum hosted several committees with the task of assessing evolving technologies and tools related to the adoption of VoiceXML. These groups investigated speech technology protocol standardization, security aspects (including biometrics), and other topics.

These areas are outside the scope of W3C standards development, yet are critical in supporting adoption and acceptance of a new standard. The role fulfilled by the VoiceXML Forum helped avoid early standards fragmentation, and fostered adoption of VoiceXML by industry.

This collaboration accelerated the cooperative effort to create the foundations of a new generation of voice applications based on public standards. In a short time an incredible sequence of Working Drafts was published, demonstrating the innovation under development. Supported by the broad involvement of stakeholders from different groups, the IVR industry began to implement these drafts as soon as they were delivered. Adoption worries were rapidly left behind in this new ecosystem.

In March 2004, after just 4 years, the first complete standards, i.e., W3C Recommendations, were released: VoiceXML 2.0 [7] for authoring a voice application; SRGS 1.0 [8] for precisely defining the syntax of speech grammars; and SSML 1.0 [9] for controlling speech synthesis (or text-to-speech). A few years later, in April/June 2007, a second round of W3C Recommendations concluded, delivering: VoiceXML 2.1 [10], adding some interesting features on top of VoiceXML 2.0; and SISR 1.0 [11], formalizing the representation of meaning within a speech grammar and complementing SRGS 1.0. The work didn't stop there. SSML 1.0 was revised to version 1.1 [12] to ease the internationalization of speech synthesis in other world regions; PLS 1.0 [13], a language to describe phonetic lexicons supporting interoperability between SRGS 1.0 and SSML 1.0/1.1, was created; and finally CCXML 1.0 [14] was developed as a real time language to implement call control in a voice browser platform. In the rest of this chapter, these languages will be briefly introduced and other aspects of this revolution will be highlighted.

It is worth noting that the entire industry took the advent of the W3C VBWG standards as a change to be immediately adopted. Vendors implemented the standards in their solutions, so that the IVR and telephony application world started to speak the VoiceXML standard language in a matter of a few years. The use of VoiceXML enforced a clean separation between the IVR platforms and the hosting of voice applications accessed by HTTP/HTTPS. Voice applications were at first

[1] All W3C Recommendations include a reference to an Implementation Report document to assess the implementability of the proposed standard. For instance, the VoiceXML 2.0 Implementation Report [44] was very important in showing how to implement a procedure to automate most of the tests.

static "pages" stored on a Web server, and then progressively became dynamic as the rest of the Web evolved. A side effect of this adoption of the web architecture by VoiceXML was that many web-related technologies became available to IVR applications. For example, VoiceXML 2.1 added the `<data>` element to take advantage of emerging AJAX access to web services. Infrastructure elements like web caches and load balancers were immediately useful within IVR deployments. Several books and articles describing VoiceXML were published. A review of the language is presented in [15] from the W3C VBWG Chair Jim Larson. For a discussion of VoiceXML in the broader context of Spoken Dialogue Systems see [16]. Many start-ups sprang up to offer voice browsers, tools, hosting and, as in all industry sectors, the larger companies filled out their offerings by acquiring these start-ups.

Over time this process slowed down, indicating that the revolution had occurred, but it also meant that continued change was becoming more difficult. There was a definite and very ambitious attempt to re-write VoiceXML to be extensible and modular, simplifying the incorporation of future advances, but this radical re-formulation stopped after producing the first Working Draft (VoiceXML 3.0 [17]). The last advance was to lay down and complete SCXML 1.0 [18], described in another chapter of this book. This language offers to both the IVR world and the world of multimodal applications a clean and powerful way to encode the interaction of different components thorough Harel's state-charts [19]. Other activities that were not completed included a standard for speaker verification/identification from voice prints and statistical language models [20].

The W3C VBWG had completed its role, developing the solid foundation of an open and powerful generation of standards, so it closed its activities in October of 2015 [21]. The VoiceXML Forum remains active in sustaining the VoiceXML 2.0/2.1 ecosystem, most notably with work in Conformance and Developer Certification.

As with most technologies, these standards will remain for a long time in the core of these industries, but there may also be opportunities to reuse them in novel ways and to fuel new advances and revolutions. The variegated world of multimodal interfaces and, more generally, of the Internet of Things (IoT), will greatly benefit from the work of the W3C VBWG, especially for a speech modality. Voice continues to prove to be the most powerful means for humans to control and influence the world around us. Section 2.4 describes the role of voice standards in these additional domains, while Sect. 2.2 is devoted to explaining the standards developed by W3C VBWG and Sect. 2.3 complements the discussion with related IETF protocols.

2.2 Quick Tour of W3C VBWG Major Standards

This section presents a quick tour of the W3C VBWG standards. The review will be limited to a brief introduction with some highlights of the major benefits of each standard.

Although VoiceXML is the most visible of the VBWG standards, there are a number of related languages that work together with VoiceXML to provide a complete facility for creating IVR applications. The individual languages have also in some cases been used independently for other purposes as well, such as:

- The grammar language, SRGS 1.0, for processing text input instead of speech.
- The speech synthesis language, SSML 1.0/1.1, for talking books, or assistive applications.
- The pronunciation lexicon language, PLS 1.0, in language training applications.
- The call control language, CCXML 1.0, for managing calls over IP.
- The state-chart language, SCXML 1.0, for generalized interactions in a multimodal interface.

The following is a brief description of these standards developed by W3C VBWG.

2.2.1 VoiceXML 2.0

The Voice Extensible Markup Language (VoiceXML), version 2.0. [7], is the flagship and most relevant standard produced by the W3C VBWG. It is an XML markup language specialized to declaratively describe a dialog interaction between a caller and an automated application. The language leverages all the advantages of the Web: an application is stored in a Web Application server; it might be statically or dynamically generated; a specialized user agent, called a Voice Browser, downloads and interprets a VoiceXML application, together with scripts, audio prompts, and grammars; the syntax is enforced by an XML Schema or a DTD; the application might be in any human language and declares an appropriate encoding; and so on.

The standard was built on top of the initial VoiceXML 1.0 [5] proposal made by the VoiceXML Forum [4]. The original Forum members were from AT&T, IBM, Lucent Technologies, and Motorola. In the W3C VBWG, a much larger number of people from many companies and organizations participated in the joint effort to transform the proposal into a widely accepted standard. This promise was realized in March 2004, when VoiceXML 2.0 was declared a W3C Recommendation, with nine companies[2] presenting an Implementation Report to demonstrate interest in promoting this standard in the industry. The Implementation Report was based on a test suite of over 600 test assertions defined in a special language to facilitate its automation. Based upon this test suite, the VoiceXML Forum delivered a Platform Certification program [22] with at least 27 platforms certified to date.

[2] The companies which submitted an Implementation Report [44] for VoiceXML 2.0 were nine: Comverse, Genesys, Loquendo, Motorola, PublicVoiceXML Consortium, Tellme Networks, Vocalocity, VoiceGenie Technologies, and Voxpilot.

A VoiceXML application is made of dialogs whose building blocks are the `<menu>` and `<form>` elements. The former is used for designing simple menu-based IVR applications, while the latter is used for extending the interaction to form filling, where an algorithm called the Form Interpretation Algorithm (FIA, described in Appendix C of the specification [7]) is used. The FIA precisely describes how the filling of different `<field>`s is performed. VoiceXML 2.0 also supports extending the interaction to a mixed-initiative dialog, where additional flexibility allows a caller to say more complex sentences, like: "I'd like to travel from Boston to Detroit in First Class next Monday around noon." In this modality, several `<field>`s are filled at the same time to take maximum advantage of the compactness and flexibility of natural language.

A novelty of VoiceXML 2.0 was to delegate[3] the definition of speech grammars and of synthesized prompts to two interoperable standards. These standards are: SRGS 1.0 [8] for grammars and SSML 1.0 [9] for prompts, described below. This choice allowed the standards to be developed in parallel, but more importantly it promoted the reuse of speech recognition and of speech synthesis in other application contexts, such as multimodal interfaces, or appliances. The `<field>` element may include one or more `<prompt>s` to solicit the caller to say or type the expected information and several <grammar>s to model callers' sentences, while the `<filled>` element is triggered if the information items are collected either by voice or DTMF, upon which they are automatically stored in a variable associated with the `<field>` itself. The filling of field values will restart the collection cycle as described in the FIA algorithm.

Besides the `<field>` element, other form items are supported by VoiceXML 2.0 to enrich the dialog interaction. For instance:

- `<block>` element to declare prompts and to perform a block of computations;
- `<record>` element to record the caller's voice and provide access to the stored audio;
- `<transfer>` element to transfer a call to another party either by a "bridge" or a "blind" transfer;
- `<subdialog>` element to pause the current interaction, spawn the processing of another context to complete a task, and then return to the calling environment with the results;
- `<object>` element to allow for new functionality extensions; this was used for extending the capabilities of a voice browser to allow additional features, for instance, the inclusion of voice biometrics capabilities in a VoiceXML application.

Data are handled and processed by an ECMAScript processor with elements in the language to declare and assign variables (`<var>` and `<assign>` elements) or load scripts (`<script>` element). The variables are organized into different

[3] The XML Schema of VoiceXML 2.0 includes the references to: SRGS 1.0 and SSML 1.0 XML Schemas, see Appendix O of VoiceXML 2.0 specification [7].

scope levels: "application" for the sharing of data across different VoiceXML documents, "document" for variables that need to be visible across a single document, "dialog" for variables active only inside a single <form> or <menu>, and the internal context of an inner element. Above these scopes, there is an additional one called "session" that contains read-only variables related to that specific session. For instance, the session scope provides access to telephony information (e.g., ANI, DNIS, etc.). Finally, each recognition step allows the browser to access information related to the most recently occurring recognition. Some examples include the input modality (either "voice" or "dtmf"), a numeric value for the confidence of the results, the recognized/keyed text and the meaning of that interaction.

The FIA describes the flow of the interaction inside a dialog element. To transition to the next dialog a <goto> element contains a URI attribute that points either to a dialog in the same document or to another VoiceXML document to which to transition. The <submit> element is used to upload data collected during the dialog interaction to the Web Application. The result is a new VoiceXML page for continuing the interaction. Moreover, an event handling mechanism is present to allow firing predefined events: "help," "repeat," but also "noinput" and "nomatch" to indicate a missing response from the caller, or that the input was not properly recognized. It is also possible to throw (<throw> element) user defined events that will trigger the execution of a handler defined via the <catch> element. In this way the application can deal with predefined and unexpected behaviors by continuing within the same dialog under the FIA, or by transitioning to another one, or even by closing the interaction. An application can be explicitly terminated by the <exit> element or closed via the <disconnect> element. The latter hangs up the call if necessary.

The VoiceXML specification had a terrific impact on the IVR industry, being widely adopted even before the language was completed. The traditional IVR platform vendors had to change their architectures by including either a homegrown VoiceXML browser or one obtained by acquiring a newly formed start-up company. Other vendors opted for hosting the VoiceXML browser and providing a Web-based development environment to create, test, and deploy voice applications. Some examples include HeyAnita, BeVocal, TellMe, and Voxeo, all since acquired by major players. Other companies specialized in tools or development environments, and they were progressively acquired as well. The VoiceXML Forum provided user groups, newsletters, journals, and events to sustain the VoiceXML ecosystem. A critical contribution from the Forum was the development of two Certification Programs, the previously mentioned Platform Certification and a Developer Certification program, both still active at present.

2.2.2 VoiceXML 2.1

Although the VoiceXML 2.0 specification was immediately implemented and rapidly became the major standard for voice and DTMF applications, a follow-on effort added a limited number of extensions. This key collection of extensions was published in June 2007 as VoiceXML 2.1 [10]. It includes eight additional features, including a means to dynamically reference grammars and scripts, a new `<foreach>` iteration element for dynamically composing prompts or executing computations on list of objects, and a `<data>` element to allow a VoiceXML application to dynamically load data from a server using the equivalent of an XML HTTP Request. These extensions were all motivated by the need to reduce the number of (expensive) VoiceXML page transitions. With VoiceXML 2.1 a single running VoiceXML page was able to adapt to external or dynamic conditions. In addition, the `<transfer>` element was extended with a new "consultation" mode to allow the interaction to be suspended while a transfer was attempted, resuming it if the transfer was not possible. The recording capabilities were extended to be active during the recognition to enable fetching of both the audio and the recognition results. Finally, the `<disconnect>` element was extended to return a list of results.

The VoiceXML 2.1 extensions were also widely implemented, and the VoiceXML Forum Platform Certification Program was extended to additionally certify both VoiceXML 2.1 and the grammar language described in the next section.

2.2.3 SRGS 1.0

Speech recognition greatly benefits by knowing in advance what a caller might say. A speech grammar is a compact way to declaratively describe the admissible sentences. The W3C VBWG was very successful in clearly defining the syntax of speech grammars. SRGS 1.0, the Speech Recognition Grammar Specification [8], defines two different formats for encoding a grammar: an XML format, called GRXML, and a textual one, called ABNF (cfr Augmented Backus-Naur Form). The two formats are homologous, with very irrelevant differences. A grammar defines sequences of words/phrases or alternatives to be legally accepted by the speech recognizer. The grammar is organized into rules, where only a few are accessible from the outside (declared "public"), while all the others are hidden (declared "private") to enforce modularity and a clean composition among different grammar files.

If words define the admissible sentences, a grammar also provides a way to compose a result to be returned to the application. This is done by the execution of small scripts contained in the `<tag>` element, with the following permitted as a return value: numbers "123" when the caller speaks "one hundred twenty three," date and time expressions, telephone numbers, or arbitrary key-value pairs.

The SRGS 1.0 specification immediately became the format supported by all speech recognition engines, allowing them to interoperate within a VoiceXML platform.

2.2.4 SISR 1.0

The production of a result remained undefined in the SRGS 1.0 specification, having been delegated to a subsequent specification, "Semantic Interpretation for Speech Recognition" (SISR 1.0 [11]), which was released as a W3C Recommendation in April 2007. SISR 1.0 formally defines the content of the `<tag>` element in SRGS 1.0 grammars to be an ECMA-327 [23] script. While ECMA-327 is a constrained version of ECMAScript, the goal was to gain computational efficiency to enable more extensive speech recognition engine processing.

The language defines how rules produce results and how they return them when they are referenced. This process allows the final result of a grammar to be composed progressively. Attention was paid to allow both a sequential and parallel execution of the result composition.

The presence of scripting capabilities inside a grammar helped move application-dependent normalization inside of the grammar and to clearly separate the application needs from the need for a natural way of expressing the caller's expected language.

2.2.5 SSML 1.0 and 1.1

Another effort was to define how to control a speech synthesis, or text-to-speech, engine. The controls help the engine to render the textual prompt in the most accurate way. The XML markup language for this purpose is the Speech Synthesis Markup Language (SSML 1.0 [9]).

SSML 1.0 includes elements that describe the structure of the text to be spoken (`<p>` element for paragraphs, and `<s>` element for sentences), text normalization and phonetic input (`<sub>` element for textual substitutions and `<phoneme>` for pronunciations), prosodic features such as pauses (`<break>` element), speed and rate (`<prosody>` element), and how to change the speaking voice (`<voice>` element).

An extension of SSML 1.0, SSML 1.1 [12], was a continuation of the standardization effort to promote the use of SSML to more international languages, in particular Asian and Indian languages.

2.2.6 PLS 1.0

Both speech grammars and synthesized prompts might require customizing the pronunciation for a specific application domain. This is often done by adding a reference to a user lexicon. The Pronunciation Lexicon Specification (PLS 1.0 [13]) was created to allow for the definition of a standard lexicon fully interoperable with SRGS 1.0 and SSML 1.0/1.1. The lexicon is a container of entries, `<lexeme>` elements, with a textual part described by the `<grapheme>` element and textual replacements provided by the `<alias>` element or phonetic transcriptions by `<phoneme>` elements.

PLS 1.0 documents support the expansion of abbreviations and acronyms, addressing both multiple orthographies and multiple pronunciations. PLS 1.0 became a W3C Recommendation in October 2008.

2.2.7 CCXML 1.0

Another language defined by the W3C VBWG focused on programming the call control of a voice browser in an innovative way. An XML markup language was developed to define handlers for telephony events generated by a telephone connection or a VoIP SIP interaction. The Voice Browser Call Control (CCXML 1.0 [14]) language was designed to allow a very efficient implementation completely based upon events and handlers to avoid creating any latency that might impact the underlying signaling.

A CCXML engine is also able to send and receive events through an HTTP/HTTPS connector, which allows for the generation of outbound calls from a web application and for the monitoring of calls and conferences via a web interface.

During the definition of CCXML 1.0 the W3C VBWG decided to start another effort to define a state-chart language to generalize the ideas behind CCXML 1.0. This new specification was State Chart XML (SCXML): State Machine Notation for Control Abstraction (SCXML 1.0 [18]) described in another chapter of this volume, and it can be used as the key component to control a generalized interaction in a multi-modal interface.

2.3 IETF, Companion Protocols

The VoiceXML revolution wouldn't have been possible without the presence of several other standards, especially protocols. For instance, in a Voice Platform the application documents, which might include VoiceXML 2.0/2.1 pages, SRGS 1.0 grammars, audio files, ECMAScript scripts, and PLS 1.0 lexicons, are accessed through HTTP/HTTPS protocols, as in any other Web user agent. Many of the web browser/web server related protocols apply equally well to voice browsers as well. For example, voice browsers respect content-types, cookies, and cache control directives, as used by web browsers.

A new requirement for IVR platforms is standardization of the communication between the VoiceXML browser platform and the servers providing speech resources. Historically based on proprietary APIs and formats, speech recognition resources require grammars and audio, returning recognition results to the IVR platform. Similarly, text to speech resources require text that is to be rendered, and then return audio to the IVR platform. With the definition of SSML, SRGS, and SISR, the high-level interaction with the speech resources became standardized. A standard network level protocol was then defined for speech resources by the IETF [24]. Media Resource Control Protocol (MRCP), whose initial draft was proposed by CISCO, Nuance and SpeechWorks in April 2006 was standardized as MRCPv1 (RFC 4463 [45]). MRCPv1 is based on Real time Protocol (RTP) for media transport and on Real time Streaming Protocol (RTSP) for controlling speech resources such as speech synthesizers and speech or DTMF recognizers. The MRCP protocol defines the requests, responses, and events to control the processing inside resource servers. For a detailed description of the MRCP protocol in relationship with W3C VBWG standards see [26].

The introduction of MRCPv1 allowed voice platforms to be implemented with a distributed and scalable architecture, and was hence immediately adopted by all the IVR platforms. In the meantime, the standardization process continued with the definition of MRCPv2, becoming an IETF standard in November 2012 (RFC 6787 [25]). MRCPv2 is based on Session Initiation Protocol (SIP) for signaling and Session Description Protocol (SDP) for exchanging and negotiating capabilities. Moreover, MRCPv2 was extended to access new resources for recording and speaker verification and identification, and to support encryption.

2.4 Current Trends and Future Evolutions

Although the W3C VWBG is now closed [21], the influence of its standards is still broadly felt across many sectors. In the IVR/Customer Care world, the presence of VoiceXML and related standards is ubiquitous, and there are no signs this will change in the near future. There is an established industry in place, so drastic changes are very unlikely. In less than 15 years the W3C VBWG standards moved from an idea to mandatory requirements for a whole industry—a remarkable outcome.

Other speech languages developed outside of the W3C VBWG include XHTML + Voice [28] from IBM and Opera Software to allow a direct integration of VoiceXML in an XHTML document, and Speech Application Language Tags (SALT) [29] from a consortium led by Microsoft, to integrate speech into a web application. More recently, Google started a separate effort to develop the Web Speech API in a W3C Community Group to enable web developers to incorporate speech recognition and synthesis into their web pages; the Final Report is available at [30].

There are several innovations and new application domains that might require these standards when ready to integrate a voice or textual interaction. A quick review of the trends and evolutions that are happening are quickly described in the following sections.

2.4.1 IVR in the Multi-Channel World

The IVR world today is experiencing change due to the proliferation of channels available to a customer for seeking support or gathering information. In the recent past the only available way was a phone call, and while these days the phone call is still predominant, the contribution of other channels is increasingly evident. For instance, textual chats may be offered to a user during a web session, often with the presence of dedicated agents, and sometimes including even some degree of automation. Moreover, a web site often provides more than web search inside its content, for instance, a text interaction to intelligently search among FAQs or even a limited capability to provide precise answers to customer requests. Finally, social media is becoming a place for seeking support as well, and can be used to express very polarized—and highly visible—opinions on the company of interest. For example, a high-profile complaint on twitter will often result in a rapid response to a problem, perhaps out of proportion to the original issue.

Given all of these options, users may switch between channels (multi-channel) if they run into challenges, or will often use multiple channels at once (omni-channel). Consequently, most vendors in the customer experience field must support a number of different channels in their solutions. In order to integrate and correlate interactions taking place over multiple channels, possibly over disparate time spans (consider a voice call vs. an SMS exchange), the standards specific to a particular channel (for example, VoiceXML for voice, HTML for web) must be combined with the ability to receive and send events and data, manage state, and coordinate activities over multiple channels. SCXML (more fully described in a separate chapter) fulfills these requirements, providing the ability to coordinate between channels. For example, an inbound voice call can be connected to a VoiceXML session under SCXML control. Once the call is complete, the SCXML session can schedule a follow-up SMS message to the original caller as a reminder (perhaps days later) or as a transaction summary (immediately).

An omni-channel session could, after an inbound voice caller is identified, take advantage of the fact the caller was browsing the company web site when they decided to call. This could lead to a co-browse web session with the caller still on the phone, helping to complete a transaction or solve a problem.

There may be opportunities for the use of VBWG standards in other ways as well. For example, representation of meaning, or extraction of meaning from textual inputs—gathered using chat or SMS channels—might be enabled using SRGS or SISR. In another example, the SSML standard could be used to improve the rendering of text to speech in web-based interfaces. In the areas of multi-channel and omni-channel communications, the W3C VBWG standards can perhaps extend their role beyond voice interactions to be exploited with other channels.

2.4.2 Virtual Assistants

Since the deployment of the Siri virtual assistant on Apple's iPhone and iPad in 2011, interest in Virtual Assistants (VAs) in general has increased. The VAs can take the aspect of avatars, or, like Siri, be just a speaking voice, allowing users to ask open-ended questions, and typically using cloud resources to determine intent and return results. This new kind of voice activated VA is moving beyond mobile phones to cars, and to home appliances—e.g., Amazon Echo and the home robot Jibo. A distinctive characteristic of this kind of VA is the ability to speak and understand user commands, sometimes with amusement and/or irony in the answers and with personality.

In this area there is still a strong need to extend the capabilities of the interaction, but certainly voice is the most natural means with which to interact with these Virtual Assistants, allowing multiple requests to be packed into a single sentence.

The standards produced by the VBWG have less applicability in this realm. VoiceXML is best suited for applications that are system-directed rather than user-directed as is seen in typical VAs. However, the supporting standards may have a role to play. SCXML can be used as described in Sect. 2.4.1 to coordinate multiple channels and interactions, where the VA can be viewed as another channel. SISR and the Extensible Multi-Modal Annotation (EMMA 1.0 [27]) specification can be used to exchange information related to input, intent, and output across different channels. Multimodal interfaces and EMMA are further detailed in the following section.

2.4.3 Multimodal Interfaces

A richer style of interaction can be offered by a multimodal interface, which integrates not only a voice modality, but also gesture, text, haptic, or other kind of input. A multimodal interface is able to integrate requests given by different complementary and supplementary modalities. The W3C MMIWG [3] is responsible for developing standards in this area, but the cross-collaboration with the W3C VBWG was very well maintained. For instance, the main specification developed by the W3C MMIWG is an MMI architecture framework [31] whose key components are an Interaction Manager and one or more Modality Components. Among them, voice input is dealt with by a specific Modality Component that can be directly modeled using W3C VBWG standards. For instance, SRGS 1.0 can be used for speech recognition, SSML 1.0 for speech synthesis, and VoiceXML 2.1 for modular dialogs.

A core aspect of multimodal interfaces is the need to integrate meanings from different modalities. This is made possible by the Extensible Multi-Modal Annotation (EMMA 1.0 [27]) specification. EMMA 1.0 was designed to encode meaning representations produced by SRGS 1.0, where semantic interpretation by SISR 1.0

can produce results in EMMA 1.0 format. This volume describes in a dedicated chapter the recent extensions to the EMMA specification (EMMA 2.0 [32]) to extend its role from representing input only, to also cover a variety of outputs. Another important contribution of W3C VBWG standards is the use of SCXML 1.0 [18] to direct the Interaction Manager inside the MMI Architecture Framework. This direction is proposed by many authors (cf. [33, 34]), and a related workshop has been active since 2014. The "EICS Workshop on Engineering Interactive Computer Systems with SCXML" [35] demonstrates the interest of the research community in this topic.

Finally, the W3C MMIWG produced a standard for describing emotions expressed by face, voice, or other modalities. The EmotionML 1.0 [36] standard is a good candidate to be integrated with SISR 1.0 to encode emotions recognized in human voice. Similarly, EmotionML 1.0 could be used with SSML 1.0 to instruct a speech synthesis engine to express emotions. A detailed description of EmotionML 1.0 is present in another chapter of this volume.

2.4.4 Internet of Things

The Web continues to evolve and expand, now with the theoretical inclusion of all objects interconnected by a network interface, often called Internet of Things (IoT). These objects can be anything in the surrounding environment including, for example, home appliances, cars, hand-held devices, televisions, etc. In literature there are many examples where both the SCXML 1.0 and VoiceXML 2.1 standards are proposed for a variety of IoT applications. These range from the SmartHome [37, 38], Ambient Assisted Living [39], Semantic Sensor Network [40] to more general Pervasive Environments [41, 42], and embedded applications like automotive ones [43]. In those projects, the control of interactions allowed by SCXML 1.0 is fundamental and often inside the MMI Architecture previously described. Moreover, a user can take advantage of a voice interaction within this multifaceted world. The W3C VBWG speech related standards offer the basis for implementing this voice interaction. Examples are voice commands from a mobile device to inquire about the status of home appliances and also give commands to remotely activate these appliances.

2.5 Conclusion

This chapter has described the development and evolution of speech-related standards, and how they have impacted the IVR industry and voice applications. The shift from proprietary to standards-based technologies provided many important benefits, including:

- Conversion of an industry from fragmented and proprietary development to environments that more easily interoperate and support application portability. Application portability prior to standards availability generally meant a complete reimplementation. Now, it may be as simple as completion of testing. An ancillary to this is the education of a workforce that is more portable as well.
- The ability to leverage widely available web-related technologies within speech environments, supporting common techniques of scaling, architecture, and development practices. And although voice user interface design and telephony are important skills, many of the other skills required to implement IVR infrastructure are now more easily available due to the migration to a web-based architecture. This can also mean one less silo in an organization, reducing costs.
- A separation of interface presentation from business logic. It is now common to use the same business-level web services to support web, voice, and other channels.

These benefits, along with advancements in speech technology, have allowed the construction of more powerful voice applications while improving portability, maintainability, and interoperability. Although some of these advancements may have occurred without the development of speech related standards, it is likely that voice application development would have remained as a separate silo within the organization, requiring niche skills across the breadth of an implementation. There are also some useful lessons that may be taken from the W3C VBWG standards development experience:

- The standards themselves are important, but shouldn't be developed in a vacuum. The involvement of industry from the beginning ensured a set of standards that would meet real-world needs. The VoiceXML ecosystem provided important support for the acceptance of the standards developed within the W3C.
- A modular collection of standards can possibly support changes in technology over a longer period of time. While VoiceXML itself is modeled around a particular type of interaction (and is limited by the FIA in this regard), the supporting standards (SCXML, SRGS, and SSML) have provided value for other communication channels and interaction types. However, it is more difficult to advance multiple specifications simultaneously, and to ensure completion of a complete set meeting the original need.

The development of speech-related standards by the W3C, in combination with wide support—both through the W3C and the VoiceXML Forum—led to a transformation of the Interactive Voice Response industry. This transformation remains an important component in overall contact center modernization, and has aided in the advancement of voice application usage and usability.

References

1. W3C (1998). Voice Browsers, W3C Workshop, Cambridge, MA. https://www.w3.org/Voice/1998/Workshop/. Accessed 1 Mar 2016.
2. W3C (2016). Voice Browser Working Group. https://www.w3.org/Voice/. Accessed 1 Mar 2016.
3. W3C (2016). Multimodal Interaction Working Group. https://www.w3.org/2002/mmi/. Accessed 1 Mar 2016.
4. VoiceXML Forum (2016). http://www.voicexml.org/. Accessed 1 Mar 2016.
5. VoiceXML Forum (2000). Voice eXtensible Markup Language (VoiceXML) version 1.0. https://www.w3.org/TR/voicexml/. Accessed 1 Mar 2016.
6. VoiceXML Forum (2016). e-zine. http://www.voicexml.org/voicexml-review-archive/. Accessed 15 Mar 2016.
7. McGlashan, S., Burnett, D. C., Carter, J., Danielsen, P., Ferrans, J., Hunt, A., et al. (2004). Voice Extensible Markup Language (VoiceXML) version 2.0, W3C Recommendation. https://www.w3.org/TR/voicexml20/. Accessed 1 Mar 2016.
8. Hunt, A., & McGlashan, S. (2004). Speech Recognition Grammar Specification Version 1.0, W3C Recommendation. https://www.w3.org/TR/speech-grammar/. Accessed 1 Mar 2016.
9. Burnett, D. C., Walker, M. R., & Hunt, A. (2004). Speech Synthesis Markup Language (SSML) Version 1.0, W3C Recommendation. https://www.w3.org/TR/speech-synthesis/. Accessed 1 Mar 2016.
10. Oshry, M., Auburn, R. J., Baggia, P., Bodell, M., Burke, D., Burnett, D. C., et al. (2007). Voice Extensible Markup Language (VoiceXML) 2.1, W3C Recommendation. https://www.w3.org/TR/voicexml21/. Accessed 1 Mar 2016.
11. van Tichelen, L., & Burke, D. (2007). Semantic Interpretation for Speech Recognition (SISR) Version 1.0, W3C Recommendation. https://www.w3.org/TR/semantic-interpretation/. Accessed 1 Mar 2016.
12. Burnett, D. C., & Shuang, Z. W. (2010). Speech Synthesis Markup Language (SSML) Version 1.1, W3C Recommendation. https://www.w3.org/TR/speech-synthesis11/. Accessed 1 Mar 2016.
13. Baggia, P. (2008). Pronunciation Lexicon Specification (PLS) Version 1.0, W3C Recommendation. https://www.w3.org/TR/pronunciation-lexicon/. Accessed 1 Mar 2016.
14. Auburn, R. J. (2011). Voice Browser Call Control: CCXML Version 1.0, W3C Recommendation. https://www.w3.org/TR/ccxml/. Accessed 1 Mar 2016.
15. Larson, J. A. (2007). W3C speech interface language: VoiceXML. *IEEE Signal Processing Magazine, 4*(3), 126–130.
16. Jokinen, K., & McTear, M. (2009). *Spoken dialogue systems*. Princeton, NJ: Morgan & Claypool.
17. McGlashan, S., Burnett, D. C., Akolkar, R., Auburn, R. J., Baggia, P., Barnett, J., et al. (2010). Voice Extensible Markup Language (VoiceXML) Version 3.0, W3C Working Draft. https://www.w3.org/TR/voicexml30/. Accessed 1 Mar 2016.
18. Barnett, J., Akolkar, R., Auburn, R. J., Bodell, M., Carter, J., McGlashan, S., et al. (2015). State Chart XML (SCXML): State Machine Notation for Control Abstraction, W3C Recommendation. https://www.w3.org/TR/scxml/. Accessed 1 Mar 2016.
19. Harel, D. (1987). StateCharts: A visual formalism for complex systems. *Journal Science of Computer Programming, 8*(3), 231–274.
20. Brown, M. K., Kellner, A., & Raggett, D. (2001). Stochastic Language Models (N-Gram) Specification, W3C Working Draft. https://www.w3.org/TR/ngram-spec/. Accessed 1 Mar 2016.
21. Burnett, D. C. (2015). ALL: Thoughts and thanks as the VBWG comes to a close. W3C Mailing List Archive. https://lists.w3.org/Archives/Public/www-voice/2015JulSep/0029.html. Accessed 1 Mar 2016.

22. VoiceXML Forum (2016). VoiceXML Platform Certification Program. http://www.voicexml.org/certification-programs/voicexml-platform-certification-program/. Accessed 1 Mar 2016.
23. ECMA (2001). ECMAScript 3rd Edition Compact Profile. http://www.ecma-international.org/publications/files/ECMA-ST-WITHDRAWN/Ecma-327.pdf. Accessed 1 Mar 2016.
24. The Internet Engineering Task Force (IETF) (2016). https://www.ietf.org/. Accessed 1 Mar 2016.
25. Burnett, D., & Shanmugham, S. (2012). Media Resource Control Protocol Version 2 (MRCPv2), RFC 6787—Internet Standard. http://www.rfc-base.org/txt/rfc-6787.txt. Accessed 1 Mar 2016.
26. Burke, D. (2007). *Speech processing for ip networks: Media resource control protocol (MRCP)*. New York, NY: Wiley.
27. Johnston, M., Baggia, P., Burnett, D. C., Carter, J., Dahl, D. A., McCobb, G., et al. (2009). EMMA: Extensible MultiModal Annotation markup language, W3C Recommendation. https://www.w3.org/TR/emma/. Accessed 1 Mar 2016.
28. Axelsson, J., Cross, C., Lie, H. W., McCobb, G., Raman, T. V., Wilson, L. (2001). XHTML +Voice Profile 1.0, W3C Note. https://www.w3.org/TR/xhtml+voice/. Accessed 1 Mar 2016.
29. Microsoft Corporation, Speech Application Language Tags (SALT) (2003). Technical article. https://msdn.microsoft.com/en-us/library/ms994629.aspx. Accessed 1 Mar 2013.
30. Shires, G., & Wennborg, H. (2012). Web Speech API Specification, W3C Community Group Final Report. https://dvcs.w3.org/hg/speech-api/raw-file/tip/speechapi.html. Accessed 1 Mar 2016.
31. Barnett, J., Bodell, M., Dahl, D., Kliche, I., Larson, J., Porter, B., et al. (2012). Multimodal Architecture and Interfaces, W3C Recommendation. https://www.w3.org/TR/mmi-arch/. Accessed 15 Mar 2016.
32. Johnston, M., Dahl, D., Denney, T., & Kharidi, N. (2015). EMMA: Extensible MultiModal Annotation markup language Version 2.0, W3C Working Draft. https://www.w3.org/TR/emma20/. Accessed 15 Mar 2016.
33. Kistner, G., & Neurenberger, C. (2004). Developing user interfaces using SCXML statecharts. In *Proceedings of the 1st EICS Workshop on Engineering Interactive Computer Systems with SCXML*, pp. 5–11. http://tuprints.ulb.tu-darmstadt.de/4053/.
34. Almeida, N., Silva, S., & Teixeira, A. (2004). Multimodal multi-device application supported by an SCXML state chart machine. In *Proceedings of the 1st EICS Workshop on Engineering Interactive Computer Systems with SCXML*. pp. 12–17. http://tuprints.ulb.tu-darmstadt.de/4053/.
35. Schnelle-Walka, D., Radomski, S., Lager, T., Barnett, J., Dahl, D., Mühlhäuser, M. (Eds.) (2014). *Proceedings of the 1st EICS Workshop on Engineering Interactive Computer Systems with SCXML*. Darmstadt: TU Darmstadt.
36. Burkhardt, F., Schröder, M., Baggia, P., Pelachaud, C., Peter, C., & Zovato, E. (2014). Emotion Markup Language (EmotionML) 1.0, W3C Recommendation. https://www.w3.org/TR/emotionml/. Accessed 15 Mar 2016.
37. Schnelle-Walka, D., Radeck-Arneth, S., & Striebinger, J. (2015). Multimodal dialog management in a smart home context with SCXML. In *Proceedings 2nd Workshop on Engineering Interactive Systems with SCXML*, Duisburg, DE.
38. López, G., Peláez, V., González, R., & Lobato, V. (2011). *Voice control in smart homes using distant microphones: A VoiceXML-based approach, in ambient intelligence*. Lecture Notes in Computer Science (Vol. 7040) (pp. 172–181). Berlin/Heidelberg: Springer.
39. Teixeira, A., Almeida, N., Pereira, C., & Oliveira, M. (2013). *W3C MMI architecture as a basis for enhanced interaction for ambient assisted living*. New York, NY: W3C Workshop on Rich Multimodal Application Development.
40. Sigüenza, A., Blanco, J. L., Bernat, J., & Hernández, L. A. (2010). Using SCXML for semantic sensor networks. In *Proceedings of the 3rd International Workshop on Semantic Sensor Networks (SSN10)*. Workshop at the 9th International Semantic Web Conference (ISWC2010) - ISWC 2010 Workshops Volume V, Shanghai, China, pp. 33–48. http://ceur-ws.org/Vol-668/.

41. Radomski, S., & Schnelle-Walka, D. (2012). VoiceXML for pervasive environments. *International Journal of Mobile Human Computer Interaction, 4*(2), 18–36.
42. Schnelle-Walka, D., Radomski, S., & Mühlhäuser, M. (2015). Modern standards for VoiceXML in pervasive multimodal applications. In J. Lumsden (Ed.), *Emerging perspectives on the design, use, and evaluation of mobile and handheld devices*. IGI Global: http://www.igi-global.com/book/emerging-perspectives-design-use-evaluation/125520
43. Bühler, D., & Hamerich, S. W. (2005). Towards VoiceXML compilation for portable embedded applications in ubiquitous environments. In *Proceedings of Interspeech 2005*, Lisbon, PT, pp. 3397–3400. http://www.isca-speech.org/archive/interspeech_2005/i05_3397.html; http://www.isca-speech.org/archive/interspeech_2005/index.html.
44. Oshry, M., Adeeb, R., Baggia, P., Blackman, A., Bodell, M., Burke, D., et al. (2004). VoiceXML 2.0 Implementation Report. https://www.w3.org/Voice/2004/vxml-ir/. Accessed 1 Mar 2016.
45. Shanmugham, S., Monaco, P., & Eberman, B. (2006). A Media Resource Control Protocol (MRCP), RFC 4463—Informational. https://tools.ietf.org/html/rfc4463. Accessed 1 Mar 2016.

Chapter 3
Extensible Multimodal Annotation for Intelligent Interactive Systems

Michael Johnston

Abstract Multimodal interactive systems enabling combination of natural modalities such as speech, touch, and gesture make it easier and more effective for users to interact with applications and services, whether on mobile devices, or in smart homes or cars. However, building these systems remains a complex and highly specialized task, in part because of the need to integrate multiple disparate and distributed system components. This task is further hindered by proprietary representations for input and output to different types of modality processing components such as speech recognizers, gesture recognizers, natural language understanding components and dialog managers. The W3C EMMA standard addresses this challenge and simplifies multimodal application authoring by providing a common representation language for capturing the interpretation of user inputs and system outputs and associated metadata. In this chapter, we describe the EMMA markup language and demonstrate its capabilities through presentation of a series of illustrative examples.

3.1 Introduction

Multimodal interfaces that allow users to provide input using different modes such as speech, touch, and gesture and support system responses combining speech, graphics, and other modes enable the creation of more natural and effective interactive systems for accessing information and services [1–3]. The critical property of these systems is that, since different modes offer different affordances for expressing content, through multimodality the user (or system) is empowered to use the mode or combination of modes best suited to the specific information to be conveyed or task to be completed [4, 5]. Having more than mode available also allows the user to switch among modes to overcome recognition problems, such as errorful speech recognition in a noisy environment [6], or to adapt to the particular physical or social environment. For example, the user might switch from voice to typing for entering sensitive information such as a credit card number or user ID.

M. Johnston (✉)
Interactions Corporation, New York, NY, USA
e-mail: mjohnston@interactions.com

D.A. Dahl (ed.), *Multimodal Interaction with W3C Standards*,
DOI 10.1007/978-3-319-42816-1_3

Interaction with mobile devices such as smart phones and tablets has always been a central use case for multimodal interaction and one that has become even more important given the widespread availability of these devices and the ubiquity of high speed mobile data networks. Increasingly though, multimodality has a broader range of applicability which extends to the connected car, control of smart devices in the home (and the Internet of Things (IOT) more broadly), interaction with media systems, wearable computers, control of virtual environments, and interaction with social and assistive robots.

Numerous multimodal prototypes have been developed across the years [7–19], and we are now starting to see deployment of truly multimodal systems [20, 21]. However, despite the increasing demand for and applicability of multimodal systems, building and maintaining them remains a complex and highly specialized task. One of the key challenges is that these systems typically involve integrating multiple different subcomponents, such as speech recognition, gesture recognition, natural language understanding, multimodal fusion, dialog modeling, and natural language generation. Communication among these components is not standardized and frequently involves a combination of ad hoc and proprietary protocols. As a result, it is difficult or impossible to plug-and-play components from different vendors or research sites, limiting the ability of authors to rapidly pull components together to prototype and develop effective solutions. The advent of cloud-based APIs for some of the technologies including speech recognition and natural language understanding is lowering the barrier of entry for development of spoken and multimodal interactive systems but further illustrates the integration problem as all of the APIs currently provide results in different formats with different encoding of metadata.

The extensible multimodal annotation (EMMA) markup language described in this chapter addresses this problem by providing a standardized common language for representing the processing of inputs to (and outputs from[1]) spoken and multimodal interactive systems. EMMA provides a representation language for encapsulating and annotating the interpretation of inputs to a multimodal system. Critically, the EMMA language does not standardize the semantic representation itself. For example, there is no standard markup for particular commands or intents, e.g., <search_flight/> for an airline query. The semantic representation remains application specific. EMMA instead provides a standard set of containers for different possible interpretations and sequences and groups of interpretations. The language also provides a set of attributes and elements for capturing common metadata regarding inputs and their interpretation such as timestamps, confidence measures, type of media involved, reference to the location of a signal file, and description of the device used to capture input. In essence, EMMA provides the "glue" that connects together the components of a spoken or multimodal interactive system. We present a multimodal architecture here, but note that EMMA also has

[1] The W3C recommendation EMMA 1.0 only addresses inputs. Proposals for EMMA 2.0 extend the standard to represent output processing.

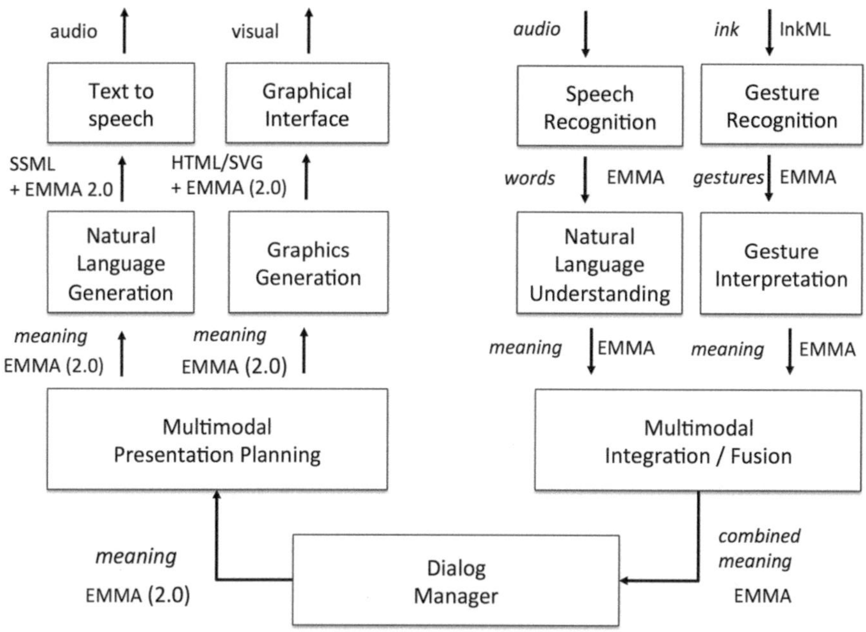

Fig. 3.1 Role of EMMA in a multimodal interactive system

significant utility for unimodal architectures as they also involve multiple components. Figure 3.1 lays out the common components of an interactive multimodal system and indicates where EMMA can be used for communication among components.

The EMMA language became a W3C Recommendation in 2009 [22]. The standardized form of EMMA is XML. In ongoing work, which we discuss later in the chapter, JavaScript Object Notation (JSON) formulations of the EMMA standard are being developed. Also, the EMMA language has provisions for carrying non-XML semantic payloads. While the initial formulation of EMMA only addressed input, in more recent proposals for an EMMA 2.0 [23], extensions to the language include support for using EMMA for expressing the stages of processing of output from spoken and multimodal systems. We also discuss this in Sect. 3.5 below.

EMMA is the language used to represent the interpretation of inputs by modality components in the W3C multimodal architecture specification [24]. The `DoneNotification` event in the multimodal architecture embeds an EMMA document in the `<mmi:Data>` element.

Historically for telephony-based spoken language systems there has been some degree of standardization. Natural Language Semantic Markup Language (NLSML) [25] was developed and published as a W3C working draft. NLSML provides for a basic set of metadata including confidence scores indication of grammar used, timestamps, and mode. NLSML was focused on speech input and has limited capability for expressing ambiguity and capturing multimodal inputs and more detailed metadata. It was adopted as the required result format for the Media Resource Control Protocol (MRCP) [26]. MRCP in turn has become the dominant standard in the space of interactive voice response (IVR) systems for accessing speech resources such as speech recognition servers. EMMA was developed within W3C as the direct replacement for NLSML and offers the capabilities needed for results for spoken language systems and for multimodal systems. The second version of MRCP, MRCP V2 [27] specifies that results may be returned in either NLSML or in the EMMA markup language.

In Sect. 3.2 below, we outline the basics of the EMMA language and provide illustrative examples showing the key metadata that can be captured. In Sect. 3.3 we focus specifically on the capabilities of the language for capturing the uncertainty and ambiguity that is characteristic of input through natural modalities to interactive systems. Here we discuss representations of N-best lists and lattices and describe the scope of annotations over more complex EMMA documents that express multiple different possible hypotheses. In Sect. 3.4, we discuss the capabilities of the language for expressing sequences of inputs and grouping inputs, including examples of multimodal inputs, and the built-in mechanisms in the language for expressing and cross-referencing multiple stages of processing of input within a single EMMA document or across several EMMA documents. In Sect. 3.5, we present recent work on extending the EMMA language to support representation of the processing and realization of system output. In Sect. 3.6, we discuss recent work on alternative formats such as JSON for the EMMA language. Section 3.7 concludes the chapter.

3.2 The Basics of EMMA

The EMMA language is an XML markup that provides mechanisms for capturing and annotating the results of the various stages of processing of user inputs. While in EMMA 2.0 the language is extended to system output, in the rest of Sects. 3.2, 3.3, and 3.4 we limit the discussion to input as specified in the EMMA 1.0 recommendation [22].

There are two key aspects to the language: a series of elements (e.g., <emma:interpretation>, <emma:group>, <emma:one-of>, <emma:sequence>, <emma:lattice>) that are used as containers for interpretations of the user's inputs, and a series of annotation attributes and elements which are used to provide standardized access to various pieces of common metadata associated with those inputs. These annotations include timestamps (emma:start, emma:end) and confidence score values (emma:confidence). Annotations with simple values, and for which there can only be one per interpretation such as timestamps and confidence scores are captured as attributes. Other annotations such as <emma:derived-from>, <emma:model>, and <emma:grammar> for which there may be more than one per interpretation or which have internal structure are captured as elements.

Given the broad range of input types to be supported, a critical design feature of EMMA is that it does not attempt to standardize the semantic representation assigned to inputs; rather it provides a series of standardized containers for mode and application specific markup, and a set of standardized annotations for common metadata. The language enables extensibility through the <emma:info> element, which is a container for application or vendor specific annotations on inputs. Note that individual EMMA documents are not intended to be authored directly by humans; rather they are generated automatically by system components such as speech recognizers, image recognizers, natural language understanding components, and multimodal fusion engines. They are, however, intended to be manipulated and read by humans in logging and annotation tools.

To make this more concrete, we consider an illustrative example. The EMMA document in Fig. 3.2 is an example of EMMA markup that might be produced by a natural language understanding component in an interactive spoken or multimodal system for making travel reservations. In this use case, the user is interacting with a speech-enabled mobile application and has requested information about flights from Boston to Denver.

```
<emma:emma version="2.0"
    xmlns:emma="http://www.w3.org/2003/04/emma"
    xmlns:xsi="http://www.w3.org/2001/XMLSchema-instance"
    xsi:schemaLocation="http://www.w3.org/2003/04/
    emma http://www.w3.org/TR/2009/REC-emma-20090210/emma.xsd"
    xmlns="http://www.example.com/example">
  <emma:interpretation id="int1"
     emma:medium="acoustic" emma:mode="voice"
     emma:function="dialog" emma:verbal="true"
     emma:start="1241035886246" emma:end="1241035889306"
     emma:duration="3060" emma:lang="en-US"
     emma:tokens="flights from boston to denver"
     emma:confidence="0.75"
     emma:source="smm:platform=iPhone-6s-ios9.2"
     emma:signal="smm:file=audio-416120.amr"
     emma:signal-size="4902"
     emma:media-type="audio/amr; rate=8000"
     emma:process="smm:type=asr&version=asr_eng2.4"
     emma:grammar-ref="gram1"
     emma:model-ref="model1">
      <flight>
         <orig>Boston</orig>
         <dest>Denver</dest>
      </flight>
  </emma:interpretation>
  <emma:info>
   <session>E50DAE19-79B5-44BA-892D</session>
  </emma:info>
  <emma:grammar id="gram1" ref="smm:grammar=flights"/>
  <emma:model id="model1" ref="smm:file=flights.xsd"/>
</emma:emma>
```

Fig. 3.2 EMMA example: voice input

All EMMA documents start with the root element <emma:emma>. This has an attribute indicating the version of EMMA along with standard XML namespace and schema declarations. To simplify presentation, in the rest of the examples in the chapter the namespace and schema information will be left out. Within <emma:emma> we find the core of an EMMA document: a tree of container elements <emma:one-of>, <emma:group>, and <emma:sequence>, terminating in a number of <emma:interpretation> elements indicating individual interpretations of the input.

In this initial example, we present the most basic case where there is a single <emma:interpretation> element containing application specific markup resulting from processing a single user input. In this example, the EMMA document is produced by a natural language understanding (NLU) component. The semantic representation assigned by the NLU consists of an XML element <flight> containing two sub elements, <orig/> and <dest/>, specifying the origin and destination cities, respectively, for a flight query, in this case "Boston" and "Denver."

Interpretation elements are required to have an <id> attribute that can be used to refer to the specific interpretation. In addition to the <id> attribute there are a series of annotation attributes which provide detailed metadata associated with the input and its interpretation. Some of these are required, such as emma:medium and emma:mode, while others such as emma:signal and emma:confidence are optional.

The first set of these annotation attributes provide a simple classification of the input. The attributes emma:medium and emma:mode provide a classification of the user input medium and the specific modality, respectively. The values for emma:medium are from the closed set of values "acoustic," "tactile," and "visual." In this example, the emma:medium value is "acoustic" and the specific modality is "voice." The values for mode are from an open set including the values: "voice," "ink," "gui," "keys," "video," and "photograph." Multimodal inputs will have multiple values, space separated, within their medium and mode attributes. The emma:function attribute differentiates interactive dialog "dialog" from other uses such as recording and verification. The attribute emma:verbal takes a boolean value indicating whether the input is verbal language. This is used to distinguish, for example, handwriting (letters) from freehand drawing (lines, areas) in pen input. Spoken language input, as in this case, will be emma:verbal="true."

Timestamps are a key type of metadata for different modes and are frequently used as a constraint an multimodal fusion strategies. EMMA provides both absolute and relative timestamp mechanisms. In the example, the attributes emma:start and emma:end are absolute timestamps indicating the start and end of the user input signal in standard Unix timestamps (milliseconds from Jan 1 1970). Temporal information can also be specified relative to other signals or to, for example, the beginning of an interaction or session. emma:time-ref-uri specifies what the time is relative to and emma:offset-to-start the offset in milliseconds. In the case where the reference item is an interval, emma:time-ref-anchor-point is used to indicate whether the offset is from the "start" or "end." There is also an emma:duration attribute which specifies the duration of an input in milliseconds, 3060 ms in our example.

The language of the input is captured using the emma:lang attribute which takes as values the standard language identifiers defined in IETF BCP 47 [28]. In the example, "en-US" indicates that this spoken input was in US English. If instead the input was in French language ("des vols de Boston á denver"), the value would be "fr". Note that emma:lang indicates the language of the actual input signal. For example, if the interpretation was the result of machine translation from French to English, the emma:lang would remain "fr" because that was the language of the original signal. This is in contrast to xml:lang which is used to indicate the language of characters within the element it appears on.

The attribute emma:tokens indicates the strings of tokens that were processed in determining the interpretation. In our example of spoken input, these tokens are words. In other modalities, such as gesture, the token stream might contain symbols representing gestures. The emma:confidence attribute is used to annotate the degree of certainty assigned to the interpretation by the processor generating the interpretation. emma:confidence takes a value between 0 and 1.[2]

The next bundle of annotations provide information about the signal processed. The emma:source attribute contains a Uniform Resource Identifier (URI) which provides a characterization of the source of the input, such as the device, or microphone, or camera used to capture the input signal. In this example, the device and platform is specified. emma:signal is used to provide a pointer to the input signal file. In our case this is an audio file. In the case of a visual input the signal might be an image or video file. The size of the signal can be specified using the emma:signal-size attribute those value indicates the size of the signal file in 8-bit octets. The emma:media-type attribute contains the MIME type of the signal, and provides a convenient location for specifying the codec and sampling rate (AMR encoded audio at 8000 Hz in this case).

The emma:process attribute is URI valued and provides a description of the process which resulted in the current interpretation. In Fig. 3.2, it indicates that the process was speech recognition (asr) and specifies the recognizer version.

The next two attributes, emma:grammar-ref and emma:model-ref, are used to provide references to the annotation elements <emma:grammar> and <emma:model>. The <emma:grammar> element, which appears as a child of <emma:emma>, is used to provide a reference or inline specification of a grammar used in generating the interpretation. Similarly the element <emma:model> is used for inline specification or reference to the data model of the semantic representation within the <emma:interpretation>. Both grammars and models are handled in this way with an emma:*-ref attribute and an element in order to support multiple models or grammars being associated with an

[2] The EMMA language does not require the confidence score to be a probability and there is no expectation or requirement that confidence values are comparable across different producers of EMMA other than that values closer to 1 are higher in confidence while values closer to 0 are lower in confidence.

```
<emma:emma version="2.0"
  <emma:interpretation id="int2"
     emma:medium="visual"
     emma:mode="photograph"
     emma:function="dialog"
     emma:verbal="false"
     emma:start="1241035896246"
     emma:end="1241035896246"
     emma:confidence="0.90"
     emma:source="smm:platform=iPhone-5s-ios9.3"
     emma:signal="smm:file=image123.jpg"
     emma:signal-size="38264"
     emma:media-type="image/jpeg"
     emma:process="smm:type=image_classification"
     emma:model-ref="model2">
      <image>
         <type emma:confidence="0.95">ball</type>
         <color emma:confidence="0.8">green</color>
      </image>
  </emma:interpretation>
  <emma:info>
   <width unit="px">300</width>
   <height unit="px">200</height>
  </emma:info>
  <emma:model id="model2" ref="smm:file=image.xsd"/>
</emma:emma>
```

Fig. 3.3 EMMA example: visual input

interpretation. This also supports use cases where there are multiple interpretations and each is associated with specific models and grammars.

In addition to the standardized set of annotations in EMMA, there is also a mechanism for adding vendor and application specific metadata. These are placed within the `<emma:info>` element which may appear either under `<emma:emma>` or within a specific `<emma:interpretation>` or other container. In our example, `<emma:info>` is used to introduce a vendor specific session identifier. In the proposal for EMMA 2.0, an `emma:info-ref` attribute is introduced to enable association of `<emma:info>` elements with specific interpretations by reference.

We now consider another example where EMMA is used instead for representation of the output of an image recognizer (Fig. 3.3). In this case, the camera on a mobile device has been used to take a picture as part of an interactive application. The medium is "`visual`" and mode "`photograph`." Function is "`dialog`" as this is an input to an interactive system, and `emma:verbal` is "`false`." The signal in this case is a jpg file. Instead of speech recognition, in this case the process is image classification. In this example, `<emma:info>` is being used to provide vendor specific annotations of the width and height in pixels of the image being processed. The interpretation in this case is application specific XML specifying that the result was classification of the image as a ball of color green. The overall confidence for the interpretation is given in the `emma:confidence` on the `<emma:interpretation>` element. We illustrate here an additional

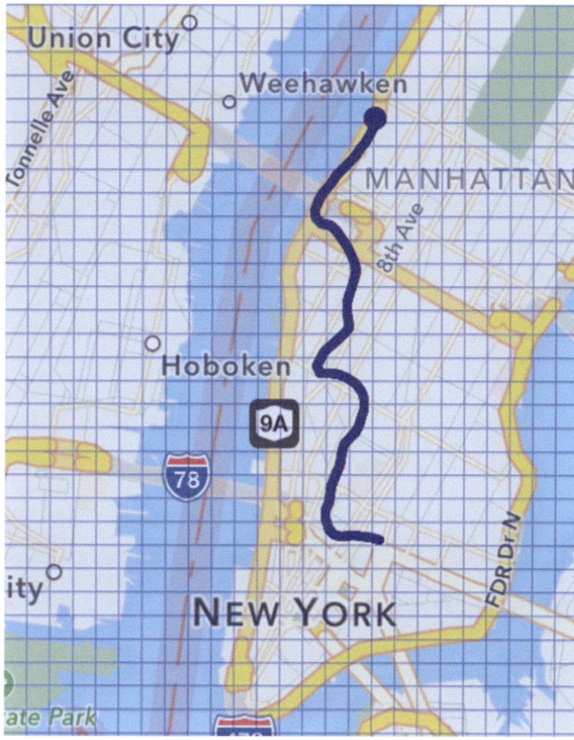

Fig. 3.4 Example of ink input on map display

capability of the mechanism in that `emma:confidence` can also appear directly on elements in the application specific markup, in this case indicating the confidence of the classification of the overall type of object and the confidence in the color classification. Another attribute which can apply directly to application markup is `emma:tokens`, enabling specification of the specific words that resulted in particular parameters in the semantic representation.

As a third and final example we consider input in the tactile medium. In this case the user draws a route on a map on a touch screen with their finger as part of interaction with a mobile application (Fig. 3.4). The EMMA interpretation is the result of gesture classification of the ink trace (Fig. 3.5).

In this example, the mode is "`ink`" and function is "`dialog`." A gesture line is non-verbal so `emma:verbal`="`false`." In the case of handwritten words, `emma:verbal` would be `true`. The signal in this example is an XML file containing InkML representation of the input ink trace made on the map [29]. InkML has a specific registered MIME type captured in `emma:media-type`. In our example interpretation, there is application specific XML markup indicating that the recognition result is of type "`line`" and the specific set of coordinates. In this case, the confidence in the interpretation is fairly low "`0.20`".

```
<emma:emma version="2.0"
  <emma:interpretation id="int3"
     emma:medium="tactile"
     emma:mode="ink"
     emma:function="dialog"
     emma:verbal="false"
     emma:start="1241035906246"
     emma:end="1241035909246"
     emma:duration="3000"
     emma:confidence="0.20"
     emma:source="smm:platform=iPhone-6s-ios9.2&device=touchscreen"
     emma:signal="smm:file=ink_3.xml"
     emma:signal-size="382"
     emma:media-type="application/inkml+xml"
     emma:process="smm:type=gesture_recognition"
     emma:model-ref="model3">
      <location>
        <type>line</type>
        <coords>[(30,34),(32,38),(34,42)…</coords>
      </location>
  </emma:interpretation>
  <emma:model id="model3" ref="smm:file=location.xsd"/>
</emma:emma>
```

Fig. 3.5 EMMA example: tactile input

EMMA provides a couple of specific annotations for cases with missing or uninterpretable input. An EMMA result can signal the absence of input with emma:no-input="true" on the EMMA interpretation:

```
<emma:interpretation id="id4" emma:medium="acoustic" emma:mode="voice" emma:no-input="true"/>.
```

Cases where there is an input but it is considered to be uninterpretable are marked as follows:

```
<emma:interpretation id="id5" emma:medium="tactile" emma:mode="ink" emma:uninterpreted="true"/>.
```

Interpretations marked as emma:uninterpreted="true" may result from, for example, the inability of a parser to assign a parse to string (in the case of natural language parsing or understanding), or in other cases where there is a result but its confidence falls below some pre-defined threshold. For example, if for the gesture recognizer producing the EMMA in Fig. 3.5 there was a confidence threshold on gesture recognition of "0.25" the recognizer would instead return an empty <emma:interpretation> with emma:uninterpreted="true."

3.3 Capturing Ambiguity and Non-determinacy

One of the key characteristics and challenges of input modes such as speech, gesture, and computer vision, employed in multimodal interactive systems, is that despite dramatic performance improvements in recent years these natural input modes remain a source of error and furthermore user inputs may be ambiguous and uncertain. One aspect of EMMA that addresses this is availability of confidence annotations both at the interpretation level and within specific semantic markup. These scores can be used by a dialog management component to drive rejection and confirmation strategies.

However, it is often the case that errors resulting from taking the top scoring interpretation of a mode can be resolved later either through fusion with content from other modes [3, 8, 20, 30] or through use of dialog and contextual information to rescore multiple hypotheses [31]. Furthermore, spoken dialog management and multimodal interface techniques can utilize multiple hypotheses in order to clarify uncertain inputs directly with the user. For example, in a multimodal interface for a task such as recognition of a spoken name, multiple hypotheses can be presented to the user in list form allowing them to select the desired result [32]. In addition, in voice search applications multiple recognition results can be used to retrieve search results. See [33] for one illustration of the use of multiple ASR results in a question answering task.

Given these motivations, a critical component in the design of the EMMA language is support for representation of multiple possible interpretations of user input. The `<emma:one-of>` container provides the core mechanism for representation of disjunction of possible interpretations. The language also supports direct representation of lattices, graphs which compactly represent multiple different possible interpretations of the input. We present first the use of `<emma:one-of>` to represent N-best lists of possible interpretations. Like `<emma:interpretation>`, `<emma:one-of>` is a container element and can appear as a child of `<emma:emma>`. In turn, `<emma:one-of>` itself contains a list of other container elements. In the example in Fig. 3.6, we return to the speech input use case from Sect. 3.2 of input to a flight information system, and show it with N-best results. Each separate result is in an `<emma:interpretation>` under `<emma:one-of>`. This captures the fact that there were three competing interpretations of the spoken input, a flight query from Boston to Denver, a flight query from Austin to Denver, or a query to search for stores in downtown Austin.

Much of the metadata associated with the input, such as the classification into medium and mode, the timing information, signal, media type, language, and process is more about the signal and its handling and is common to all of the interpretations. To avoid having to repeat annotations that apply across multiple interpretations, EMMA allows for annotations to appear on the `<emma:one-of>` directly and defines a scoping relation on annotations. Annotations on an `<emma:one-of>` element are assumed to apply to all of the interpretations within the `<emma:one-of>`. Certain annotations are not shared and are specific

```
<emma:emma version="2.0">
  <emma:one-of id="r1"
     emma:medium="acoustic" emma:mode="voice"
     emma:function="dialog" emma:verbal="true"
     emma:start="1241035886246"
     emma:end="1241035889306"
     emma:source="smm:platform=iPhone-6s-ios9.2"
     emma:signal="smm:file=audio-416120.amr"
     emma:signal-size="4902"
     emma:process="smm:type=asr&version=asr_eng2.4"
     emma:media-type="audio/amr; rate=8000"
     emma:lang="en-US">
    <emma:interpretation id="int1"
    emma:confidence="0.75"
    emma:tokens="flights from boston to denver">
      <search_flight><orig>Boston</orig>
                     <dest>Denver</dest></search_flight>
    </emma:interpretation>
    <emma:interpretation id="int2"
      emma:confidence="0.68"
      emma:tokens="flights from austin to denver">
       <search_flight><orig>Austin</orig>
                     <dest>Denver</dest></search_flight>
    </emma:interpretation>
    <emma:interpretation id="int3"
      emma:confidence="0.28"
      emma:tokens="florists in Austin downtown">
       <search_store><category>florists</orig>
                    <loc>downtown austin</loc></search_store>
    </emma:interpretation>
  </emma:one-of>
</emma:emma>
```

Fig. 3.6 EMMA representation of N-best lists: <emma:one-of>

to individual interpretations. For example, for speech recognition results these are the confidence scores and <emma:tokens>. In certain applications, emma:lang might also be specific to some interpretations if the system in question recognizes multiple languages. If a confidence measure is present the interpretations must be listed in order of confidence, best first.

The contents of <emma:one-of> can also be the other container elements such as <emma:group> and <emma:sequence>. <emma:one-of> elements can be nested to structure a set of interpretations based on the source of the ambiguity. For example, an N-best list of recognition results where individual ASR strings have multiple possible interpretations can be represented as a list of <emma:one-of> elements embedded within an <emma:one-of>. Figure 3.7 contains an example from an interactive mobile assistant where the string is recognized as either “action movies tonight” or “action movies tomorrow” and the natural language understanding (NLU) assigns each one an interpretation as either a search for movies in theaters or a search for movies on television. The attribute “disjunction-type” on <emma:one-of> can be used to annotate the source of the ambiguity at each level of the embedding.

In some use cases, these representations of multiple hypotheses will be consumed by a dialog manager and contextual information may be used in order to

```
<emma:emma version="2.0">
  <emma:one-of id="r1" disjunction-type="recognition"
      emma:medium="acoustic" emma:mode="voice"
      emma:function="dialog" emma:verbal="true"
      emma:start="1241035886246"
      emma:end="1241035889306"
      emma:process="smm:type=asr&version=asr_eng2.4"
      emma:lang="en-US">
      <emma:one-of id="d1" disjunction-type="understanding"
        emma:tokens="action movies tonight"
        emma:process="smm:type=nlu&version=3.7">
        <emma:interpretation id="int1"
           emma:confidence="0.75">
           <search_tv><genre>action movies</genre>
                  <time>tonight</time></search_tv>
        </emma:interpretation>
        <emma:interpretation id="int2"
           emma:confidence="0.68">
           <search_theaters><genre>action</genre>
                  <time>tonight</time></search_theaters>
        </emma:interpretation>
      </emma:one-of>
      <emma:one-of id="d2" disjunction-type="understanding"
        emma:tokens="action movies tomorrow"
        emma:process="smm:type=nlu&version=3.7">
        <emma:interpretation id="int3"
           emma:confidence="0.9">
           <search_tv><genre>action movies</genre>
                  <time>tomorrow</time></search_tv>
        </emma:interpretation>
        <emma:interpretation id="int4"
           emma:confidence="0.6">
           <search_theaters><genre>action</genre>
                  <time>tomorrow</time></search_theaters>
        </emma:interpretation>
      </emma:one-of>
  </emma:one-of>
</emma:emma>
```

Fig. 3.7 EMMA embedded <emma:one-of>

select among hypotheses. For example, considering Fig. 3.7, knowledge of the preceding conversation, such as whether the user has been searching for television vs. planning an evening out can be used to select among the hypotheses. N-best hypotheses can also be used to drive clarification strategies, for example, in speech if given the context we assume this is a television programming search, given the remaining ambiguity on the time value, the system might construct a targeted clarification question [34, 35] such as "action movies *when*?". In a multimodal context, the system could instead decide to present multiple options to the user visually. In the example in Fig. 3.8, the <emma:one-of> contains a series of different recognition results of different person names and an associated language understanding (tagging of first and last name). These different options can be presented directly to the user as a visual menu as in Fig. 3.9 with an accompanying prompt "Who were you trying to call?". This kind of multimodal confirmation is

```
<emma:emma version="2.0">
 <emma:one-of id="one-of1"
  emma:medium="acoustic" emma:mode="voice"
  emma:function="dialog" emma:verbal="true" emma:lang="en-US"
  emma:start="1241641821513" emma:end="1241641823033">
  <emma:interpretation id="nbest1"
    emma:confidence="1.00" emma:tokens="jon smith">
    <fn>jon</fn><ln>smith</ln></emma:interpretation>
  <emma:interpretation id="nbest2"
    emma:confidence="0.99" emma:tokens="john smith">
    <fn>john</fn><ln>smith</ln></emma:interpretation>
  <emma:interpretation id="nbest3"
  emma:confidence="0.99" emma:tokens="joann smith">
    <fn>joann</fn><ln>smith</ln></emma:interpretation>
  <emma:interpretation id="nbest4"
  emma:confidence="0.98" emma:tokens="joan smith">
    <fn>joan</fn><ln>smith</ln></emma:interpretation>
 </emma:one-of>
</emma:emma>
```

Fig. 3.8 N-best lists of names represented in <emma:one-of>

Fig. 3.9 Visual interface for disambiguation

particularly effective on mobile devices with a touch display where a spinnable list can easily be used to pick among even quite a long list of alternatives [32].

In addition to N-best lists, EMMA also provides a mechanism for direct representation of lattices or confusion networks. Lattices and confusion networks provide a more compact representation and are particularly useful when the range of alternatives is large. A lattice is a directed graph which represents a series of possible interpretations of the input signal. A lattice has a start state and one or more end states, and symbols such as words label transitions between intervening states. Each path through the lattice represents a possible recognition result.

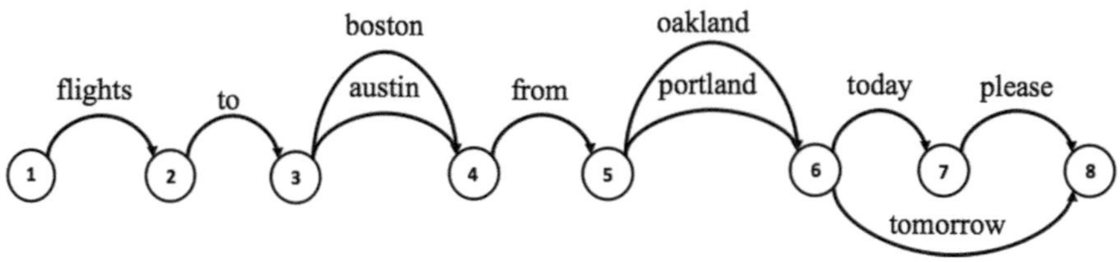

Fig. 3.10 Speech recognition lattice example

```
a. flights to boston from portland today please
b. flights to austin from portland today please
c. flights to boston from oakland today please
d. flights to austin from oakland today please
e. flights to boston from portland tomorrow
f. flights to austin from portland tomorrow
g. flights to boston from oakland tomorrow
h. flights to austin from oakland tomorrow
```

Fig. 3.11 N-best list of strings captured by lattice

A confusion network (or word confusion network) is a special case of a lattice that presents alternatives word by word. The alternatives in lattices and confusion networks are often weighted.

Lattices are a commonly available output from speech recognizers, and are also useful for other modes including representation of the possible interpretations of gesture input. Consider first an example from speech recognition. If the user has said "flights to boston from Portland tomorrow," a possible resulting speech recognition lattice can be represented graphically as a finite state automaton as in Fig. 3.10.

This representation compactly represents a list of eight strings, enumerated in Fig. 3.11, each a different path through the lattice. Lattices are encoded in EMMA using the `<emma:lattice>` element, which includes `initial` and `final` attributes to capture the start and end states of the lattice. Each arc in the lattice is captured in an `<emma:arc>` element with "`from`" and "`to`" attributes giving the origin state and destination state for the arc. Arcs can also be weighted using the `emma:cost` attribute. The example in Fig. 3.12 shows the EMMA representation of the lattice in Fig. 3.10.

The lattice representation capability also has important applications for other modes such as gestures made with a stylus or hand motion. We present here an example drawn from Johnston and Bangalore [36] showing the representation of the possible different interpretations of ink input to a mobile application as a finite state machine. After gesture and handwriting recognition, the ink trace in Fig. 3.13 (left) could be assigned multiple different interpretations. These are represented in the finite state automaton in Fig. 3.13 (right). The path "G hw 0" represents recognition of the input as a handwritten digit "0". The path "G area sel 2 rest SEM(r12,r15)"

```
<emma:emma version="2.0">
  <emma:interpretation id="interp1"
    emma:medium="acoustic" emma:mode="voice">
    <emma:lattice initial="1" final="8">
      <emma:arc from="1" to="2">flights</emma:arc>
      <emma:arc from="2" to="3">to</emma:arc>
      <emma:arc from="3" to="4">boston</emma:arc>
      <emma:arc from="3" to="4">austin</emma:arc>
      <emma:arc from="4" to="5">from</emma:arc>
      <emma:arc from="5" to="6">portland</emma:arc>
      <emma:arc from="5" to="6">oakland</emma:arc>
      <emma:arc from="6" to="7">today</emma:arc>
      <emma:arc from="7" to="8">please</emma:arc>
      <emma:arc from="6" to="8">tomorrow</emma:arc>
    </emma:lattice>
  </emma:interpretation>
</emma:emma>
```

Fig. 3.12 EMMA representation of speech recognition lattice with <emma:lattice>

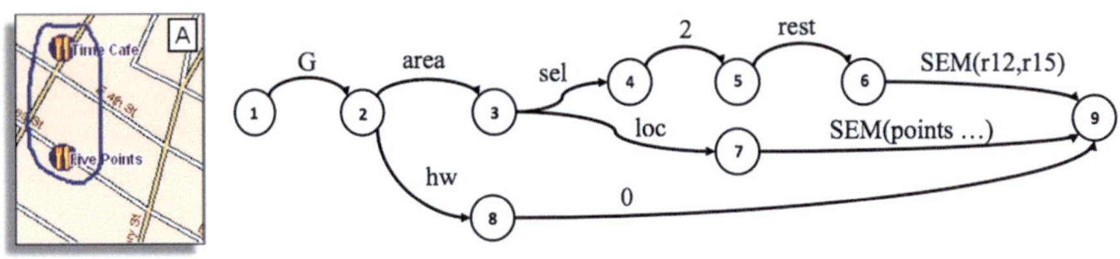

Fig. 3.13 Example of ink input and lattice representation

represents the interpretation of this ink gesture as a selection of two restaurants. The remaining path "G area loc SEM(points..)" represents recognition of the gesture as an area which indicates a particular geographic extent on the map. In a multimodal system with combined multimodal commands, the restaurant selection path might combine with a speech command such as "compare these restaurants" while the location path might combine with "zoom in here." For more details on this representation and its use to support interactive multimodal systems see Johnston and Bangalore [36]. This representation of ink gesture can be captured in EMMA markup using the <emma:lattice> element as in Fig. 3.14.

3.4 Groups, Sequences, and Derivations

Groupings of inputs and sequences of inputs can be represented using the remaining container elements <emma:group> and <emma:sequence> each of which itself can contain multiple container elements.

```
<emma:emma version="2.0">
  <emma:interpretation id="interp2"
    emma:medium="tactile" emma:mode="ink">
    <emma:lattice initial="1" final="9">
      <emma:arc from="1" to="2">G</emma:arc>
      <emma:arc from="2" to="3">area</emma:arc>
      <emma:arc from="3" to="4">sel</emma:arc>
      <emma:arc from="4" to="5">2</emma:arc>
      <emma:arc from="5" to="6">rest</emma:arc>
      <emma:arc from="6" to="9">SEM(r12,r15)</emma:arc>
      <emma:arc from="3" to="7">loc</emma:arc>
      <emma:arc from="7" to="9">SEM(points..)</emma:arc>
      <emma:arc from="2" to="8">hw</emma:arc>
      <emma:arc from="8" to="9">0</emma:arc>
    </emma:lattice>
  </emma:interpretation>
</emma:emma>
```

Fig. 3.14 EMMA representation of ink lattice

One key use case for `<emma:group>` is for containing bundles of multimodal input, which in turn may then be passed to a multimodal integration component in order to determine their combined interpretation. In the example in Fig. 3.15 there has been a spoken input "traffic along this route" and the user has drawn a line on a visual display that is captured as an ink trace (for an example of this kind of ink input see Fig. 3.4 above). This package of multimodal input can be captured in `<emma:group>` as a grouping, in this case of two N-best lists in `<emma:one-of>`. The `<emma:group-info>` element can be used to provide information, in application specific markup, regarding the criteria by which the inputs were grouped. In this case, the `<emma:group-info>` indicates that the inputs are grouped based on their respective time intervals overlapping.

The `<emma:sequence>` container is essentially a special case of `<emma:group>` in which the inputs represented in the containers inside the `<emma:sequence>` are required to be in temporal order. For example, a sequence of two point gestures made on a 2D visual display could be represented as in Fig. 3.16.

A key characteristic of input processing in interactive multimodal systems is that there are often multiple different stages of processing. For example, a speech signal may be processed by a speech recognizer yielding an EMMA result containing speech recognition hypotheses. These may then be consumed by a natural language understanding component yielding a list of possible semantic representations for the input. That in turn may serve as input to a multimodal integration component where the speech interpretation is combined with possible interpretations of the gesture, in order to yield a multimodal interpretation (see, for example, [37] for one example of a declarative approach to multimodal integration). The output of multimodal integration can then be consumed by a dialog manager or interaction management component that controls the overall flow of the interaction.

Each stage of processing yields a different EMMA document. The EMMA language provides a mechanism for representing the connections between these

```
<emma:emma version="2.0">
  <emma:group id="g1">
    <emma:one-of id="d1"
      emma:medium="acoustic" emma:mode="voice"
      emma:function="dialog" emma:verbal="true" emma:lang="en-US"
      emma:start="1241035886246" emma:end="1241035889306"
      emma:process="smm:type=asr&version=asr_eng2.4"
      disjunction-type="recognition">
      <emma:interpretation id="s1"
          emma:tokens="traffic along this route"
          emma:confidence="0.75">
          <search_traffic><loc><type>line</type></loc></search_traffic>
      </emma:interpretation>
      <emma:interpretation id="s2"
          emma:tokens="hotels around here"
          emma:confidence="0.65">
          <search_hotels><loc><type>area</type></loc></search_hotels>
      </emma:interpretation>
    </emma:one-of>
    <emma:one-of id="d2"
      emma:medium="tactile" emma:mode="ink"
      emma:function="dialog" emma:verbal="false"
      emma:start="1241035885246" emma:end="1241035887002"
      emma:process="smm:type=gesture_recognition&version=3.7"
      disjunction-type="recognition">
      <emma:interpretation id="g1" emma:confidence="0.9">
         <loc><type>area</type><coords> … </coords></loc>
      </emma:interpretation>
      <emma:interpretation id="g2" emma:confidence="0.6">
         <loc><type>line</type><coords> … </coords></loc>
      </emma:interpretation>
    </emma:one-of>
  </emma:group>
  <emma:group-info><type>temporal_overlap</type></emma:group-info>
</emma:emma>
```

Fig. 3.15 EMMA representation of multimodal input using <emma:group>

```
<emma:emma version="2.0">
  <emma:sequence id="s1">
    <emma:interpretation id="g1"
      emma:medium="tactile" emma:mode="ink">
        <location type="point" x="2300" y="1200"/>
    </emma:interpretation>
    <emma:interpretation id="g2"
      emma:medium="tactile" emma:mode="ink">
        <location type="point" x="3000" y="2400"/>
    </emma:interpretation>
  </emma:sequence>
</emma:emma>
```

Fig. 3.16 EMMA sequence container example

derivations or processing steps in handling input to interactive systems. The <emma:derived-from> element is an empty element which can appear inside <emma:interpretation> or other containers and provides a pointer to the

<emma:interpretation> or other container element that was input to the process indicated in emma:process in order to generate that interpretation. The resource attribute on <emma:derived-from> is URI valued and provides the reference to the container the interpretation is derived from. The other attribute composite indicates whether this is an interpretation derived by combining multiple modalities of input. In one case of a multimodal combination there will be two or more <emma:derived-from> elements pointing to the different modes being combined. The references can be to other documents, or alternatively, the previous stages of processing can be held in an element <emma:derivation> under <emma:emma> and referenced within the document, enabling a whole chain of processing to be captured in a single document. This format is particularly useful for logging.

We return to the multimodal example from Fig. 3.15. Assuming that the document containing the EMMA group is processed by a multimodal fusion component, the multimodal combined result and the earlier stages of processing can be represented in a single document as in Fig. 3.17. The combined interpretation appears directly under <emma:emma>. In this case there are two possible combined interpretations and each has two <emma:derived-from> elements inside referencing the unimodal interpretations, held in the <emma:derivation> container lower in the document that were combined to derive those multimodal interpretations.

The EMMA mechanisms we have described so far are all part of the EMMA 1.0 specification that was established as a W3C recommendation in 2009 [22]. In the next two sections, we present some key extensions and enhancements to the language that are being developed as part of the enhanced EMMA 2.0 language [23] and related work.

3.5 Extending EMMA to System Outputs

All of the markup described so far is directed specifically at the representation and annotation of input such as speech commands, drawings made on touch displays, and free-form hand/arm gesture (the right hand side of Fig. 3.1). Much the same set of concerns and requirements for containing representatives and providing metadata apply also to the outputs generated by interactive spoken and multimodal systems. As in the case of input, often there are multiple stages of processing. For example, a dialog manager may generate a semantic representation capturing the desired next action of a system, which is then processed by a natural language generation component yielding a sequence of words, which is then passed to a speech synthesis component for rendering as audio.

As in the input side, there are also needs in the output processing pipeline for representation of multiple alternatives. For example, natural language generation models can generate multiple possible candidate utterances which are rescored or selected among based on considerations downstream, such as the quality of the

```
<emma:emma version="2.0">
  <emma:one-of
     emma:medium="acoustic tactile" emma:mode="voice ink"
     emma:function="dialog" emma:verbal="true"
     emma:start="1241035885246" emma:end="1241035889306"
     emma:process="smm:type=multimodal_fusion"
     emma:lang="en-US">
     <emma:interpretation id="mm1" emma:confidence="0.585">
       <search_hotels><loc><type>area</type>
                     <coords> … </coords></loc></search_hotels>
       <emma:derived-from resource="#s2" composite=true/>
       <emma:derived-from resource="#g1" composite=true/>
      </emma:interpretation>
     <emma:interpretation id="mm2" emma:confidence="0.45">
       <search_traffic><loc><type>line</type>
                      <coords> … </coords></loc></search_traffic>
     </emma:interpretation>
     <emma:derived-from resource="#s1" composite=true/>
     <emma:derived-from resource="#g2" composite=true/>
  </emma:one-of>
  <emma:derivation>
    <emma:one-of id="d1"
      emma:medium="acoustic" emma:mode="voice"
      emma:function="dialog" emma:verbal="true" emma:lang="en-US"
      emma:start="1241035886246" emma:end="1241035889306"
      emma:process="smm:type=asr&version=asr_eng2.4"
      disjunction-type="recognition">
      <emma:interpretation id="s1"
          emma:tokens="traffic along this route"
          emma:confidence="0.75">
          <search_traffic><loc><type>line</type></loc></search_traffic>
      </emma:interpretation>
      <emma:interpretation id="s2"
          emma:tokens="hotels around here"
          emma:confidence="0.65">
          <search_hotels><loc><type>area</type></loc></search_hotels>
      </emma:interpretation>
     </emma:one-of>
     <emma:one-of id="d2"
      emma:medium="tactile" emma:mode="ink"
      emma:function="dialog" emma:verbal="false"
      emma:start="1241035885246" emma:end="1241035887002"
      emma:process="smm:type=gesture_recognition&version=3.7"
      disjunction-type="recognition">
       <emma:interpretation id="g1"
          emma:confidence="0.9">
          <loc><type>area</type><coords> … </coords></loc>
       </emma:interpretation>
       <emma:interpretation id="g2"
          emma:confidence="0.6">
          <loc><type>line</type><coords> … </coords></loc>
       </emma:interpretation>
     </emma:one-of>
  </emma:derivation>
</emma:emma>
```

Fig. 3.17 Multimodal combination with representation of derivation using `<emma:derived-from>`

text-to-speech rendering. In these cases, <emma:one-of> can be used to capture multiple alternative outputs. System output can also be multimodal. <emma:group> can be used to contain contributions for multiple modes being generated by a multimodal presentation planning component which takes a semantic representation from a dialog manager and distributes it across the available modalities such as speech and graphical displays.

In order to address this asymmetry and extend the utility of EMMA, EMMA 2.0 includes an <emma:output> element, the output variant of <emma:interpretation> for containment of planned system output. <emma:output> is a container element and can appear directly under <emma:emma> or as a child of the various containers, <emma:one-of>, <emma:group>, etc. To make this more concrete, consider the example in Fig. 3.18. In this case, assume the user has asked which airlines have flights from Denver to Boston on September 20th. The dialog manager conducts a search and generates a semantic representation, the <inform> element here indicates information to be conveyed to the user. In this example, the <time> element contains an expression in TIMEML [38].

Note that for <emma:output> many of the same annotation attributes and elements are relevant but in some cases have a subtly different semantics. emma:confidence in this case is not a recognition confidence, but rather the confidence of the dialog manager in the decision to make this response. The emma:medium and emma:mode and the emma:lang are not the actual classification of an input but rather the desired medium and mode and language that the output should be rendered in, in this case voice output in US English. The annotation emma:media-type can be used to indicate the desired codec and sampling rate for audio output. Other annotations are more complex, for example, at the point that a response is specified by the dialog manager, timestamps indicate not the actual timing of input but the desired or proposed timing of output. Generally a dialog manager may not be able to specify the absolute timing of output. Relative timestamps can potentially be used to capture the desired temporal relation between

```
<emma:emma version="2.0">
  <emma:output emma:confidence="0.9" id=sem1"
     emma:medium="acoustic"
     emma:mode="voice"
     emma:lang="en-US"
     emma:process="http://example.com/dialog_engine">
       <inform><flights>
                <src>DEN</src>
                <dest>BOS</dest>
                <airline>United,Delta</airline>
                <time><timex tid="t1" type="DATE" value="XXXX-9-20"/>
                </time>
               </flights></inform>
  </emma:output
</emma:emma>
```

Fig. 3.18 EMMA representation for output from dialog manager

```
<emma:emma version="2.0">
  <emma:one-of id="nlg_options1"
      emma:medium="acoustic" emma:mode="voice" emma:lang="en-US"
      emma:process="http://example.com/nlg"
      emma:result-format="application/ssml+xml">
      <emma:output emma:confidence="0.9" id="nlg1">
         <speak version="1.0" xmlns=http://www.w3.org/2001/10/synthesis
          xml:lang="en-US">There are several flights to Boston from
              Denver on United and Delta on September twentieth.</speak>
      </emma:output>
      <emma:output emma:confidence="0.7" id="nlg2">
         <speak version="1.0" xmlns=http://www.w3.org/2001/10/synthesis
          xml:lang="en-US">For September twentieth, both United and
              Delta have flights from Denver to Boston.</speak>
      </emma:output>
      <emma:derived-from resource="http://example.com/logs/log12#sem1"/>
  </emma:one-of>
</emma:emma>
```

Fig. 3.19 EMMA representation for output from natural language generation component

multiple chunks or modes of system output, e.g., spacing between multiple spoken utterances or coordination of speech and visual presentation. Absolute timestamps potentially could be added by later stages of processing and captured in logs indicating the actual time of playback of the audio.

The EMMA output representation in Fig. 3.18 would be consumed by a natural language generation component yielding a new EMMA document (Fig. 3.19) capturing multiple possible different surface realizations of the semantic representation into English text. Here we see the use of `<emma:one-of>` to represent indeterminacy in system output generation with confidence scores associated with each possible output. The payload in these examples is in SSML [39] a W3C standard language for specifying speech to be rendered by text-to-speech engines. The MIME type for the content of `<emma:output>` is specified in `emma:result-format`. Here again you see the EMMA scope mechanism at work, the annotations, `emma:medium`, `emma:mode`, `emma:process`, etc., appear on `<emma:one-of>` and are assumed to apply to all of the enclosed `<emma:output>` elements. Assuming the document in Fig. 3.18 is accessible by URI, the `<emma:derived-from>` element provides a reference to the `<emma:output>` that was processed by the natural language generation component in order to generate the `<emma:one-of>` in Fig. 3.19.

In a multimodal system, the same semantic representation in Fig. 3.18 could be processed by a multimodal presentation planner creating an `<emma:group>` with two `<emma:output>` elements, one with graphical content in HTML5 and another with SSML text to be rendered by TTS. In this case, `<emma:group>` is used as a container for the results of multimodal fission. Relative timestamps can be used in order to specify the temporal synchronization of the graphical element of the response and the speech element of the response.

3.6 JSON Representation for EMMA

When the EMMA language was initially developed, XML representations were dominant in the standards community both for document markup and for messaging among components. In that setting, it was natural to adopt XML for EMMA. Since then, along with the increase in the capabilities and use of both client side and server side JavaScript, JSON [40] has become prevalent as a format for communication among web services, an interchange format among different programming languages, and also for specification of content to be rendered by various client side frameworks. Here I discuss some previous and ongoing work in EMMA to integrate with and support JSON.

In EMMA 2.0 [23] there are proposals for an extension to the language to enable specification of non-XML payloads for `<emma:interpretation>` and `<emma:output>`. The `emma:result-format` attribute can be used to specify the MIME type of the content in the payload and the XML character data mechanism (CDATA) can be used to embed non-XML content directly inline. For example, if the result of an NLU processor was in JSON format, instead of having to convert to XML for inclusion in an EMMA document, the content could be contained directly by specifying the `emma:result-format` as "`application/json`" as in the example in Fig. 3.20. In this case, the spoken input "I want a medium coke and a large pepperoni pizza with mushrooms" is assigned a semantic representation in JSON by a natural language understanding component. The representation could be built, for example, by SISR attachments [41] to an SRGS grammar [42]. The `<emma:literal>` element is in EMMA 1.0 and is used to support string literals as the contents of an `<emma:interpretation>`. In this example it contains the CDATA.

In ongoing work, a JSON specific variant of EMMA is being developed. Many of the details of this representation are still under discussion, but here we present a possible representation and discuss some of the issues. Figure 3.21 is an example of possible JSON format for an N-best list with two recognition results and their interpretations.

JSON consists of sets of attributes and values captured in { } with":" between the attributes and values. Values may themselves be sets of attributes and values, atomic strings, or lists in square brackets []. The distinction between elements and attributes that there is in XML is not present in JSON. As a result there is no way to place annotations directly on a container such as a `one-of`. The assumption we make here instead is that annotations applying to a container appear as sisters of the attribute introducing its contents, in this case `"emma:one-of."` In order to allow for multiple interpretations, `"emma:one-of"` is list valued. The semantic representation in each interpretation is itself encoded in JSON. Annotations that apply to the `"emma:interpretation"` appear as sister attributes in that array of attributes and values. Although JSON has no inbuilt mechanism similar to namespaces in XML, we maintain the `"emma:"` prefix on EMMA containers and annotations here in order to differentiate the EMMA language from the JSON representation of the application specific content.

```
<emma:emma version="2.0">
   <emma:interpretation id="json_ex1"
     emma:confidence="0.75"
     emma:medium="acoustic"
     emma:mode="voice"
     emma:verbal="true"
     emma:function="dialog"
     emma:tokens="I want a medium coke and a
                  large pepperoni pizza with mushrooms"
     emma:result-format="application/json"
     emma:process="smm:nlu_v3">
       <emma:literal>
         <![CDATA[
                   {"drink": {"liquid":"coke",
                             " size":"medium"},
                    "pizza": {"size":"large",
                              "topping":["pepperoni","mushrooms"]}}
         ]]>
       </emma:literal>
   </emma:interpretation>
 </emma:emma>
```

Fig. 3.20 EMMA XML with JSON semantic representation

```
{"id":"d1",
 "emma:medium":"acoustic",
 "emma:mode":"voice",
 "emma:function":"dialog",
 "emma:verbal":"true",
 "emma:start":"1241035886246",
 "emma:end":"1241035889306",
 "emma:source":"smm:platform=iPhone-6s-ios9.2",
 "emma:process":"smm:type=asr&version=asr_eng2.4",
 "emma:media-type":"audio/amr; rate=8000",
 "emma:lang":"en-US",
 "emma:one-of":[{"emma:interpretation":{"flight":{"orig":"BOS",
                                                   "dest":"DEN"}},
                 "id":"int1" ,
                 "emma:confidence":"0.75",
                 "emma:tokens":"flights from boston to denver"},
                {"emma:interpretation":{"flight":{"orig":"AUS",
                                                   "dest":"DEN"}},
                 "id":"int2",
                 "emma:confidence":"0.68",
                 "emma:tokens":"flights from austin to denver"}]
}
```

Fig. 3.21 EMMA JSON representation

Many issues remain to be worked out, but a native JSON version of EMMA would potentially be extremely beneficial to developers building multimodal interactive systems in an environment where they are primarily loading, processing, and generating JSON rather than XML.

3.7 Conclusion

In recent years, driven by the widespread availability of mobile devices, in-car systems, and smart home technologies with touch screens, cameras, and voice input and output, there is increasing demand for the creation of spoken and multimodal interactive systems. These systems remain complex to author in part as they involve integration of multiple different systems components that have to work in concert to provide a robust and compelling user experience. In this chapter, we have described and illustrated the EMMA language, a recent W3C standard for representation of the interpretation of input to and output from spoken and multimodal interactive systems.

EMMA provides a series of containers and annotations that facilitate the representation of common metadata required for representation of inputs in different modalities. Key features of the language include mechanisms for representation of the ambiguous and uncertain input resulting from recognition and interpretation of natural input modalities and mechanisms supporting representation of grouping and temporal sequences of inputs—important capabilities for capturing composite inputs distributed over multiple input modes. The language also provides flexibility so that multiple levels of processing can be captured incrementally within a single growing document, or distributed across multiple EMMA documents with cross-references capturing the derivation of the final interpretation from the initial input signal.

The use of XML as a language for representing user inputs facilitates the generation and parsing of EMMA documents by EMMA producers and consumers, since tools for XML parsing and querying are readily available in almost all programming environments. The rich representation of input in EMMA and its extensibility facilitate the capture of detailed interaction logs for system analysis, debugging, and training models.

In recent work, proposals have been developed for the extension of the EMMA language from input representation to system output, completing the circle and enabling use of a common representation for the majority of communication among components of a spoken or multimodal system. In ongoing work, the EMMA language is being extended to other formats, providing support for embedding of JSON and other representations, along with evolving proposals for a native JSON version of EMMA. Other future directions for EMMA include support for incremental results from speech recognizers and other modality processing components, extensions to enable use of EMMA for capture not just of machine interpretation of inputs but also metadata added by human annotators, and support for new metadata types including more detailed specification of the location in which input is received.

Acknowledgements I would like to acknowledge the many contributors to the EMMA standard from around the world including Deborah Dahl, Kazuyuki Ashimura, Paolo Baggia, Roberto Pieraccini, Dan Burnett, Dave Raggett, Stephen Potter, Nagesh Kharidi, Raj Tumuluri,

Jerry Carter, Wu Chou, Gerry McCobb, Tim Denney, Max Froumentin, Katrina Halonen, Jin Liu, Massimo Romanelli, T. V. Raman, and Yuan Shao.

References

1. Hauptmann, A. (1989). Speech and gesture for graphic image manipulation. In *Proceedings of CHI'89*, Austin, TX, pp. 241–245.
2. Nishimoto, T., Shida, N., Kobayashi, T., & Shirai, K. (1995). Improving human interface in drawing tool using speech, mouse, and keyboard. In *Proceedings of the 4th IEEE International Workshop on Robot and Human Communication, ROMAN95*, Tokyo, Japan, pp. 107–112.
3. Oviatt, S. L. (1999). Mutual disambiguation of recognition errors in a multimodal architecture. In *Proceedings of the Conference on Human Factors in Computing Systems: CHI'99*, Pittsburgh, PA, pp. 576–583.
4. Cohen, P. R. (1992). The role of natural language in a multimodal interface. In *Proceedings of the 5th Annual ACM Symposium on User Interface Software and Technology*, Monterey, CA, ACM Press, New York, NY, pp. 143–149
5. Rudnicky, A., & Hauptman, A. (1992). Multimodal interactions in speech systems. In M. Blattner & R. Dannenberg (Eds.), *Multimedia interface design* (pp. 147–172). New York: ACM Press.
6. Oviatt S., & VanGent, R. (1996). Error resolution during multimodal human-computer interaction. In *Proceedings of International Conference on Spoken Language Processing (ICSLP)*, Philadelphia, PA, USA, pp. 204–207.
7. Allgayer, J., Jansen-Winkeln, R. M., Reddig, C., & Reithinger, N. (1989). Bidirectional use of knowledge in the multi-modal NL access system XTRA. In *Proceedings of IJCAI*, Detroit, MI, USA, pp. 1492–1497.
8. Bangalore, S., & Johnston, M. (2000). Tight-coupling of multimodal language processing with speech recognition. In *Proceedings of the International Conference on Spoken Language Processing*, Beijing, pp. 126–129.
9. Bolt, R. A. (1980). "Put-That-There": Voice and gesture at the graphics interface. *Computer Graphics, 14*(3), 262–270.
10. Chai, J., Hong, P., & Zhou, M. (2004). A probabilistic approach to reference resolution in multimodal user interfaces. In *Proceedings of 9th International Conference on Intelligent User Interfaces (IUI)*, Madeira, pp. 70–77.
11. Cohen, P. R., Johnston, M., McGee, D., Oviatt, S. L., Pittman, J., Smith, I., et al. (1997). Multimodal interaction for distributed interactive simulation. In *Proceedings of Innovative Applications of Artificial Intelligence Conference*. Menlo Park: AAAI/MIT Press.
12. House, D., & Wirn, M. (2000). Adapt—A multimodal conversational dialogue system in an apartment domain. In *Proceedings of the International Conference on Spoken Language Processing (ICSLP)*, Beijing, pp. 134–137.
13. Johnston, M., Bangalore, S., Vasireddy, G., Stent, A., Ehlen, P., Walker, M., et al. (2002). MATCH: An architecture for multimodal dialog systems. In *Proceedings of the Association of Computational Linguistics Annual Conference*, Philadelphia, PA, pp. 376–383.
14. Koons, D. B., Sparrell, C. J., & Thorisson, K. R. (1993). Integrating simultaneous input from speech, gaze, and hand gestures. In M. T. Maybury (Ed.), *Intelligent multimedia interfaces* (pp. 257–276). Cambridge, MA: AAAI Press/MIT Press.
15. Neal, J. G., & Shapiro, S. C. (1991). Intelligent multi-media interface technology. In J. W. Sullivan & S. W. Tyler (Eds.), *Intelligent user interfaces* (pp. 45–68). New York: Addison Wesley.

16. Sharma, R., Yeasin, M., Krahnstoever, N., Rauschert, I., Cai, G., Brewer, I., MacEachren, A. M., & Sengupta, K. (2003). Speech-gesture driven multimodal interfaces for crisis management. *Proceedings of the IEEE, 91*(9), 1327–1354.
17. Wahlster, W. (2002). SmartKom: Fusion and fission of speech, gestures, and facial expressions. In *Proceedings of the 1st International Workshop on Man-Machine Symbiotic Systems*, Kyoto, pp. 213–225.
18. Waibel, A., Vo, M., Duchnowski, P., & Manke, S. (1996). Multimodal interfaces. *AI Review Journal, 10*:299–319.
19. Wauchope, K. (1994). Eucalyptus: Integrating natural language input with a graphical user interface. Naval Research Laboratory, Report NRL/FR/5510-94-9711.
20. Cohen, P. R., Kaiser, E. C., Buchanan, C. M., & Lind, S. (2015). Sketch-Thru-Plan: A multimodal interface for command and control. *Communications of ACM, 58*(4), 56–65.
21. Johnston, M., & Ehlen, P. (2010). Speak4it^{SM}: Multimodal interaction in the wild. In *Proceedings of IEEE Spoken Language Technology Workshop*, Berkeley, CA, pp. 59–60.
22. Johnston, M., Baggia, P., Burnett, D. C., Carter, J., Dahl, D. A., McCobb, G., et al. (2009). EMMA:Extensible MultiModal Annotation markup language. W3C Recommendation. https://www.w3.org/TR/2009/REC-emma-20090210/.
23. Johnston, M., Dahl, D. A., Denney, T., & Kharidi, N. (2015). EMMA: Extensible MultiModal Annotation markup language Version 2.0. W3C Public working draft. https://www.w3.org/TR/emma20/. Accessed Sept 2015.
24. Barnett, J., Bodell, M., Dahl, D., Kliche, I., Larson, J., Porter, B., et al. (2012). Multimodal architecture and interfaces. W3C Recommendation. https://www.w3.org/TR/mmi-arch/.
25. Dahl, D. (2000). Natural language semantics markup language for the speech interface framework. W3C Working Draft. https://www.w3.org/TR/2000/WD-nl-spec-20001120/.
26. Shanmugham, S., Monaco, P., & Eberman, B. (2006). A media resource control protocol (MRCP). IETF RFC 4463. https://tools.ietf.org/html/rfc4463.
27. Burnett, D., & Shanmugham, S. (2012). Media resource control protocol Version 2 (MRCPv2). IETF RFC 6787. https://tools.ietf.org/html/rfc6787.
28. Phillips, A., & Davis, M. (2006). Tags for the Identification of Languages, IETF. http://www.rfc-editor.org/rfc/bcp/bcp47.txt.
29. Chee, Y.-M., Franke, K., Froumentin, M., Madhvanath, S., Magana, J. A., Pakosz, G., et al. (2011). Ink markup language (InkML). W3C Recommendation. https://www.w3.org/TR/InkML/.
30. Kaiser, E., Olwal, A., McGee, D., Benko, H., Corradini, A., Li, X., et al. (2003). Mutual disambiguation of 3D multimodal interaction in augmented and virtual reality. In *Proceedings of the 5th International Conference on Multimodal Interfaces (ICMI)*, Vancover, BC, Canada, pp. 12–19.
31. Jonson, R. (2006). Dialog context-based re-ranking of ASR hypotheses. In *Proceedings of IEEE Spoken Language Technology Workshop*, Palm Beach, Aruba, pp. 174–177.
32. Johnston, M. (2009). Building multimodal applications with EMMA. In *Proceedings of the ICMI Conference*, Boston, MA, USA, pp. 47–54.
33. Mishra, T., & Bangalore, S. (2011). Finite-state models for speech-based search on mobile devices. *Natural Language Engineering, 17*(2), 243–264.
34. Stoyanchev, S., & Johnston, M. (2015). Localized error detection for targeted clarification in a virtual assistant. In *Proceedings of International Conference on Acoustics, Speech, and Signal Processing* (ICASSP), Bribane, Australia, pp. 5241–5245.
35. Stoyanchev, S., Liu, A., & Hirschberg, J. (2014). Towards natural clarification questions in dialogue systems. In *Proceedings of AISB*, London, England.
36. Johnston, M., & Bangalore, S. (2009). Robust understanding in multimodal interfaces. *Computational Linguistics, 35*(3), 345–397.
37. Johnston, M. (1998). Unification-based multimodal parsing. In *Proceedings of the Association for Computational Linguistics Annual Conference (ACL)*. Montreal, pp. 624–630.

38. Pustejovsky, J., Castaño, J., Ingria, R., Saurí, R., Gaizauskas, R., Setzer, A., et al. (2003). TimeML: Robust specification of event and temporal expressions in text. In *Proceedings of Fifth International Workshop on Computational Semantics (IWCS-5)*, Tilburg, The Netherlands.
39. Burnett, D., Walker, M. R., & Hunt, A. (2004). Speech synthesis markup language (SSML) Version 1.0. W3C Recommendation. https://www.w3.org/TR/speech-synthesis/.
40. ECMA International (2013). The JSON data interchange format. Standard ECMA-404. http://www.ecma-international.org/publications/files/ECMA-ST/ECMA-404.pdf. Accessed 14 May 2016.
41. Van Tichelen, L., & Burke, D. (2007). Semantic interpretation for speech recognition (SISR) Version 1.0. W3C Recommendation. https://www.w3.org/TR/semantic-interpretation/.
42. Hunt, A., & McGlashan, S. (2004). Speech recognition grammar specification Version 1.0. W3C Recommendation. https://www.w3.org/TR/speech-grammar/.

Chapter 4
EmotionML

Felix Burkhardt, Catherine Pelachaud, Björn W. Schuller, and Enrico Zovato

Abstract EmotionML is a W3C recommendation to represent emotion related states in data processing systems. Given the lack of agreement in the literature on the most relevant aspects of emotion, it is important to provide a relatively rich set of descriptive mechanisms. It is possible to use EmotionML both as a standalone markup and as a plug-in annotation in different contexts. Emotions can be represented in terms of four types of descriptions taken from the scientific literature: categories, dimensions, appraisals, and action tendencies, with a single `<emotion>` element containing one or more of such descriptors. EmotionML provides a set of emotion vocabularies taken from the scientific and psychology literature. Whenever users have a need for a different vocabulary, however, they can simply define their own custom vocabulary and use it in the same way as the suggested vocabularies. Several applications have already been realized on the basis of EmotionML.

4.1 Introduction

EmotionML is a W3C recommendation to represent emotion related states in data processing systems. It was developed by a subgroup of the W3C MMI (Multimodal Interaction) Working Group chaired by Deborah Dahl in a first version from

F. Burkhardt (✉)
Telekom Innovation Laboratories, Winterfeldstr. 21, 10785 Berlin, Germany
e-mail: Felix.Burkhardt@telekom.de

C. Pelachaud
LTCI, CNRS, Télécom ParisTech, Université Paris-Saclay, Paris, France
e-mail: catherine.pelachaud@telecom-paristech.fr

B.W. Schuller
Imperial College, London, UK

University of Passau, Chair CIS, Passau, Germany
e-mail: schuller@ieee.org

E. Zovato
Nuance, Turin, Italy
e-mail: ezovato@gmail.com

D.A. Dahl (ed.), *Multimodal Interaction with W3C Standards*,
DOI 10.1007/978-3-319-42816-1_4

approximately 2005 until 2013, most of this time the development was led by Marc Schröder.

In the scientific literature on emotion research, there is no single agreed description of emotions, not even a clear consensus on the use of terms like affect, emotion, or other related phenomena. For a markup language representing emotional phenomena it therefore appears important to allow the representation of their most relevant aspects in the wider sense. Given the lack of agreement in the literature on the most relevant aspects of emotion, it is inevitable to provide a relatively rich set of descriptive mechanisms.

It is possible to use EmotionML both as a standalone markup and as a plug-in annotation in different contexts. Emotions can be represented in terms of four types of descriptions taken from the scientific literature: categories, dimensions, appraisals, and action tendencies, with a single `<emotion>` element containing one or more of such descriptors.

This chapter starts with a report on the analysis of use cases and requirements in Sect. 4.2. Section 4.3 gives a brief overview on previous attempts to formalize emotional vocabularies. In the following we discuss the elements that constitute EmotionML in Sect. 4.4. Section 4.5 motivates the suggested emotion vocabularies that were published as a W3C Note. Lastly, Sect. 4.6 introduces applications that were realized on the basis of EmotionML. The chapter closes with conclusions and outlook in Sect. 4.7.

4.2 Use Cases and Requirements

Although there are differences in emotion modeling theories, there is a sort of consensus on the essential components of emergent emotions. Among these the most important are feelings, appraisal, action tendencies, and emotion expressions (for example expressed in facial expressions, body gestures, and vocal cues). A shared formalism of encoding these aspects of emotions is a key point to allow inter-operability in technological environments.

The goal of representing emotions in a markup language was addressed with a bottom-up approach. Use cases were gathered from contributors with different expertise in the field of emotion in technology and research. The working group iteratively extracted requirements on the markup language from a number of 39 collected use cases. Based on these requirements, a syntax for EmotionML has been produced.

The collected use cases can be classified into three broader categories:

- Data Annotation
- Emotion Recognition
- Emotion Generation

Data annotation class includes many use cases that deal with human annotation of the emotion contained in some material, for example by means of a simple text, or a node in an XML tree, representing pictures or voice recordings. The main requirement is defining the scope of the annotated emotion together with the emotion annotation itself. Annotation of emotions can be an important factor in training data for emotion classification. In data annotation it is sometimes necessary to keep track of the evolution of the emotion over time. An important requirement is in fact the possibility to trace events over time, in order to annotate the "dynamic" aspects of emotions, beyond the static ones.

The Emotion Recognition use cases deal with the detection of low level features from human–human or human–machine interaction. These features could be, for example, speech prosodic features [1] or facial action parameters [2]. Recognition can be unimodal or multimodal, where individual modalities can be merged. In this case timing and synchronization mechanisms are required. In emotion recognition also confidence measures have to be taken into consideration.

The Emotion Generation use cases deal mainly with the generation of face and body gestures and the generation of emotional speech. Emotion eliciting events trigger the generation of emotional behavior, through a mapping mechanism in which certain states are associated with specific behavioral actions.

These use cases showed that there is a big variety of information that is passed to and received from an emotion processing system. As a consequence, the emotion markup language had among its requirements a flexible method of receiving and sending data. The definition of which kind of information can be handled had to be specified as well.

The analysis of use cases showed that one of the primary requirements for the emotion markup language was the definition of a way of interfacing with external representations of data involved in the processes. In other words this markup language should not try to represent the expression of emotion itself, for example facial expressions or sensor data. Other languages should be used for this purpose. Also the system oriented concepts of input and output are replaced by more specific concepts like "experiencer," "trigger," and "observable behavior."

The emotion markup language has among its requirements the possibility to describe emotions both in terms of categories, defined in specific sets (see Sect. 4.5), and of dimensions, for example "evaluation," "activation," and "power." Beyond categories and dimensions the emotion markup language includes the possibility to specify "action tendencies" which are tendencies linked to specific emotions (e.g., anger can be linked to a tendency to attack). Also emotion "appraisals" can be described. These are important in cases where emotions play a role in driving behavior, i.e. when emotions are used to model the behavior of a person.

EmotionML has also to deal with mixed emotions and has to include the possibility to define a value in a well-defined scale for each of the component emotions. Temporal aspects are also important. The emotion markup language has to specify absolute times, relative times, and interval durations. As previously mentioned an important requirement is the specification of the time evolution of a dynamic scale value, through a tracing mechanism.

Meta information has to be also taken into account, such as the specification of the degree of confidence or probability that the represented emotion is correct.

The modality through which an emotion is expressed is also useful information. A general mechanism to specify meta information has been added in EmotionML allowing the definition of key-value pairs, in a similar fashion as in the EMMA markup language (@todo: link to EMMA chapter).

A mechanism to link the emotion description to external source of data is mandatory, since many use cases rely on media representations like video or audio files, pictures, documents, etc. Also the semantics of the external link has to be specified, and EmotionML fulfills this requirement by means of a specific attribute ("role"), that indicates whether the referenced link points to an observable behavior expressing the emotion, to the subject experiencing the emotion, to an emotion eliciting event, or to an object towards which an emotion-related action is directed.

4.3 Previous Work

The representation of emotions and related states has been part of several activities. For example, as part of their Virtual Human Markup Language, the Curtin University of Technology introduced in the Interface project EML, Emotion Markup Language, a vocabulary used to control the emotional expression of virtual characters.

In the area of labelling schemes, maybe the most thorough attempt to propose an encompassing labelling scheme for emotion-related phenomena has been the work on the HUMAINE database [3].

The relevant concepts were identified in prose, and made available as a set of configuration files for the video annotation tool Anvil [4]. A formal representation format was not proposed in this work. Markup languages including emotion-related information were defined mainly in the context of research systems generating emotion-related behaviour of ECAs (Embodied Conversational Agent).

The expressive richness is usually limited to a small set of emotion categories, possibly an intensity dimension, and in some cases a three-dimensional continuous representation of activation-evaluation-power space (see [5] for a review).

For example, the Affective Presentation Markup Language APML [6] provides an attribute "affect" to encode an emotion category for an utterance (a "performative") or for a part of it:

```
<performative type="inform" affect="afraid">
Do I have to go to the dentist?
</performative>
```

An interesting contribution to the domain of computerized processing and representation of emotion-related concepts is A Layered Model of Affect, ALMA [7]. Following the OCC model [8], ALMA uses appraisal mechanisms to trigger emotions from events, objects, and actions in the world. Emotions have an intensity varying over time. Each individual emotion influences mood as a longer-term affective state. ALMA uses an XML-based markup language named AffectML in two places: to represent the antecedents to emotion, i.e. the appraisals leading to emotions, or to represent the impact that emotions and moods have on a virtual agent's behaviour.

4.4 Emotion Markup Language Elements

Based on the requirements, a syntax for EmotionML has been produced in a sequence of steps.

The following snippet exemplifies the principles of the EmotionML syntax.

```
<sentence id="sent1">
  Do I have to go to the dentist?
</sentence>
<emotion xmlns="http://www.w3.org/2009/10/emotionml"
category-set="http://.../xml#everyday-categories">
  <category name="afraid" value="0.4"/>
  <reference role="expressedBy" uri="#sent1"/>
</emotion>
```

The following properties can be observed:

- The emotion annotation is self-contained within an <emotion> element.
- All emotion elements belong to a specific namespace.
- It is explicit in the example that emotion is represented in terms of categories.
- It is explicit from which category set the category label is chosen.
- The link to the annotated material is realized via a reference.

EmotionML is conceived as a plug-in language, with the aim to be usable in many different contexts. Therefore, proper encapsulation is essential. All information concerning an individual emotion annotation is contained within a single <emotion> element. All emotion markup belongs to a unique XML namespace.

EmotionML differs from many other markup languages in the sense that it does not enclose the annotated material. In order to link the emotion markup with the annotated material, either the reference mechanism in EmotionML or another mechanism external to EmotionML can be used.

A top-level element emotionml enables the creation of stand-alone EmotionML documents, essentially grouping a number of emotion annotations together, but also providing document-level mechanisms for annotating global meta data and for defining emotion vocabularies (see below). It is thus possible to use EmotionML both as a standalone markup and as a plug-in annotation in different contexts.

4.4.1 Representations of Emotion

Emotions can be represented in terms of four types of descriptions taken from the scientific literature [9]: <category>, <dimension>, <appraisal>, and <action-tendency>. An <emotion> element can contain one or more of these descriptors; each descriptor must have a name attribute and can have a value attribute indicating the intensity of the respective descriptor. For <dimension>, the value attribute is mandatory, since a dimensional emotion description is always a position on one or more scales; for the other descriptions, it is possible to omit the value to only make a binary statement about the presence of a given category, appraisal or action tendency.

The following example illustrates a number of possible uses of the core emotion representations.

```
<category name ="affectionate "/>
<dimension name ="valence " value="0.9"/>
<appraisal name ="agent-s e l f "/>
<action-tendency name ="approach"/>
```

4.4.2 Mechanism for Referring to an Emotion Vocabulary

Since there is no single agreed-upon vocabulary for each of the four types of emotion descriptions, EmotionML provides a mandatory mechanism for identifying the vocabulary used in a given <emotion>. The mechanism consists in attributes of <emotion> named category-set, dimension-set, etc., indicating which vocabulary of descriptors for annotating categories, dimensions, appraisals, and action tendencies are used in that emotion annotation. These attributes contain a URI pointing to an XML representation of a vocabulary definition. In order to verify that an emotion annotation is valid, an EmotionML processor must retrieve the vocabulary definition and check that every name of a corresponding descriptor is part of that vocabulary.

Some vocabularies are suggested by the W3C [10]. Users are encouraged to use them to make EmotionML documents interoperable.

4.4.3 Meta-Information

Several types of meta-information can be represented in EmotionML.

First, each emotion descriptor (such as <category>) can have a confidence attribute to indicate the expected reliability of this piece of the annotation. This can reflect the confidence of a human annotator or the probability computed by a machine classifier. If several descriptors are used jointly within an <emotion>, each descriptor has its own `confidence` attribute. For example, it is possible to have high confidence in, say, the arousal dimension but be uncertain about the pleasure dimension:

```
<emotion dimension-set="http://www.w3.
org/ TR /emotion- voc/xml # pad -dimensions">
   <dimension name ="arousal " value ="0.7" confidence ="0.9"/>
   <dimension name ="pleasure " value ="0.6" confidence ="0.3"/>
</emotion>
```

Each <emotion> can have an `expressed-through` attribute providing a list of modalities through which the emotion is expressed. Given the open-ended application domains for EmotionML, it is naturally difficult to provide a complete list of relevant modalities. The solution provided in EmotionML is to propose a list of human-centric modalities, such as `gaze`, `face`, `voice`, etc., and to allow arbitrary additional values. The following example represents a case where an emotion is recognized from, or to be generated in, face and voice:

```
<emotion category-set="http: //... / xml
  # everyday-categories " expressed-through="face voice">
   <category name ="s a t i s f a c t i o n "/>
</emotion>
```

For arbitrary additional meta data, EmotionML provides an <info> element which can contain arbitrary XML structures. The <info> element can occur as a child of <emotion> to provide local meta data, i.e. additional information about the specific emotion annotation; it can also occur in standalone EmotionML documents as a child of the root node <emotionml> to provide global meta data, i.e. information that is constant for all emotion annotations in the document. This can include information about sensor settings, annotator identities, situational context, etc.

4.4.4 References to the "Rest of the World"

Emotion annotation is always *about* something. There is a subject "experiencing" (or simulating) the emotion. This can be a human, a virtual agent, a robot, etc. There is observable behavior expressing the emotion, such as facial expressions, gestures, or vocal effects. With suitable measurement tools, this can also include physiological changes such as sweating or a change in heart rate or blood pressure. Emotions are often caused or triggered by an identifiable entity, such as a person, an object, an event, etc. More precisely, the appraisals leading to the emotion are triggered by that entity. And finally, emotions, or more precisely the emotion-related action tendencies, may be directed towards an entity, such as a person or an object.

EmotionML considers all of these external entities to be out of scope of the language itself; however, it provides a generic mechanism for referring to such entities. Each `<emotion>` can use one or more `<reference>` elements to point to arbitrary URIs. A `<reference>` has a `role` attribute, which can have one of the following four values: `expressedBy` (default), `experiencedBy`, `triggeredBy`, and `targetedAt`. Using this mechanism, it is possible to point to arbitrary entities filling the above-mentioned four roles; all that is required is that these entities be identified by a URI.

4.4.5 *Time*

Time is relevant to EmotionML in the sense that it is necessary to represent the time during which an emotion annotation is applicable. In this sense, temporal specification complements the above-mentioned reference mechanism.

Representing time is an astonishingly complex issue. A number of different mechanisms are required to cover the range of possible use cases. First, it may be necessary to link to a time span in media, such as video or audio recordings. For this purpose, the `<reference role="expressedBy">` mechanism can use a so-called Media Fragment URI to point to a time span within the media [11]. Second, time may be represented on an absolute or relative scale. Absolute time is represented in milliseconds since 1 January 1970, using the attributes `start`, `end`, and `duration`. Absolute times are useful for applications such as affective diaries, which record emotions throughout the day, and whose purpose is to link back emotions to the situations in which they were encountered. Other applications require relative time, for example time since the start of a session. Here, the mechanism borrowed from EMMA is the combination of `time-ref-uri` and `offset-to-start`. The former provides a reference to the entity defining the meaning of time 0; the latter is time, in milliseconds, since that moment.

4.4.6 *Representing Continuous Values and Dynamic Changes*

A mentioned above, the emotion descriptors `<category>`, `<dimension>`, etc. can have a `value` attribute to indicate the position on a scale corresponding to the respective descriptor. In the case of a dimension, the value indicates the position on that dimension, which is mandatory information for dimensions; in the case of categories, appraisals, and action tendencies, the value can be optionally used to indicate the extent to which the respective item is present.

In all cases, the `value` attribute contains a floating-point number between 0 and 1. The two end points of that scale represent the most extreme possible values, for example the lowest and highest possible positions on a dimension, or the complete absence of an emotion category vs. the most intense possible state of that category.

The `value` attribute thus provides a fine-grained control of the position on a scale, which is constant throughout the temporal scope of the individual `<emotion>` annotation. It is also possible to represent changes over time of these scale values, using the `<trace>` element which can be a child of any `<category>`, `<dimension>`, `<appraisal>`, or `<action-tendency>` element. This makes it possible to encode trace-type annotations of emotions as produced.

4.5 Vocabularies

As described above, EmotionML takes into account a number of key concepts from scientific emotion research [5]. Four types of descriptions are available: categories, dimensions, appraisals, and action tendencies. These types correspond to the four main existing representation scheme of emotions.

Depending on the tradition of emotion research and on the use case, it may be appropriate to use any single one of these representations; alternatively, it may also make sense to use combinations of descriptions to characterize more fully the various aspects of an emotional state that are observed: how the value attributed to an appraisal caused the emotion; how an emotion can be described in terms of a category and/or a set of dimensions; and the potential actions an individual may be executing when an emotion is triggered. Insofar, EmotionML is a powerful representational device.

This description glosses over one important detail, however. Whereas emotion researchers may agree to some extent on the types of facets that play a role in the emotion process (such as appraisals, feeling, expression, etc.), there is no general consensus on the representation schema nor on the descriptive vocabularies that should be used. Which set of emotion categories is considered appropriate varies dramatically between the different traditions, and even within a tradition such as the Darwinian tradition of emotion research, there is no agreed set of emotion categories that should be considered as the most important ones (see, e.g., [12]). Similarly, emotion theoreticians do not agree on the number nor the type of dimensions to consider. Similar remarks can be made for appraisals.

For this reason, any attempt to enforce a closed set of descriptors for emotions would invariably draw heavy criticism from a range of research fields. Given that there is no consensus in the community, it is impossible to produce a consensus annotation in a standard markup language. The obvious alternative is to leave the choice of descriptors up to the users; however, this would dramatically limit interoperability.

The solution pursued in EmotionML is of a third kind. The notion of an 'emotion vocabulary' is introduced: any specific emotion annotation must be specific about the vocabulary that is being used in that annotation. This makes it possible to define in a clear way the terms that make sense in a given research tradition. Components that want to interoperate need to settle on the emotion vocabularies to use; whether a given piece of EmotionML markup can be meaningfully interpreted by an EmotionML engine can be determined.

The specification includes a mechanism for defining emotion vocabularies. It consists of a '`<vocabulary>`' element containing a number of '`<item>`' elements. A vocabulary has a 'type' attribute, indicating whether it is a vocabulary for representing categories, dimensions, appraisals, or action tendencies. A vocabulary item has a 'name' attribute. Both the entire vocabulary and each individual item can have an '`<info>`' child to provide arbitrary metadata.

A W3C Working Group Note [13] complements the specification to provide EmotionML with a set of emotion vocabularies taken from the scientific and psychology literature. When the user considers them suitable, these vocabularies rather than other arbitrary vocabularies should be used in order to promote interoperability.

Whenever users have a need for a different vocabulary, however, they can simply define their own custom vocabulary and use it in the same way as the vocabularies listed in the Note. This makes it possible to add any vocabularies from scientific research that are missing from the pre-defined set, as well as application-specific vocabularies.

In selecting emotion vocabularies, the group has applied the following criteria. The primary guiding principle has been to select vocabularies that are either commonly used in technological contexts, or represent current emotion models from the scientific and psychology literature. A further criterion is related to the difficulty to define mappings between categories, dimensions, appraisals, and action tendencies.

For this reason, groups of vocabularies were included for which some of these mappings are likely to be definable in the future.

The following vocabularies are defined. For categorical descriptions, the "big six" basic emotion (often referred to as universal) vocabulary by Ekman [4], an everyday emotion vocabulary by Cowie et al. [14], and three sets of categories that lend themselves to mappings to, respectively, appraisals, dimensions, and action tendencies: the 22 OCC labels of emotion categories [8], the 24 labels used by Fontaine et al. [15] that are further defined in terms of 4 dimensions, and the 12 categories that are linked to actions tendency as introduced by Frijda [16].

Three dimensional vocabularies are provided, the pleasure-arousal-dominance (PAD) vocabulary by Mehrabian [17], the four-dimensional vocabulary proposed by Fontaine et al. [15], and a vocabulary providing a single 'intensity' dimension for such use cases that want to represent solely the intensity of an emotion without any statement regarding the nature of that emotion. For appraisal, three vocabularies are proposed: the appraisals defined in the OCC model of emotions [8], Scherer's Stimulus Evaluation Checks which are part of the Component Process

Model of emotions [18], and the appraisals used in the computational model of emotions EMA [19]. Finally, for action tendencies, only a single vocabulary is currently listed, namely that proposed by Frijda [10]. The following example represents the call for an emotion described by its four dimensions based on Fontaine et al. [16]:

```
<emotion dimension-set="http://www.w3. org/ TR /emotion- voc/xml
   #fsre -dimensions">
<dimension name ="valence " value="0.8"/> <! high positive
valence -->
<dimension name ="arousal " value="0.5"/> <! average arousal -->
<dimension name ="potency" value="0.2"/> <! very low potency -->
<dimension name ="unpredictability " value="0.4"/> <! some-
how lower than average predictable event -->
</emotion>
```

While these vocabularies should provide users with a solid basis, it is likely that additional vocabularies or clarifications about the current vocabularies will be requested. Due to the rather informal nature of a W3C Note, it is rather easy to provide future versions of the document that provide the additional information required.

4.6 Applications

In the following, a range of applications is named that provide EmotionML support on the input and/or output side. While a complete listing is not possible due to the increasing usage of the standard, a selection has been made on the criteria to (a) cover for the different use-cases as listed above and (b) based on popularity and spread of the solutions to solve these. Some actually delivered implementation reports to the recommendation.[1] The applications are grouped.

4.6.1 Data Annotation

GTrace provides a tool for the annotation of continuous dimensional emotion in the sense of "traces" whether the emotion primitive such as arousal or valence is rising or falling over time [20]. This is done via a mouse (or joystick, etc.) in a 1D window that is shown side-by-side with the material to be annotated. Various pre-specified

[1] https://www.w3.org/2002/mmi/2013/emotionml-ir/.

scales are provided and one can customize these or add new ones. The program is fully compatible with EmotionML.

The Speechalyzer by Deutsche Telekom Laboratories is an open source project for analysis, annotation, and transcription of speech files [21]. It can be used to rapidly judge large numbers of audio files emotionally, an automatic classification is integrated. The Speechalyzer was part of a project to identify disgruntled customers in an automated voice service portal [22] with two use cases in mode: (a) transfer angry users to a trained human agent, and (b) gain some statistic insight on the number of angry customers at the end of each day. It utilizes EmotionML as an exchange format to import and export emotionally annotated speech data.

iHEARu-PLAY is a gamified crowd-sourcing platform [23] that supports audio, image, and video annotation for emotion supporting EmotionML. It also allows for crowd-sourced recording of data. At present, dynamic active learning abilities are integrated.

Further examples of annotation software supporting EmotionML include DocEmoX, a tool to annotate documents in 3D emotion space [24],

4.6.2 Emotion Recognition

The openSMILE tool first developed during the European SEMAINE project supports extraction of large audio feature spaces in real time for emotion analysis from audio and video [25] and has also been used for other modalities such as physiological data or CAN-Bus data in the car. It is written in C++ and has been ported to Android for mobile usage. The main features are the capability of on-line incremental processing and high modularity that allows for feature extractor components to be freely interconnected for the creation of custom features via a simple configuration file. Further, new components can be added to openSMILE via an easy plugin interface and a comprehensive API. openSMILE is free software licensed under the GPL license. The toolkit has matured to a standard in the field of Affective Computing also due to its usage in a broad range of competitions in the field including AVEC 2011–2016, Interspeech ComParE 2009–2016, EmotiW, and MediaEval.

The openEAR extension of openSMILE provides pre-trained models for emotion recognition and a ready-to-use speech emotion recognition engine [26].

The EyesWeb platform was enhanced in the ASC-Inclusion European project enabling it to send text messages containing emotion Markup Language messages to give information about recognized emotions from body gestures [27]. EyesWeb is an open platform that supports a wide number of input devices including motion capture systems, various types of professional and low cost video cameras, game interfaces (e.g., Kinect, Wii), multichannel audio input (e.g., microphones), and analog inputs (e.g., for physiological signals). Supported outputs include multichannel audio, video, analog devices, and robotic platforms.

4.6.3 Emotion Generation

MARY TTS is an open-source, multilingual text-to-speech synthesis platform that includes modules for expressive speech synthesis [28]. Particularly the support for both categorical and dimensional representations of emotions by EmotionML is important to Mary's expressive speech synthesis. These categories and dimensions are implemented by modifying the predicted pitch contours, pitch level, and speaking rate.

Using this approach, expressive synthesis is most effective when using HMM-based voices, since the statistical parametric synthesis framework allows appropriate prosody to be realized with consistent quality. Expressive unit selection voices support EmotionML best if they are built from multiple-style speech databases [29], which preserve intonation and voice quality better than when applying signal manipulation to conventional unit-selection output.

Greta is a real-time 3D embodied conversational agent with a 3D model of an agent compliant with MPEG-4 animation standard [30]. It is able to communicate using a rich palette of verbal and nonverbal behaviours. Greta can talk and simultaneously show facial expressions, gestures, gaze, and head movements. Besides the standard XML languages FML and BML that allow to define the communicative intentions and behaviours, EmotionML support was added and used in a range of European projects such as SEMAINE, TARDIS, and ARIA-VALUSPA.

4.6.4 Platforms and Projects

The SEMAINE platform [31] stems from the European Semaine-Project. It provides a free to use virtual agent system including full audio/visual input analysis (e.g., via openSMILE) and output generation (via MARY TTS and Greta) as well as a dialogue manager. The communication between modules is based on EmotionML.

Finally, a range of projects use the standard. Examples are the above named SEMAINE and ARIA-VALUSPA European projects, both dealing with audiovisual emotionally intelligent chatbots. Further the ASC-Inclusion and De-ENIGMA projects aiming to help children on the autism spectrum to learn about emotions, the TARDIS European project that provides serious gaming to young individuals to prepare for their first job interviews, the MixedEmotions and SEWA European projects focusing on sentiment analysis as well as a range of national projects such as the Finnish "Detecting and visualizing emotions and their changes in text" project [32, 33].

4.7 Conclusions

We presented EmotionML, a W3C recommendation to represent emotion related states in data processing systems.

It is possible to use EmotionML both as a standalone markup and as a plug-in annotation in different contexts. Emotions can be represented in terms of four types of descriptions taken from the scientific literature: categories, dimensions, appraisals, and action tendencies, with a single `<emotion>` element containing one or more of such descriptors.

A W3C Working Group Note complements the specification to provide EmotionML with a set of suggested emotion vocabularies taken from the scientific and psychology literature. Whenever users have a need for a different vocabulary, however, they can simply define their own custom vocabulary and use it in the same way as the vocabularies listed in the Note.

Several applications have already been realized on the basis of EmotionML but we're still far away from widespread use. This of course reflects the fact that emotion processing systems are still in their technological infancy and up to now are more research topic than product feature.

Nonetheless we believe that with the spreading of user interfaces that are more natural than keyboard typing, as, for example, speech interfaces, wearables or physical sensors, emotional processing will become a necessity for such systems to be able to interact in a natural and intuitive manner.

Another technology trend that pushes emotional processing is the renaissance of artificial intelligence, as intelligence and emotions are strongly connected concepts.

We hope this article encourages the reader to use EmotionML in her/his own projects and give feedback to the W3C to pave the way towards EmotionML version 2.0. Some topics that came up in the group's discussions have been left out in the first version, for the sake of simplicity. For example, the blend of emotions, emotion regulation, or a direct link to RDF (Resource Description Framework) for semantic annotation was dropped at some point. Another issue that could be pursued are requirements that resolve from use cases concerned with sentiment analysis, which is an important topic given the automatic analysis of user generated content.

References

1. Devillers, L., Vidrascu, L., & Lamel, L. (2005). Challenges in real-life emotion annotation and machine learning based detection. *Neural Networks, 18*(4), 407–422 (2005 special issue).
2. Tekalp, A. M., & Ostermann, J. (2000). Face and 2-D mesh animation in MPEG-4. *Image Communication Journal, 15*, 387–421.
3. Douglas-Cowie, E., Cowie, R., Sneddon, I., Cox, C., Lowry, O., McRorie, M., et al. (2007). The HUMAINE database: Addressing the collection and annotation of naturalistic and induced

emotional data. In *Proceedings of Affective Computing and Intelligent Interaction*, Lisbon, Portugal (pp. 488–500).

4. Kipp, M. (2014). ANVIL: a universal video research tool. In J. Durand, U. Gut, & G. Kristofferson (Eds.), *Handbook of corpus phonology*, pp. 420–436. Oxford: Oxford University Press.
5. Schröder, M., Pirker, H., Lamolle, M., Burkhardt, F., Peter, C., & Zovato, E. (2011). Representing emotions and related states in technological systems. In P. Petta, R. Cowie, & C. Pelachaud (Eds.), *Emotion-oriented systems – The humaine handbook* (pp. 367–386). Berlin: Springer.
6. de Carolis, B., Pelachaud, C., Poggi, I., & Steedman, M. (2004). APML, a markup language for believable behavior generation. In H. Prendinger & M. Ishizuka (Eds.), *Life-like characters* (pp. 65–85). New York: Springer.
7. Gebhard, P. (2005). ALMA - A layered model of affect. In *Proceedings of the Fourth International Joint Conference on Autonomous Agents and Multiagent Systems (AAMAS-05)*, Utrecht.
8. Ortony, A., Clore, G. L., & Collins, A. (1988). *The cognitive structure of emotion*. Cambridge, UK: Cambridge University Press.
9. Schröder, M., Pirker, H., Lamolle, M, Burkhardt, F., Peter, C., & Zovato, E. (2011). Representing emotions and related states in technological systems. In *Emotion-oriented systems - The humaine handbook* (pp. 367–386). Berlin: Springer.
10. Frijda, N. H. (1986). *The emotions*. Cambridge, UK: Cambridge University Press.
11. Troncy, R., Mannens, E., Pfeiffer, S., & van Deursen, D. (2012, March 15). Media fragments URI 1.0: W3c proposed recommendation.
12. Cowie, R., & Cornelius, R. R. (2003). Describing the emotional states that are expressed in speech. *Speech Communication, 40*(1–2), 5–32.
13. Schröder, M., Pelachaud, C., Ashimura, K., Baggia, P., Burkhardt, F., Oltramari, A., et al. (2011). Vocabularies for emotionml. http://www.w3.org/TR/emotion-voc/
14. Cowie, R., Douglas-Cowie, E., Appolloni, B., Taylor, J., Romano, A., & Fellenz, W. (1999). What a neural net needs to know about emotion words. In N. Mastorakis (Ed.), *Computational intelligence and applications* (pp. 109–114). Singapore: World Scientific & Engineering Society Press.
15. Fontaine, J. R. J., Scherer, K. R., Roesch, E. B., & Ellsworth, P. C. (2007). The world of emotions is not two-dimensional. *Psychological Science, 18*(12), 1050–1057.
16. Frijda, N. H. (1986). *The emotions*. Cambridge, UK: Cambridge University Press.
17. Mehrabian, A. (1996). Pleasure-arousal-dominance: A general framework for describing and measuring individual differences in temperament. *Current Psychology, 14*(4), 261–292.
18. Scherer, K. R. (1999). Appraisal theory. In T. Dalgleish & M. J. Power (Eds.), *Handbook of cognition & emotion* (pp. 637–663). New York: Wiley.
19. Gratch, J., & Marsella, S. (2004). A domain-independent framework for modeling emotion. *Cognitive Systems Research, 5*(4), 269–306.
20. Cowie, R., Sawey, M., Doherty, C., Jaimovich, J., Fyans, C., & Stapleton, P. (2013). Gtrace: General trace program compatible with emotionml. In *2013 Humaine Association Conference on Affective Computing and Intelligent Interaction (ACII)* (pp. 709–710). New York: IEEE.
21. Burkhardt, F. (2011). Speechalyzer: A software tool to process speech data. In *Proceedings of the ESSV, Elektronische Sprachsignalverarbeitung*.
22. Burkhardt, F., Polzehl, T., Stegmann, J., Metze, F., & Huber, R. (2009). Detecting real life anger. In *Proceedings ICASSP*, Taipei, Taiwan (Vol. 4).
23. Hantke, S., Appel, T., Eyben, F., & Schuller, B. (2015). iHEARu-PLAY: Introducing a game for crowd sourced data collection for affective computing. In *Proceedings of 1st International Workshop on Automatic Sentiment Analysis in the Wild (WASA 2015)*, Xi'an, P.R. China (pp. 891–897). New York: IEEE.

24. Kouroupetroglou, G., Tsonos, D., & Vlahos, E. (2009). Docemox: A system for the typography-derived emotional annotation of documents. In *Universal Access in Human-Computer Interaction. Applications and Services* (pp. 550–558). New York: Springer.
25. Eyben, F., Weninger, F., Groß, F., & Schuller, B. (2013). Recent developments in openSMILE, the Munich open-source multimedia feature extractor. In *Proceedings of the 21st ACM International Conference on Multimedia, MM 2013*, Barcelona, Spain (pp. 835–838). New York: ACM.
26. Eyben, F., Wöllmer, M., & Schuller, B. (2009, September). openEAR – Introducing the Munich open-source emotion and affect recognition toolkit. In *Proceedings 3rd International Conference on Affective Computing and Intelligent Interaction and Workshops, ACII 2009*, Amsterdam, The Netherlands (Vol. I, pp. 576–581). HUMAINE Association. New York: IEEE.
27. Piana, S., Staglianò, A., Camurri, A., & Odone, F. (2013). A set of full-body movement features for emotion recognition to help children affected by autism spectrum condition. In *IDGEI International Workshop*.
28. Charfuelan, M., & Steiner, I. (2013). Expressive speech synthesis in mary tts using audiobook data and emotionml. In *Proceedings of Interspeech*.
29. Steiner, I., Schröder, M., & Klepp, A. (2013). The PAVOQUE corpus as a resource for analysis and synthesis of expressive speech. *Proceedings of Phonetik & Phonologie* (Vol. 9).
30. Bevacqua, E., Prepin, K., Niewiadomski, R., de Sevin, E., & Pelachaud, C. (2010). Greta: Towards an interactive conversational virtual companion. In *Artificial Companions in Society: Perspectives on the Present and Future* (pp. 143–156).
31. Schröder, M., Bevacqua, E., Cowie, R., Eyben, F., Gunes, H., Heylen, D., et al. (2012). Building autonomous sensitive artificial listeners. *IEEE Transactions on Affective Computing, 3*(2), 165–183.
32. Munezero, M., Kakkonen, T., & Montero, C. S. (2011). Towards automatic detection of antisocial behavior from texts. In *Sentiment analysis where AI meets psychology (SAAIP)* (p. 20).
33. Burkhardt, F., Becker-Asano, C., Begoli, E., Cowie, R., Fobe, G., & Gebhard, P. (2014). Application of emotionml. In *Proceedings of the 5th International Workshop on Emotion, Sentiment, Social Signals and Linked Open Data (ES3LOD)*.

Chapter 5
Introduction to SCXML

Jim Barnett

Abstract SCXML is a control flow language based on Harel State Charts. It offers powerful, application-independent control constructs, along with a plug-in capability that allows platforms to customize the language for specific domains. This paper offers an overview of the language along with examples of its use.

5.1 Overview

SCXML [1] is a modality-independent control flow language that is adaptable to a wide range of tasks. It was developed in the W3C's Voice Browser Working Group as part of an attempt to separate control flow from the user interaction. The specific motivation for SCXML was the observation that in VoiceXML [2], control flow and user interaction are not cleanly separated, and, as a result, it could be difficult to re-use VoiceXML markup. For example, a company might develop a VoiceXML routine to collect a caller's credit card number as part of a retail application, and then want to re-use it for a customer service application. At an abstract level, the logic for collecting the credit card number is the same in the two applications, but the relevant voice interaction markup in the retail application is tangled up with control flow logic that is specific to that application (for example, deciding where to go next in the retail application once we have gathered the credit card number). Thus re-using the credit card collection routine was not a simple cut-and-paste operation, but a complex re-write that attempted to disentangle the user interaction logic from the control flow logic. The Voice Browser Group therefore decided to develop SCXML as a modality-independent control flow language. Although the Group's primary interest was centered on interactive voice dialogs, the fact that SCXML is a pure control flow language means that it can be used for a wide variety of modalities and tasks.

J. Barnett (✉)
Department of Architecture Team, Genesys, Daly City, CA, USA
e-mail: jim.barnett@genesys.com

D.A. Dahl (ed.), *Multimodal Interaction with W3C Standards*,
DOI 10.1007/978-3-319-42816-1_5

SCXML is a state machine language, and is based on Harel State Charts [3]. Professor David Harel developed a powerful, compact, graphical state machine notation, which is capable of expressing anything from simple event handlers to complex, loosely coupled processes operating in parallel. Harel's goal was to model reactive systems, namely those that respond to their environment. It is possible to use Harel State Charts (or SCXML) to model a fixed, predictable process flow—do x then 2 s later do y then immediately do z—but their expressiveness is wasted in such applications. They are best suited to situations where we don't know exactly what will happen next or when it will happen, where we have to react to whatever happens whenever it happens. This uncertainty is a basic feature of user interface applications; in voice applications we don't know when the user will speak or what he will say. Multimodal applications add an additional layer of uncertainty, since we don't know which modality the user will select—will he speak, click on an icon, or do both at once? Such applications are interactive, rather than simply reactive, since we can guide users by offering them alternatives or prompting them for needed information. Harel State Charts (and SCXML) turn out to be equally well suited to interactive applications and purely reactive ones.

Another advantage of Harel State Charts is that they are part of UML [4], so they have been widely used for design of a broad variety of applications. However Harel State Charts and UML are purely graphical languages, used only at the design time. The task for SCXML is to define the desired run-time behavior.

The definition of the run-time consists of specifying the behavior of the SCXML interpreter, describing how it interprets SCXML markup, and, in particular, how it processes and generates events. The bulk of the SCXML specification is devoted to providing these definitions. Furthermore, as part of the W3C standardization process, the Voice Browser Working Group produced a suite of tests that demonstrated that there were multiple interoperable implementations of specification [5]. The specification also includes a pseudo-code algorithm that shows how one might implement the language, but implementers are not required to follow it (as long as their behavior matches the requirements of the specification).

One of the most important parts of the run-time specification consists in defining the interpreter's interactions with underlying platform. SCXML is a pure control flow language so it can't actually *do* anything without hooks to interact with its environment. Furthermore, these hooks must be extensible to allow the language to be adapted to different domains. SCXML therefore is based on a pluggable design consisting of the core language plus platform-specific plugins. The specification defines some plugins that the Voice Browser Group thought would be commonly useful, but platforms are free to add their own.

SCXML offers four kinds of plugability:

- Invocation. An SCXML script can invoke other scripts. For example, an SCXML state machine representing a voice user interface might invoke VoiceXML scripts to interact with the user.
- Executable Content. An SCXML interpreter can extend the language with new primitives that expose capabilities of the underlying platform. For example, if

SCXML was being used to control a robot, the platform might offer primitives to rotate the robot, to move its arm, and to grasp an object.

- Data Model. An SCXML interpreter may define one or more data plus scripting languages that can be used to store values and perform calculations. Since ECMAScript is such an obvious candidate for this purpose, the SCXML specification defines an ECMAScript data model, but platforms are free to define others if they chose.
- Event I/O Processor. An SCXML interpreter may offer various event transport mechanisms that provide scripts the ability to exchange events/messages with other components in the environment. As was the case with ECMAScript, the Voice Browser Group considered HTTP to be widely useful for event transport, so the SCXML specification defines a basic HTTP transport mechanism, but platforms may define other transports (or none at all, in the case of applications where no communication with outside components is needed).

SCXML is defined so that these various plugins can be used without changing the semantics of the core state machine language. The result is a powerful control flow language that can be customized for wide variety of applications by plugging in customized components that are suitable for the domain in question.

5.2 Basic State Machine Concepts

At the heart of state machines is the concept of a *state*, which represents the condition of an object or process. For example, consider a light which has two states: on and off. When the light goes from on to off or vice versa, we say it takes a *transition* between those states. Transitions are triggered by *events* which represent actions in the world. In the case of the light, the relevant event is toggling its switch, which moves it from state "`on`" to state "`off`" or from state "`off`" to state "`on`." This simple pair of states is represented in SCXML as follows:

```
<state id="on">
  <transition event="toggle" target="off"/>
  </state>
<state id="off">
  <transition event="toggle" target="on"/>
</state>
```

The "`event`" attribute in `<transition>` specifies the event which triggers the transition, and the "target" attribute specifies the state that the system will be in after the transition is taken. If an event does not trigger a transition, it is discarded. Thus in the example above, events other than "`toggle`" will simply be ignored because they do not trigger any transitions.

To turn this pair of states into a full state machine, we would add a <scxml>...</scxml> wrapper element which specifies, among other things, whether the state machine starts in state "on" or state "off."

5.3 Extensions to Basic State Machines

5.3.1 Data Model

The simple state machine defined in the previous section consisting of states, events, and transitions is a finite state automation such as you will encounter in a course on formal language theory. Harel enriched this basic notation with additional concepts that increase the expressiveness of the language. One such feature is the data model, which allows application authors to declare variables and assign values to them.[1] The following state machine skeleton declares a variable called "counter" and initializes it to 0.[2]

```
  <scxml.....>
  <datamodel>
     <data id="counter" expr="0"/>
  </datamodel>
..........
</scxml>
```

In addition to initializing the value of a variable, we can assign new values to it at run-time using the <assign> element. We also have <onentry> and <onexit> elements as children of <state> that contain operations (such as <assign>) that are executed whenever the state is entered or exited. Expanding our skeleton example, we add an "enterPassword" state and increment the "counter" variable each time we enter it (in this and subsequent examples, we use italics to highlight the relevant parts of the state machine):

```
<scxml.....>
  <datamodel>
    <data id="counter" expr="0"/>
  </datamodel>
  <state id="enterPassword">
    <onentry>
```

[1] The term "data model" is the SCXML equivalent for what Harel called an "action language."

[2] An SCXML implementation will support one or more data model languages, for example, ECMAScript or XPath. To avoid the syntactic details of specific languages, the examples in this paper use a neutral notation whose meaning should be intuitively clear.

```
      <assign location="counter" expr="counter + 1"/>
    </onentry>
  </state>
..........
</scxml>
```

The "counter" variable is thus keeping track of the number of times we enter the "enterPassword" state. This is useful because the data model can be used to place conditions on transitions, via the "cond" attribute of the <transition> element. We will complete our example to represent a typical login sequence in which the user either succeeds or is locked out after five unsuccessful attempts. We add a "loggedIn" state and a "lockedOut" state, and also add an "initial" attribute to <scxml> indicating that we start off in the "enterPassword" state. We also add three transitions from the "enterPassword" state. One transition takes us to the "loggedIn" state if we get an "accepted" event (meaning that the password has been accepted). Another transition takes us back to the "enterPassword" state if we get a "rejected" event and "counter" is less than or equal to 5. (Exiting and re-entering the "enterPassword" state will increment "counter.") A third transition takes us to the "lockedOut" state when we get a "rejected" event and "counter" is greater than 5.

```
<scxml initial="enterPassword"....>
  <datamodel>
    <data id="counter" expr="0"/>
  </datamodel>
  <state id="enterPassword">
    <onentry>
      <assign location="counter" expr="counter + 1"/>
    </onentry>
    <transition event="accepted" target="loggedIn"/>
    <transition  event="rejected"  cond="counter  <=  5"
target="enterPassword"/>
    <transition event="rejected" cond="counter>=6"
target="lockedOut"/>
  </state>
  <state id="loggedIn"/>
  <state id="lockedOut"/>
</scxml>
```

5.4 Operations and Conditions

In addition to <assign>, SCXML supports an <if>, <elseif>, <else> conditional construction, and a <foreach> construction that allows iteration over the members of a collection (if the underlying data model supports arrays or lists), a <log> element that logs a message in a platform-dependent manner, and a <raise> element that generates an event. It is possible to place these operations inside <onentry>, <onexit>, or inside <transition>, where it will be executed when the transition is taken. Finally, wherever a single operation can occur, it is possible to place multiple operations, which will be executed in document order. Here is an example using the <log> tag to show the order of operations when a transition is taken:

```
<scxml initial="state1"....>
  <state id="state1">
    <transition event="e" target="state2">
      <log expr="In transition\n"/>
    </transition>
    <onexit>
      <log expr="Leaving state1/n"/>
    </onexit>
  </state>
  <state id="state2">
    <onentry>
      <log expr="Entering state2"/>
    </onentry>
  </state>
</scxml>
```

This state machine starts off in state1. When event "e" occurs, it transitions to state2. When the transition is taken, we first execute the operations in state1's <onexit> handler. Then we execute those in the <transition> itself, then those in state2's <onentry> handler. Thus when the transition completes, the log file will contain:

```
Leaving state1
In transition
Entering state2
```

Note that the order in which the log expressions are printed is different from the order in which they occur in the SCXML document. What matters is the order in which the state machine executes its operations.

Here is a somewhat contrived example showing the ordering of operations and the use of <if>:

```
<scxml initial="state1"....>
  <datamodel>
    <data id="counter" expr="0"/>
  <state id="state1">
    <onentry>
      <assign location="counter" expr="counter + 1"/>
    </onentry>
    <transition event="e" target="state2">
      <if expr="counter > 1">
        <raise event="event2"/>
      </if>
      <assign location="counter" expr="counter - 1"/>
    </transition>
    <onexit>
      <if expr="counter > 1">
        <raise event="event1"/>
      </if>
      <assign location="counter" expr="counter + 1"/>
    </onexit>
  </state>
  <state id="state2">
    <onentry>
      <if expr="counter > 1">
        <raise event="event3"/>
      <else>
        <raise event="event4"/>
      </if>
    </onentry>
  </state>
</scxml>
```

Again we start off in state1. The data model initializes variable "counter" to 0, but the <onentry> element of state1 increments it to 1. When event "e" is raised we go to state2. (In this example we assume that "e" is raised by some external entity.) We execute the <onexit> code in state1. The <onexit> code consists of an <if> statement followed by an <assign> statement. The <if> statement raises "event1" if "counter" is greater than 1. "counter" is not greater than 1, so we don't raise "event1". The following <assign> statement increments "counter" to 2, and we now execute the statements inside the <transition>. The first statement is an <if> that checks that "counter" is now greater than 1, and then raises "event2". The next statement is an <assign> that decrements "counter." Now we enter state2. Its <onentry> block consists of an <if> statement with an <else> clause. The <if> statement raises "event3" if "counter" is greater than 1. It is not, so we execute the <else> clause which raises "event4". Thus the result of taking the transition from state1 to state2 is to generate events "event2" and "event4" (but not events "event1" and "event3"). "counter" is equal to 1 when we start to take the transition, and when we finish it, but it does have the value 2 for a moment during the transition, which is why "event2" gets raised. (To be precise, in Harel's state chart model all transitions and operations are instantaneous, so the "moment" during which "counter" equals 2 has no duration. In an SCXML implementation, of course, nothing is instantaneous, so there will be a very short interval of time during with "counter" will have the value 2.)

5.5 Executable Content

These operations that can occur in <onentry>, <onexit>, and inside transitions, along with the <if> conditional and <foreach> are called "executable content." Executable content occurs in blocks consisting of one or more elements, which are executed in document order. If the execution of an element of executable content causes an error, the remaining elements in the block are not executed. This leads to one subtlety in the definition of SCXML. Multiple instances of <onexit> and <onentry> are allowed inside <state>. When multiple instances occur, the executable content they contain is executed in document order (when the state is entered for <onentry> and when it is exited for <onexit>). The only difference between a single large block of executable content and multiple smaller blocks involves error handling. Consider the case below, an <onentry> element containing three elements of executable content:

```
<onentry>
  <content1/>
  <content2/>
  <content3/>
</onentry>
```

Suppose that the execution of <content2> causes an error. The execution of that block of executable content will terminate, and <content3> will not be executed. If we want <content3> to be executed even if there is an error in an earlier element, we can put it in a separate block (i.e., in a separate <onentry> element:

```
<onentry>
  <content1/>
  <content2/>
</onentry>
<onentry>
  <content3/>
</onentry>
```

Since <content3> is in a separate block, its execution is not affected by errors in other, earlier blocks of executable content.

5.6 External Communications

One of the most useful pieces of executable content is the <send> element, which can be used to send events/messages to external entities.[3] The format of the message and the means of delivery are specified by the Event I/O Processor that is used. The SCXML specification defines a Basic HTTP Event I/O Processor that delivers messages by HTTP POST, and platforms are free to define other ones. As an example of <send>, suppose that an online store accepts orders at the URI http://somestore.com/orders. We can use <send> to place an order as follows:

```
<send event="newOrder" target="http://somestore.com/
orders" type=http://www.w3.org/TR/scxml/
#BasicHTTPEventProcessor>
  <param name="username" location="..."/>
  <param name="password" location="..."/>
  <param name="firstItem" location="..."/>
  <param name="secondItem location="..."/>
.........
</send>
```

In this example, the "event" parameter gives the name of the message, "target" gives the URI to deliver the message to, and "type" is used to specify that the Basic HTTP Event I/O Processor is to be used. (Implementations are allowed to come up with shorter, more user-friendly names for Event I/O Processors, such as type="HTTP.") The <param> elements are used to define HTTP POST parameters to include with the message.[4] The values of the "name" attribute can be anything, and depend on what the receiving end expects (so in this case, we must know that somestore.com expects the items in the order to be labelled "firstItem," "secondItem," etc.). The "location" attributes fetch the corresponding values from the data model. (The "location" attribute is also used with <assign>. Its value can be a path expression referring to any part of the data model, not just the top level variables that are declared with <data>. This

[3] <send> can also be used to raise events in the current SCXML session or in other SCXML sessions.

[4] <send> also has an optional "namelist" attribute that can be used in place of <param> to include data in the message. The value of "namelist" is a series of space-separated data model locations. The names and values at those locations will be included as key-value pairs in the message. <param> is a more general and flexible method of incorporating data into the message, but "namelist" can be a useful shorthand notation.

is useful in data models based on languages like ECMAScript or XPath, where the value of a top level variable may be a complex tree structure.)

Event I/O Processors can be used to receive events as well as send them. Suppose somestore.com replies to all orders by sending back an "orderAccepted" or "orderRejected" event. We can now send the order and use transitions to handle these replies:

```
<state id="placingOrder">
  <onentry>
    <send target=http://somestore.com/orders....>
.........
    </send>
  </onentry>
  <transition event="orderAccepted"
target="orderSuccessful"/>
  <transition event="orderRejected"
target="orderFailed"/>
</state>
```

The "orderFailed" states will contain the logic to decide whether to retry the order or to skip it altogether, while the "orderSuccessfull" state can proceed to the next stage of processing knowing that the order has been accepted. (In general, it is simpler to split the different outcomes into different states, rather than having one big messy state that attempts to handle both success and failure.)

One other useful feature of <send> is the "delay" attribute which can be used to delay the delivery of the event, along with the <cancel> element, which can be used to cancel a delayed event that hasn't been delivered yet. These features are useful primarily for setting timers. Suppose we are worried that somestore.com might not deliver any response to our order. Given the way the "placingOrder" state is defined above, it will sit there forever waiting for either the "orderAccepted" or "orderRejected" event. We can break out of that situation by sending a delayed event as a timer in the following way:

```
<state id="placingOrder">
  <onentry>
    <send target=http://somestore.com/orders....>
.........
    </send>
    <send delay="30s" event="timerExpired" id="timer1"/>
  </onentry>
  <transition event="orderAccepted"
target="orderSuccessfull">
    <cancel sendid="timer1"/>
  </transition>
  <transition event="orderRejected"
```

```
target="orderFailed">
    <cancel sendid="timer1"/>
  </transition>
  <transition event="timerExpired" target="noResponse"/>
</state>
```

Here in the <onentry>, right after we send the order to somestore.com, we send a "timerExpired" event that is to be delivered after 30 s. Since we do not specify a "target" URI for this event, the event will be raised in this session (i.e., the one in which the delayed <send> tag was executed). We now add <cancel> elements as executable content to the transitions on "orderAccepted" and "orderRejected." (The "sendid" attribute on the <cancel> elements must match the "id" element on the <send> because there might be multiple delayed <send>s in this session, and we need to know which one to cancel.) Thus if we get the "orderAccepted" or "orderRejected" event within 30 s of the time we submitted the order, the "timerExpired" event will be cancelled (meaning it will never be raised). Therefore if the "timerExpired" event occurs, we know that 30 s have expired and we have not received either "orderAccepted" or "orderRejected" from somestore.com. We now move to the "noResponse" state, which will determine what to do next. For example, if we think that the lack of response was due to an intermittent connectivity problem, the "noReponse" state could set its own timer, wait a minute or so, and then return to the "placingOrder" state to try again. (The "noResponse" state would need to use a counter, as shown above in the password example, to keep from looping back to "enterOrder" indefinitely in the case where the connectivity problem was persistent.)

5.7 Invoking Platform Resources

The <invoke> element provides another way to interact with platform resources. As its name implies, it is intended to invoke processing by such a resource. Suppose we have an SCXML application that is managing a voice-based order processing system. As part of handling an order, we need to get the customer's address. The following snippet of SCXML markup invokes a VoiceXML script to do this:

```
<state id="address">
   <invoke   type="vxml"   src="http://somestore.com/
orders/getAddress.vxml/">
</state>
```

When we enter the "address" state, the platform will start up an instance of the "getAddress.vxml" script.[5] The "type" attribute indicates the kind of resource to invoke, and the "src" attribute specifies the markup that the resource should execute.[6] We can think of the <invoke> element as delegating the activity of the "address" state to the VoiceXML interpreter. In specific, if we leave the "address" state while the VoiceXML interpreter is still running, its execution will be cancelled. On the other hand, if the VoiceXML interpreter terminates while we are still in the "address" state, the platform will automatically generate a "done.invoke" event to indicate the completion. We can add a transition triggered by the "done" event to take us to the next state in the order processing once the address has been collected:

```
<state id="address">
    <invoke type="vxml" src="http://somestore.com/
orders/getAddress.vxml/">
  <transition event="done.invoke" target="creditCard"/>
</state>
```

On the other hand, suppose that the "cancel" event indicates that order processing should be cancelled. We can add a transition for this event out of the "address" state and the platform will shut down the VoiceXML processing cleanly and deallocate any voice resources:

```
<state id="address">
  <invoke type="vxml" src="http://somestore.com/orders/
getAddress.vxml/">
  <transition event="done.invoke" target="creditCard"/>
  <transition event="cancel" target="terminateOrder"/>
</state>
```

All SCXML implementations must support SCXML scripts invoking other SCXML scripts (thus providing a simple form of state machine re-use), but otherwise the set of invocable resources is platform-specific. SCXML platforms that support voice user interfaces are likely to support invocation of VoiceXML, but they are not required to do so. When SCXML scripts invoke other SCXML scripts, they may communicate using the <send> element, and return data in the "done.invoke" event. In the case of other types of resources, the communication mechanisms will be platform-specific. In the example above, for instance, the

[5] We will see below that this is not strictly true. The platform will not invoke the VoiceXML script if it can determine that it is about to leave the "address" state immediately.

[6] The SCXML specification defines "http://www.w3.org/TR/voicexml21/" as the proper value for "type" to indicate a VoiceXML 2.1 interpreter, but platforms are allowed to support author-friendly shorthand notations such as "vxml."

platform would have to specify how the VoiceXML interpreter returned the address to the SCXML script, whether in the "`done.invoke`" event or by some other platform-specific method. (The example above is incomplete in that the address gathered by the VoiceXML script is not added to the SCXML data model.)

As in the case of `<send>`, `<param>` and "`namelist`" can be used to pass data from the SCXML session to the invoked process.

Multiple instances of `<invoke>` are permitted in a state. Suppose that we want to offer the user the choice of providing information either by voice or by typing in a GUI. We can invoke the voice user interface and the GUI at the same time. One of the two will complete and return a "`done.invoke`" event, while the other will just sit there waiting for input. In this case, though, we will want to know which one of the invocations completed since subsequent processing may depend on whether the user choose voice or GUI input. We can add an "`id`" attribute to the `<invoke>` element, and the platform will use that value as part of the done event[7]:

```
<state id="address">
  <invoke type="vxml" id="voice" src="http://somestore.
com/orders/getAddress.vxml/">
  <invoke type="html" id="GUI" src="http://somestore.
com/orders/getAddress.html/">
  <transition event="done.invoke.voice"
target="continueVoiceInput"/>
  <transition event="done.invoke.GUI"
target="continueGUIInput"
  <transition event="cancel" target="terminateOrder"/>
</state>
```

In fact, there is always an id at the end of the "`done.invoke`" event, because the platform will generate one if the author does not provide it. The reason that `<transition event="done.invoke".../>` is triggered though the event's name is "done.invoke.`<someID>`" (where `<someID>` can be any string) is that SCXML does prefix matching on event names. Specifically, an event name consists of tokens separated by the period " . ". A transition matches an event if the value of the "`event`" attribute consists of a string of tokens that is a prefix of the tokens in the event name. Thus `<transition event="done.invoke".../>` matches events named "`done.invoke.foo`," "`done.invoke.foo.bar`," and "`done.invoke.anything`". In fact, `<transition event="done".../>` also matches all of these events. However, `event="done.invoke"` does not match "`done.inv`" since "`invoke`" and "`inv`" are different tokens.

[7] In this example, we assume that the platform lets us use `<invoke>` with type "`html`" to display an HTML page. This would be platform-specific functionality and is not part of the SCXML standard.

In addition, the "event" expression in a transition may contain multiple, space separated event descriptions, and the transition matches an event if the event name matches any of the descriptions. For example, <transition event="this that the.other".../> will match events "this.now," "that.then," and "the.other.thing," but not "the" (since "the.other" is not a prefix of "the"). In general, SCXML transitions match sets of events, with the set being larger or smaller depending on how specific the "event=..." expression is.

5.8 Compound States

Consider a complex retail application that must get the user's credit card information as part of its processing. We can think of "getting credit card information" as a single state in that application, or we can break it down into sub tasks, such as "getting first name," "getting last name," "getting card type," "getting card number," etc. In Harel's model, we call "get credit card information" a compound state, namely one that contains multiple substates. In SCXML, this is represented as follows:

```
<state id="getCCInfo">
  <state id="getFirstName"...../>
  <state id="getLastName".../>
.........
  <state id="getCCNumber".../>
.........
</state>
```

When the state machine is in the "getCCInfo" state, it is always in one and only one of its substates (which one it is in will change as processing proceeds). Furthermore, the substates themselves may be complex. For example, the "getCCNumber" state might itself consist of two steps: querying for the number and then confirming it with the user. The resulting nested state structure would look like this:

```
<state id="getCCInfo">
  <state id="getFirstName"...../>
  <state id="getLastName".../>
.........
  <state id="getCCNumber"...>
    <state id="queryCCNumber"..../>
    <state id="confirmCCNumber".../>
</state>
.........
</state>
```

As a result of this nested structure, an SCXML state machine may be in multiple states at the same time. For example, if the state machine is in "confirmCCNumber," it is also in "getCCNumber" and "getCCInfo." A state with no child <state> elements is called an atomic state, while one with <state> children is called a compound state. ("confirmCCNumber" and "queryCCNumber" are atomic states, while "getCCNumber" and "getCCInfo" are compound states.)

A state machine moves among the substates of a complex state by taking transitions. For example, suppose we have a VoiceXML script that we can invoke to capture the user's first name, and that we want to ask for his last name once we have the first name. The resulting state structure would look like this:

```
<state id="getCCInfo">
  <state id="getFirstName">
    <invoke src="firstName.vxml" type="vxml"/>
    <transition event="done.invoke"
target="getLastName"/>
  </state>
  <state id="getLastName".../>
.........
  <state id="getCCNumber"...>
    <state id="queryCCNumber"..../>
    <state id="confirmCCNumber".../>
</state>
.........
</state>
```

Transitions may take atomic or compound states as targets. When considered as the target of a transition, a compound state bears a certain resemblance to a sub-routine. Suppose the retail application containing the credit card logic is being developed by multiple people. The person responsible for the high-level logic can insert a transition to "getCCInfo" without needing to know anything about the internal structure of its substates. That internal structure can be changed—for example, by making "getLastName" a compound state—without requiring modifications to any state or transition that takes the parent state "getCCInfo" as its target.

When a transition takes a compound state like "getCCInfo" as its target, the state machine needs to know which of its substates it should start in. We therefore introduce an "initial" attribute on compound states, similar to the one on <scxml>.

```
<state id="getCCInfo" initial="getFirstName">
  <state id="getFirstName">
    <invoke src="firstName.vxml" type="vxml"/>
    <transition event="done.invoke"
target="getLastName"/>
```

```
  </state>
  <state id="getLastName".../>
..........
  <state id="getCCNumber" initial="queryCCNumber">
    <state id="queryCCNumber"..../>
    <state id="confirmCCNumber".../>
  </state>
..........
</state>
```

In this example, if a transition takes "`getCCInfo`" as its target, it will start off in substate "`getFirstName`." Similarly, if a transition takes "`getCCNumber`" as its target, the state machine will start off in substate "`queryCCNumber`."[8] We refer to states specified by the "`initial`" attribute as default initial states, because they can be overridden by transitions that explicitly target a substate. For example, if a transition takes "`getLastName`" as its target, the "`initial`" attribute on "`getCCInfo`" will have no effect and the state machine will go directly to "`getLastName`."

A compound state may also have a <`final`> child element. As the name indicates, entering a <`final`> child indicates that the compound state has finished its processing (that is, moving through its child states). When the state machine enters the <`final`> child of a compound state, it automatically raises the event "`done.state.`<`id`>", where "<`id`>" is the id of the parent state. This event can serve as a signal to the rest of the state machine that it is time to leave the compound state. Expanding our example, we add a transition from "`getCCInfo`" indicating that it should go to state "`prepareShipment`" when it terminates its processing (that is, when it enters final state "`ccDone`").[9]

```
<state id="getCCInfo" initial="getFirstName">
  <transition event="done.state.getCCInfo" target="pre-
pareShipment"/>
  <final id="ccDone"/>
  <state id="getFirstName">
```

[8] The logic of default initial states is somewhat more complex than the example indicates. First of all, the value of the "`initial`" attribute can be any descendent state, not just an immediate child. Second, a state may contain a child <`initial`> element which contains a <`transition`> instead of the "`initial`" attribute, If such an <`initial`> child is present, the state machine uses that transition to determine the initial descendent state. The only practical difference between an "`initial`" attribute and an <`initial`> element is that the transition in the latter may contain executable content, which will be executed when and if the transition is used to determine the default initial state. Finally, if neither an "`initial`" attribute nor an <`initial`> child is present, the default initial state is the first child state in document order.

[9] <`final`> states may have <`onentry`> and <`onexit`> children, as well as a <`donedata`> child that specifies data to be included in the "`done.state.`<`id`>" event, but they may not contain transitions or substates.

```
    <invoke src="firstName.vxml" type="vxml"/>
    <transition event="done.invoke"
target="getLastName"/>
  </state>
  <state id="getLastName".../>
.........
  <state id="getCCNumber" initial="queryCCNumber">
    <state id="queryCCNumber"..../>
    <state id="confirmCCNumber".../>
</state>
.........
</state>
```

5.9 Selecting Transitions

SCXML documents contain a large number of `<transition>` elements. Transitions may occur in atomic or compound states, and multiple transitions can occur in any state.[10] The location of the transitions within the document is significant since the SCXML interpreter follows a specific order in deciding which transition to take. First of all, the SCXML interpreter only considers transitions in the states that it is currently in. (We refer to these as the active states.) Second, the interpreter considers the transitions in an active state in document order. Consider the following example:

```
<state id="state1">
  <transition event="e" cond="x>1" target="state2"/>
  <transition event="e" cond="x<2" target="state3"/>
</state>
```

If the state machine is in "`state1`" and event "`e`" occurs, the interpreter will consider the first transition and evaluate its condition. If "`x>1`" is true, it will take the transition and go to "`state2`," even if "`x<2`" also evaluates to true. Only if "`x>1`" evaluates to false will it consider the second transition. It will transition to "`state3`" only if "`x>1`" is false and "`x<2`" is true.

Finally, the SCXML interpreter looks for transitions first in the active atomic state, and only considers transitions in its parent (and the parent of its parent, etc.) if it does not find a matching transition in the child.[11] This is useful because

[10] Transitions may also occur in parallel states, which are discussed in the next section.

[11] In a state machine consisting only of compound and atomic states, only one atomic state can be active at a time, and when it is active, all its ancestors (parent, parent of parent, etc.) are active. If the state machine includes parallel states, which are discussed in the next section, more than one atomic state can be active.

transitions in ancestor states serve as defaults which are taken only if they are not overridden by transitions their descendent states. Returning to the credit card example, suppose that the system can raise an ‘invalidValue’ error at any point, indicating perhaps a speech recognition failure or some other problem. If we want to transfer to a human operator when this happens, we can put a single transition in the top level “getCCInfo” state:

```
<state id="getCCInfo" initial="getFirstName">
  <transition event="invalidValue" target="transfer-
ToOperator"/>
  <transition event="done.state. CCInfo" target="prepar-
eShipment"/>
  <final id="ccDone"/>
  <state id="getFirstName">
    <invoke src="firstName.vxml" type="vxml"/>
    <transition event="done.invoke"
target="getLastName"/>
  </state>
  <state id="getLastName".../>
.........
  <state id="getCCNumber" initial="queryCCNumber">
    <state id="queryCCNumber"..../>
    <state id="confirmCCNumber".../>
  </state>
.........
</state>
```

Suppose that the state machine is in the “queryCCNumber” state when “invalidValue” is raised. The interpreter will first look in that state for a transition matching the event. It won’t find one so it will next check in the parent state “getCCNumber.” It again will not find a transition matching “invalidValue”, so it will check in the parent state of “getCCNumber.” It will find a transition matching “invalidValue” in “getCCInfo,” so it will take this transition and go to “transferToOperator.” Similar logic will apply if the state machine is in any of the children of “getCCInfo” (or in any of their children, etc.).

Now suppose that we would like to treat the “invalidValue” event differently when the state machine is in “queryCCNumber.” For example, an invalid value might indicate possible fraud so the state machine will go to a special state to check for that. If we put a transition to “checkForFraud” in “queryCCNumber,” the state machine will take that transition if it is in that state when “invalidValue” is raised, but in any other state, it will still take the transition to “transferToOperator” that is in the ancestor “getCCInfo” state:

```
<state id="getCCInfo" initial="getFirstName">
  <transition event="invalidValue" target="transfer-
ToOperator"/>
  <transition event="done.state.CCInfo" target="prepar-
eShipment"/>
  <final id="ccDone"/>
  <state id="getFirstName">
    <invoke src="firstName.vxml" type="vxml"/>
  <transition event="done.invoke" target="getLastName"/>
  </state>
  <state id="getLastName".../>
.........
  <state id="getCCNumber" initial="queryCCNumber">
    <state id="queryCCNumber">
      <transition event="invalidValue" target="check-
ForFraud"/>
    </state>
    <state id="confirmCCNumber".../>
  </state>
.........
</state>
```

There are further subtleties in how SCXML selects transitions. When the state machine enters a new state, it first checks for eventless transitions (ones without an "event" attribute), then for transitions triggered by internal events (events raised by the platform or the <raise> element), and last of all for transitions triggered by external events (events generated by external entities).[12] Consider the following example:

```
<state id="state1">
  <onentry>
    <raise event="event1"/>
  </onentry>
  <transition event="event1" target="state2"/>
  <transition cond="x > 1" target="state3"/>
</state>
```

The <onentry> element of "state1" raises event "event1," and there is a transition triggered by "event1" going to "state2." However there is also an eventless transition with condition "x > 1". The SCXML interpreter will test this

[12] The <send> element can be used to raise either internal or external events.

transition first, and if x is greater than 1 the state machine will go directly to "state3" without considering transitions triggered by "event1."[13] In fact the interpreter will select transitions triggered by "event1" only when it reaches a state that has no eventless transition that can be taken (that is, a state with no transition that lacks an "event" attribute and whose "cond" attribute evaluates to "true"). This sequence of processing events and transitions is also relevant for <invoke>, since the interpreter will execute <invoke> elements only when it has processed all eventless transitions and all transitions triggered by internal events, and is waiting for an external event. In our example above, there would be no point in putting an <invoke> element in "state1" since it would never be executed. Whenever the state machine enters "state1," it goes immediately to either "state2" or "state3," so there is no point in starting an invocation that will be cancelled right away.

Another quirk involves targetless transitions, namely those that lack a "target" attribute. When a state machine takes such a transition, it stays in its current state(s). It is thus different from a transition that takes its parent state as its target. Consider the following example:

```
<state id="state1">
  <onentry>
    <assign location="var1" expr="var1 + 1"/>
  </onentry>
  <transition event="event1" target="state1"/>
  <transition event="event2">
    <assign location="var2" expr="var2 + 1"/>
  </transition>
</state>
```

When the state machine is in "state1" and "event1" occurs, it will take the transition with "target" of "state1." It will therefore exit "state1" and re-enter it, executing the <onentry> element and incrementing "var1". However if "event2" occurs, the state machine will select the targetless transition, and not exit and re-enter "state1," and thus not increment "var1". However it will execute the <assign> element that it contains, and increment "var2". A targetless transition is thus a kind of event handler, which contains executable content to be executed when a certain event occurs, without causing a change of state.[14]

[13] Note that the eventless transition to "state3" is taken even though the transition to "state2" precedes it in the document. The SCXML interpreter uses document order to choose among transitions that have the same type of trigger. So first it considers all eventless transitions in document order, then all transitions triggered by internal events in document order, then all transitions triggered by external events in document order.

[14] A targetless transition that does not contain executable content does nothing. It is harmless but useless.

5.10 Parallel States

We now come to one of the most complex and powerful features of Harel State Charts, namely parallel states. A <parallel> element will have two or more child states, which are normally themselves compound states. When the state machine enters a <parallel> state it also enters *all* of its child states. This is a bit like being in multiple state machines at the same time, but those state machines (i.e., the child states) are not isolated from each other. In particular, the child states see the same set of events, share a data model, and can check to see which states other children are in. As an example of this, consider two companies doing a joint software development project. There is a business side to the work, and a technical side, and the two are somewhat independent of each other. Suppose the business side moves through the following states: preliminary discussions, negotiation of terms, signing of the contract, and joint sales. Meanwhile, the technical side moves through states: discussion of concept, high-level architecture, detailed design, joint testing, sharing of source code, and development complete. The two sides do not move through their states in lock step. In particular, the technical side of the state machine could be in either high-level architecture, detailed design, or joint testing while the business side was in negation of terms or signing contract. To represent this, we start with a <parallel> element with child states "BusinessNegotiations" and "TechnicalWork." Their child states are the ones mentioned above.

```
<parallel id="JointDevelopment">
  <state id="BusinessNegotiations"
initial="preliminaryDiscussions">
    <state id="preliminaryDiscussions".../>
    <state id="negotiationOfTerms".../>
    <state id="signingContract".../>
    <state id="jointSales".../>
  </state>
  <state id="TechnicalWork"
initial="discussionOfConcept">
    <state id="discussionOfConcept".../>
    <state id="highLevelArchitecture".../>
    <state id="detailedDesign".../>
    <state id="jointTesting".../>
    <state id="shareSourceCode".../>
    <state id="developmentDone".../>
  </state>
</parallel>
```

When the state machine enters "JointDevelopment" it also enters both "BusinessNegotations" and "TechnicalWork." Since those two states are themselves compound states, the state machine must be in a single child state for each of them. Since "BusinessNegotiations" has four child states, and "TechnicalWork" has six, there are 24 possible pairs of child states that we could be in. The semantics of the <parallel> element allows all these combinations to exist, just as if "BusinessNegotiations" and "TechnicalWork" were separate state machines.

However, we want to place some constraints on which combinations of states can occur. Suppose, for example, that we don't want "BusinessNegotions" and "TechnicalWork" to leave their initial states ("preliminaryDiscussions" and "discussionOfConcept," respectively) until a Non-Disclosure Agreement has been signed. We can introduce an event, "ndaSigned" and place transitions for it in both the initial states:

```
<parallel id="JointDevelopment">
  <state id="BusinessNegotiations"
initial="preliminaryDiscussions">
    <state id="preliminaryDiscussions">
      <transition event="ndaSigned"
target="negotiationOfTerms"/>
    </state>
    <state id="negotiationOfTerms".../>
    <state id="signingContract".../>
    <state id="jointSales".../>
    </state>
  <state id="TechnicalWork"
initial="discussionOfConcept">
    <state id="discussionOfConcept">
      <transition event="ndaSigned"
target="highLevelArchitecture"/>
    </state>
    <state id="highLevelArchitecture".../>
    <state id="detailedDesign".../>
    <state id="jointTesting".../>
    <state id="shareSourceCode".../>
    <state id="developmentDone".../>
  </state>
</parallel>
```

When the "ndaSigned" event occurs, we select a transition for it in *each* of the active atomic states. Therefore that single event will cause a transition from "preliminaryDicussions" to "negotiationOfTerms" *and* from "discussionOfConcept" to "highLevelArchitecture." (Either one of these states could have ignored this event, of course. If only one of the active atomic states had a transition that matched this event, only that transition would have been taken. If neither had a matching transition, no transition would have been taken.) Thus the fact that the children of a <parallel> state see the same set of events permits coordinated transitions without requiring them.

Now suppose that we add another constraint, namely that source code cannot be shared until the contract has been signed. We don't know how long it will take to sign the contract, or what state inside "TechnicalWork" we will be in when that happens, so we cannot rely on shared events here. Instead we add a "contractSigned" variable in the data model and have the "BusinessNegotiations" state set it to "true" when it leaves the "signingContract" state[15]:

```
<parallel id="JointDevelopment">
  <datamodel>
    <data id="contractSigned" expr="false"/>
  </datamodel>
  <state id="BusinessNegotiations"
initial="preliminaryDiscussions">
    <state id="preliminaryDiscussions">
      <transition event="ndaSigned"
target="negotiationOfTerms"/>
    </state>
    <state id="negotiationOfTerms".../>
    <state id="signingContract">
    <onexit>
      <assign location="contractSigned" expr="true"/>
    </onexit>
    </state>
    <state id="jointSales".../>
    </state>
  <state id="TechnicalWork"
initial="discussionOfConcept">
    <state id="discussionOfConcept">
      <transition event="ndaSigned"
target="highLevelArchitecture"/>
    </state>
    <state id="highLevelArchitecture".../>
    <state id="detailedDesign".../>
    <state id="jointTesting".../>
    <state id="shareSourceCode".../>
    <state id="developmentDone".../>
  </state>
</parallel>
```

[15] The <datamodel> element may occur as a child of <scxml> or of any <state> or <parallel> element.

Now we check this flag in the transition from "jointTesting" to "shareSourceCode":

```
<parallel id="JointDevelopment">
  <datamodel>
    <data id="contractSigned" expr="false"/>
  </datamodel>
  <state id="BusinessNegotiations"
initial="preliminaryDiscussions">
    <state id="preliminaryDiscussions">
      <transition event="ndaSigned"
target="negotiationOfTerms"/>
    </state>
    <state id="negotiationOfTerms".../>
    <state id="signingContract">
      <onexit>
        <assign location="contractSigned" expr="true"/>
      </onexit>
    </state>
    <state id="jointSales".../>
  </state>
  <state id="TechnicalWork"
initial="discussionOfConcept">
    <state id="discussionOfConcept">
      <transition event="ndaSigned"
target="highLevelArchitecture"/>
    </state>
    <state id="highLevelArchitecture".../>
    <state id="detailedDesign".../>
    <state id="jointTesting">
      <transition cond="contractSigned == true"
target="shareSourceCode"/>
    </state>
    <state id="shareSourceCode".../>
    <state id="developmentDone".../>
  </state>
</parallel>
```

In a realistic state machine the condition on the transition would be more complicated, since we would need to check that the joint testing was actually complete, but this example shows how the children of `<parallel>` can communicate via the shared data model.

There is also an "`In()`" predicate that can be used in conditions to check which state another parallel substate is in. We might be tempted to use this predicate to check that the high-level architecture was complete before signing the contract, by checking that the "TechnicalWork" child had advanced to the "detailedDesign" state in the following manner:

```
<parallel id="JointDevelopment">
  <datamodel>
    <data id="contractSigned" expr="false"/>
  </datamodel>
  <state id="BusinessNegotiations"
initial="preliminaryDiscussions">
    <state id="preliminaryDiscussions">
      <transition event="ndaSigned"
target="negotiationOfTerms"/>
    </state>
    <state id="negotiationOfTerms">
      <transition cond="In('detailedDesign')
target="signingContract"/>
    </state>
    <state id="signingContract">
      <onexit>
        <assign location="contractSigned" expr="true"/>
      </onexit>
    </state>
      <state id="jointSales".../>
    </state>
    <state id="TechnicalWork"
initial="discussionOfConcept">
    <state id="discussionOfConcept">
      <transition event="ndaSigned"
target="highLevelArchitecture"/>
    </state>
    <state id="highLevelArchitecture".../>
    <state id="detailedDesign".../>
    <state id="jointTesting">
      <transition cond="contractSigned == true"
target="shareSourceCode"/>
    </state>
    <state id="shareSourceCode".../>
    <state id="developmentDone".../>
  </state>
</parallel>
```

Now the state machine will not transition from "negotiationOfTerms" to "signingContract" unless it is also in the "detailedDesign" state. However this isn't actually what we want. Suppose that the technical work progresses quickly and the state machine has moved to the "jointTesting" state by the time the business side is ready to move from "negotiationOfTerms" to "signingContract." The business side state machine will be blocked and will never move to "signingContract" because the technical side will never return to "detailedDesign." The right thing to do in this case would be to use a flag such as the "contractSigned" on shown above. This example should stand as a warning that it is important to think through the state logic carefully, particularly when using parallel states.[16]

5.11 Conclusion

SCXML was not specifically defined for multimodal applications, but it is well suited to them. First of all, it contains powerful constructs such as <onentry> and <onexit> handlers, conditions on transitions, and compound states, which can express complex logic in a compact, relatively perspicuous notation. In particular, parallel states provide a loose coordination between activities that is well suited to multimodal interfaces, where the various modalities are usually partly, but not completely, independent.

A second reason that SCXML is suitable for multimodal applications is that it offers a variety of means for interacting with the application environment. Multimodal applications need to access a broad variety of platform capabilities and to interact with other components in their environment. <send>, <invoke> and executable content provide hooks into such capabilities.

Furthermore the hooks that SCXML provides are customizable, giving platform developers the ability to modify the language to match their domain. For example, the data model is an important part of just about any application, but no single data model is right for all domains. The SCXML specification defines an ECMAScript data model, but allows platforms to define others. In fact, the specification only specifies how a data model called "ecmascript" must behave. A platform that wanted to use ECMAScript but with a different behavior could define its own data model and call it "ecmascriptTheWayWeThinkItShouldBe." That data model would not be part of the standard, the way the existing "ecmascript" data model is, but applications that used it would still be valid SCXML applications. For example, the Voice Browser Group provided a test suite for SCXML as part of the standardization process [5]. The purpose of the tests was to show that there were interoperable implementations of the language, in the sense that they passed the same set of tests. One implementation passed all the tests for the core

[16] SCXML is one of the most powerful foot-shooting weapons ever developed, and it is easy to write state machines that block or go into infinite loops. Individual implementations or development tools may include safety features such as reachability analysis or non-terminating loop detection, but the specification does not include any safeguards that might restrict the power of the language.

language using a data model that was very different from the ones defined by the Voice Browser Group.[17]

Similarly for <send> the SCXML specification defines an SCXML Event I/O Processor, which can be used to communicate with other SCXML sessions (particularly ones created by <invoke>) and a Basic HTTP Event I/O Processor, which can be used to communicate with any component that can accept or send HTTP messages. However platforms may define other Event I/O Processors to suit their needs. As an example, the Voice Brower Group defined a DOM Event I/O Processor, which can be useful when working inside a web browser, using SCXML to control an HTML page [6].[18]

Finally, platforms can use <invoke> or custom executable content to provide access a broad range of resources. <invoke> is particularly well suited to re-using existing languages such as VoiceXML.

In summary, SCXML combines Harel's powerful, application-independent state machine logic with flexible, customizable hooks into platform, and environment resources. The result is a control flow language that can be adapted to a wide variety of domains, including multimodal applications.

References

1. Barnett, J., Akolkar, R., Auburn, R., Bodell, M., Burnett, D., Carter, J., et al. (2015). State chart XML (SCXML) state machine notation for control abstraction. W3C Recommendation. https://www.w3.org/TR/scxml/.
2. Oshry, M., Auburn, R., Baggia, P., Bodell, M., Burke, D., Burnett, D., et al. (2007). Voice extensible markup language (VoiceXML) 2.1. W3C Recommendation. https://www.w3.org/TR/2007/REC-voicexml21-20070619/.
3. Harel, D., & Politi, M. (1998). *Modeling reactive systems with statecharts: The STATEMATE approach.* http://www.wisdom.weizmann.ac.il/~dharel/reactive_systems.html.
4. Object Management Group (2009). UML specification version 2.3. http://www.omg.org/spec/UML/2.3/.
5. Barnett, J. (2015). SCXML implementation report. https://www.w3.org/Voice/2013/scxml-irp/.
6. Barnett, J., Akolkar, R., Auburn, R., Bodell, M., Burnett, D., Carter, J., et al. (2015). DOM event I/O processor for SCXML. W3C Working Group Note. https://www.w3.org/TR/scxml-dom-iop/.
7. Barnett, J., Akolkar, R., Auburn, R., Bodell, M., Burnett, D., Carter, J., et al. (2015). XPath data model for SCXML. W3C Working Group Note. https://www.w3.org/TR/scxml-xpath-dm/.

[17] In addition to the "ecmascript" datamodel, the Voice Browser Group also defined an XPath data model [7]. It was removed from the final specification because the group did not receive enough implementation reports for it.

[18] Like the XPath Data Model, the DOM Event I/O Processor was removed from the final specification due to the lack of sufficient implementation reports.

Chapter 6
Dialogue Act Annotation with the ISO 24617-2 Standard

Harry Bunt, Volha Petukhova, David Traum, and Jan Alexandersson

Abstract This chapter describes recent and ongoing annotation efforts using the ISO 24617-2 standard for dialogue act annotation. Experimental studies are reported on the annotation by human annotators and by annotation machines of some of the specific features of the ISO annotation scheme, such as its multidimensional annotation of communicative functions, the recognition of each of its nine dimensions, and the recognition of dialogue act qualifiers for certainty, conditionality, and sentiment. The construction of corpora of dialogues, annotated according to ISO 24617-2, is discussed, including the recent DBOX and DialogBank corpora.

6.1 Introduction

The ISO 24617-2 annotation standard [10, 11, 30] has been designed for the annotation of spoken, written and multimodal dialogue with information about the dialogue acts that make up a dialogue, with the aim to create interoperable annotated resources. A dialogue act is a unit in the description of communicative behaviour

H. Bunt (✉)
Tilburg Center for Cognition and Communication (TiCC), Tilburg University, Tilburg, The Netherlands
e-mail: bunt@uvt.nl

V. Petukhova
Spoken Language Systems Group, Saarland University, Saarbrücken, Germany
e-mail: v.petukhova@lsv.uni-saarland.de

D. Traum
Institute fro Creative Technologies, University of Southern California, Los Angeles, CA 90094, USA
e-mail: traum@ict.usc.edu

J. Alexandersson
Department of Intelligent User Interfaces DFKI, Gernan Research Center for Artificial Intelligence, Saarbrücken, Germany
e-mail: janal@dfki.de

D.A. Dahl (ed.), *Multimodal Interaction with W3C Standards*,
DOI 10.1007/978-3-319-42816-1_6

that corresponds semantically to certain changes that the speaker wants to bring about in the information state of an addressee. ISO 24617-2 defines a dialogue act as

(1) *Communicative activity of a dialogue participant, interpreted as having a certain communicative function and semantic content.*

The communicative function of a dialogue act, such as *Propositional Question, Inform, Confirmation, Request, Apology,* or *Answer*, specifies how the act's semantic content changes the information state of an addressee upon understanding the speaker's communicative behaviour.

According to the annotation schemes that existed prior to the establishment of ISO 24617-2 and its immediate predecessor DIT^{++}, such as DAMSL; MRDA; HCRC Map Task; Verbmobil; SWBD-DAMSL; and DIT,[1] dialogue act annotation consisted of segmenting a dialogue into certain grammatical units and marking up each unit with one or more communicative function labels. The ISO 24617-2 standard supports the annotation of dialogue acts in semantically more complete ways by additionally annotating the following aspects:

Dimensions The annotation scheme supports 'multidimensional' annotation, where multiple communicative functions may be assigned to dialogue segments; different from DAMSL and other multidimensional schemes, the ISO scheme uses an explicitly defined notion of 'dimension', which corresponds to a certain type of semantic content.

Qualifiers are defined for expressing that a dialogue act is performed conditionally, with uncertainty, or with a particular sentiment.

Functional and feedback dependence relations link a dialogue act to other units in a dialogue, e.g. for indicating which question is answered by a given answer, or which utterance a speaker is providing feedback about.

Rhetorical relations may optionally be annotated to indicate, e.g. that one dialogue act motivates the performance of another dialogue act.

The following example illustrates the use of dimensions, communicative functions, qualifiers, dependence relations, and rhetorical relations (where `"#fs1"`, `"#fs2"`, and `"#fs3"` indicate the segments in P1's and P2's utterances that express a dialogue act—see Section 6.2.2 for more on segmentation).

(2) 1. P1: Is there an earlier connection?
2. P2: Ehm,.. no, unfortunately there isn't.

```
<diaml xmlns:"http://www.iso.org/diaml/">
<dialogueAct xml:id="da1" target="#fs1"
   sender="#p1" addressee="#p2"
   communicativeFunction="propositionalQuestion" dimension="task"/>
<dialogueAct xml:id="da2" target="#fs2"
   sender="#p2" addressee="#p1"
   communicativeFunction="stalling" dimension="timeManagement"/>
```

[1]See Allen and Core [2], Dhillon et al. [19], Carletta et al. [16], Jurafsky et al. [32], Alexandersson et al. [1], Bunt [4, 5].

```
<dialogueAct xml:id="da3" target="#fs2"
   sender="#p2" addressee="#p1"
   communicativeFunction="turnTake" dimension="turnManagement"/>
<dialogueAct xml:id="da4" target="#fs3"
   sender="#p2" addressee="#p1"
   communicativeFunction="answer" dimension="task" sentiment="regret"
   functionalDependence="#da1"/>
</diaml>
```

The development of ISO 24617-2 was supported by annotation experiments in which preliminary versions of the scheme were tested for their usability by human annotators and by machine-learned annotation. After its establishment as an international standard in 2012, further annotation efforts have been undertaken in applying the standard in several corpus annotation, collection, and re-annotation projects. This chapter describes the most substantial of these experiments and annotation efforts.

This chapter is organized as follows. Section 6.2 outlines the use of the ISO 24617-2 annotation scheme. Section 6.3 describes the results of experiments concerned with some of the special features of the annotation scheme. Section 6.4 presents several new and emerging corpora of dialogues, annotated with the ISO 24617-2 annotation scheme. Section 6.5 closes this chapter with concluding remarks and perspectives for future studies and applications using the ISO 24617-2 standard.

6.2 Annotating with ISO 24617-2

6.2.1 Features of the ISO 25617-2 Annotation Standard

Dimensions Utterances in dialogue often have more than one communicative function, as several authors have observed [3, 4, 8, 40, 46]. The following dialogue fragment illustrates this:

(3) 1. Anne: Henry, can you take us through these slides?
    2. Henry: Ehm... sure, just ordering my notes.

In the first utterance, Anne makes a request and assigns the next speaking turn to Henry. In the second utterance, Henry accepts the turn and stalls for time; accepts the request, and explains why he does not fulfill the request right away. The multidimensional DIT^{++} annotation scheme was designed to optimally support the annotation of multifunctional utterances [7]. This scheme is based on a well-founded notion of dimension, inspired by the observation that participation in a dialogue involves a range of communicative activities beyond those strictly related to performing the task or activity that motivates the dialogue. Dialogue participants also perform communicative activities such as giving and eliciting feedback, taking turns, stalling for time, and showing attention; moreover, they often perform several of these activities at the same time. The term 'dimension' refers to these various types of communicative activity.

The ISO 24617-2 annotation scheme inherits the following nine dimensions from the DIT^{++} scheme: (1) *Task:* dialogue acts that move the task or activity forward which motivates the dialogue; (2–3) *Feedback*, divided into *Auto-* and *Allo-Feedback*: acts providing or eliciting information about the processing of previous utterances by the current speaker or by the current addressee, respectively; (4) *Turn Management:* activities for obtaining, keeping, releasing, or assigning the right to speak; (5) *Time Management:* acts for managing the use of time in the interaction; (6) *Discourse Structuring:* dialogue acts dealing with topic management, opening and closing (sub-)dialogues, or otherwise structuring the dialogue; (7–8) *Own-* and *Partner Communication Management:* actions by the speaker to edit his current contribution or a contribution of another current speaker, respectively; (9) *Social Obligations Management:* dialogue acts for dealing with social conventions such as greeting, introducing oneself, apologizing, and thanking.

The ISO 224617-2 inventory of communicative functions consists of 56 of the 88 functions of the DIT^{++} taxonomy.[2] Some of these are specific for a particular dimension; for instance *Turn Take* is specific for Turn Management; *Stalling* is specific for Time Management, and *Self-Correction* is specific for Own Communication Management. Other functions can be applied in any dimension; for example, *You misunderstood me* is an *Inform* in the Allo-Feedback dimension. All types of question, statement, and answer can be used in any dimension, and the same is true for commissive and directive functions, such as *Offer, Suggest,* and *Request.* These functions are called *general-purpose* functions, as opposed to *dimension-specific* functions. Table 6.1 lists the communicative functions defined in ISO 24617-2.

Qualifiers The different qualifiers defined in ISO 24617-2 are applicable to different classes of dialogue acts. Sentiment qualifiers are applicable to any dialogue act with a general-purpose function (GPF); conditionality qualifiers to dialogue acts with a commissive or directive function (*Promise, Offer, Suggestion, Request*, etc.); and certainty qualifiers are applicable to dialogue acts with an 'information-providing' function' (*Inform, Agreement, Disagreement, Correction, Answer, Confirm, Disconfirm).*

Functional Dependence Relations are indispensable for the interpretation of dialogue acts that are responsive in nature, such as *Answer, Confirmation, Disagreement, Accept Apology,* and *Decline Offer*. The semantic content of these acts depends crucially on the content of the dialogue act that they respond to. Functional dependence relations connect occurrences of such dialogue acts to their 'antecedent' and correspond to links for marking up a segment not only as having the function of an answer, for example, but also indicating which question is answered.

Feedback Dependence Relations play a similar role for determining the semantic content of feedback acts, which is co-determined by the utterance(s) that the

[2]DIT^{++} has a fine-grained set of 29 feedback functions, whereas ISO 241617-2 has only 5, which are, however, more reliably annotated.

Table 6.1 ISO 24617-2 communicative functions

General-purpose communicative functions	Dimension-specific communicative functions	
	Function	Dimension
Inform	AutoPositive	Auto-Feedback
Agreement	AutoNegative	
Disagreement	AlloPositive	Allo-Feedback
Correction	AlloNegative	
Answer	FeedbackElicitation	
Confirm	Staling	Time Management
Disconfirm	Pausing	
Question	Turn Take	Turn Management
Set-Question	Turn Grab	
Propositional Question	Turn Accept	
Choice-Question	Turn Keep	
Check-Question	Turn Give	
Offer	Turn Release	
Address Offer	Self-Correction	Own Communication Man.
Accept Offer	Self-Error	
Decline Offer	Retraction	
Promise	Completion	Partner Communication Man.
Request	Correct Misspeaking	
Address Request	Init-Greeting	Social Obligations Man.
Accept Request	Return Greeting	
Decline Request	Init-Self-Introduction	
Suggest	Return Self-Introduction	
Address Suggest	Apology	
Accept Suggest	Accept Apology	
Decline Suggest	Thanking	
Instruct	Accept Thanking	
	Init-Goodbye	
	Return Goodbye	
	Interaction Structuring	Discourse Structuring

feedback is about. Feedback acts often refer to the immediately preceding utterance, but can also refer further back and to more than one utterance [39]. The ISO 24617-2 annotation scheme therefore includes links for marking up these dependences; an example occurs in (7).

Rhetorical Relations, which have been studied extensively for written texts, also occur in spoken dialogue where they occur in two different ways, illustrated in the following examples (where the participants talk about remote TV controls):

(4) 1. A: I can never find them.
 2. B That's because they don't have a fixed location.
(5) 1. A: Where would you position the buttons?
 2. A: I think that has some impact on many things

In (6.2.1) the dialogue acts expressed by A's and B's utterances are related by a *Cause* relation between their respective semantic contents: the content of the second causes the content of the first; in (6.2.1), by contrast, the second dialogue act forms a reason for performing the first, so the causal relation is between the two dialogue acts as a whole, rather than between their semantic contents. The annotation of a rhetorical relation is illustrated in example (6.2.3).

Different from functional and feedback dependences, which are an integral part of dialogue acts with a responsive function and of feedback acts, respectively, rhetorical relations give additional information about the ways in which dialogue acts are semantically or pragmatically related.

6.2.2 Multidimensional Segmentation

Dialogues are often segmented into *turns*, defined as stretches of communicative behaviour produced by one speaker, bounded by periods of inactivity of that speaker. Such a segmentation is too coarse for accurate dialogue act annotation, as example (3) above illustrates. More accurate annotation is possible by using *'functional segments'* as the units to which annotations are attached. Functional segments are defined as the *minimal stretches of communicative behaviour that have a communicative function*—'minimal' in the sense of not containing material that does not contribute to its communicative function(s). Functional segments are mostly shorter than turns, may be discontinuous, may overlap, and may have parts contributed by different speakers. Functional segments by definition have *at least one* communicative function, and possibly several. An example of the use of functional segments is shown in (6), where we see the utterance *The first train to the airport on Sunday is at…let me see… 6.16* in response to the question *What time is the first train to the airport on Sunday?* The response has parts which have a communicative function in three different dimensions: Task, Auto-Feedback (expressed by the repetition in the second utterance), and Time Management; in each of these dimensions the relevant functional segment is shown; the DiAML annotation is represented in (6.2.2).

(6) C: What time is the first train to the airport on Sunday?
I: The first train to the airport on Sunday is at…let me see… 6.16

Auto-Feedback fs2 *The first train to the airport on Sunday*
Task: fs3 *The First train to the airport on Sunday is at 6.16*
Time Man. fs4 *…let me see…*

```
<diaml xmlns:"http://www.iso.org/diaml/">
<dialogueAct xml:id="da1" target="#fs1"
  sender="#p1" addressee="#p2"
  communicativeFunction="setQuestion" dimension="task"/>
<dialogueAct xml:id="da2" target="#fs2"
  sender="#p2" addressee="#p1"
  communicativeFunction="autoPositive"
  dimension="autoFeedback" feedbackDependence="#fs1"/>
```

```
(7) <dialogueAct xml:id="da3" target="#fs3"
      sender="#p2" addressee="#p1" communicativeFunction="answer"
      dimension="task" functionalDependence="#da1"/>
    <dialogueAct xml:id="da4" target="#fs4"
      sender="#p2" addressee="#p1"
      communicativeFunction="stalling" dimension="timeManagement"/>
    </diaml>
```

6.2.3 The Dialogue Act Markup Language (DiAML)

The ISO 24617-2 standard includes the specification of the Dialogue Act Markup Language (DiAML), designed in accordance with the ISO Linguistic Annotation Framework (ISO 24612 [31]), which draws a distinction between the concepts of *annotation* and *representation*. The term 'annotation' refers to the linguistic information that is added to segments of language data, independent of the format in which the information is represented; 'representation' refers to the format in which an annotation is rendered, independent of its content [28].

This distinction is implemented in the DiAML definition following the ISO Principles for Semantic Annotation (ISO 24617-6; see also [9]). The definition specifies, besides a class of XML-based *representation structures*, also a class of more abstract *annotation structures* with a formal semantics. These components are called the *concrete* and *abstract syntax*, respectively. Annotation structures are set-theoretical structures like pairs and triples, for which the concrete syntax defines an XML-based rendering. An annotation structure is a set of *entity structures*, which contain semantic information about a functional segment, and *link structures*, which describe semantic relations between functional segments. An entity structure contains the conceptual information of a single dialogue act, and specifies: (1) a sender; (2) one or more addressees; (3) possible other participants, like an audience or side-participants; (4) a communicative function; (5) a dimension; (6) possible qualifiers for sentiment,[3] conditionality or certainty; and (7) zero, one or more functional dependence relations or feedback dependence relations.

The concrete syntax, defined following the CASCADES method (see ISO 24617-6 [31] and [9]), has a unit that corresponds to entity structures in the form of the XML element `dialogueAct`, as illustrated in (2). The question asked by participant P1 is represented by the `dialogueAct` element with identifier `da1`, which refers to the functional segment `fs1` formed by P1's utterance. Participant P2's response consists of two functional segments. First, a turn-initial *Ehm,...* which forms a multifunctional segment signalling that P2 is taking the turn and also stalls for time. The second functional segment contains the actual answer,

[3]ISO 24617-2 does not prescribe the use of any particular set of sentiment labels. See, e.g., the EmotionML language (www.w3.org/TR/emotionml) for possible choices in this respect.

which includes an expression of regret that is annotated by means of a qualifier, represented as the value of the `sentiment` attribute.

Functional dependence relations are components of a `dialogueAct` element since they form part of a dialogue act viewed as a semantic unit. The same is true for feedback dependence relations as a component of a feedback act, as illustrated in example (6). Rhetorical relations, by contrast, do not play a role in determining the meaning of a dialogue act, but provide additional information about the semantic/pragmatic relations between dialogue acts. They are represented by means of `rhetoricalLink` elements as shown in (8).

(8) 1. P4: Where would you position the buttons?
    2. P4: I think that has some impact on many things

```
<diaml xmlns:"http://www.iso.org/diaml/">
<dialogueAct xml:id="da1" target="#fs1"
   sender="#p4" addressee="#p3"
   communicativeFunction="setQuestion" dimension="task"/>

<dialogueAct xml:id="da2" target="#fs2"
   sender="#p4" addressee="#p3"
   communicativeFunction="inform" dimension="task"/>
<rhetoricalLink dact="#da2"
   rhetoRelatum="#da1" rhetoRel="cause"/>
</diaml>
```

6.3 Experiences in the Use of ISO 24617-2

6.3.1 Communicative Function Recognition

Multidimensional annotation using a rich inventory of dialogue act tags is often thought to be too difficult for human annotators as well as for automatic annotation to give reliable results. In order to investigate this, Geertzen and Bunt [24] determined the inter-annotator agreement for assigning communicative functions in the ten dimensions of DIT^{++}, nine of which are inherited by ISO 24617-2.

They observed that, when a hierarchically structured tag set is used, the popular standard kappa coefficient [17] is not an appropriate measure of agreement, since the assignment to a functional segment of two different but hierarchically related tags, like *Answer* and *Confirm*, or *Inform* and *Agreement*, does not reflect total disagreement, as the standard kappa would assume, but *partial* (dis-)agreement, since a *Confirm* act is a particular kind of *Answer*, and an *Agreement* is a particular kind of *Inform*. Instead, they defined a weighted kappa coefficient, using Cohen's weighted kappa coefficient [18] with a distance metric that takes the hierarchical structure of the tag set into account (see also [34]). The *taxonomically weighted kappa* is defined as follows:

Table 6.2 Standard and weighted kappa-scores for annotator agreement in the annotation of communicative functions, per ISO 24617-2 dimension (adapted from [24])

Dimension	Standard kappa			Weighted kappa		
	P_o	P_e	κ	P_o	P_e	κ_{tw}
Task	0.52	0.09	0.47	0.76	0.17	0.71
Auto-Feedback	0.32	0.14	0.21	0.87	0.69	0.57
Allo-Feedback	0.53	0.19	0.42	0.79	0.50	0.58
Turn Management	0.90	0.42	0.82	0.90	0.42	0.82
Time Management	0.91	0.79	0.58	0.91	0.79	0.58
Own Communication Management	1.00	0.50	1.00	1.0 0	0.95	1.00
Partner Communication Management	1.00	1.00	–	1.00	1.00	–
Dialogue structuring	0.87	0.48	0.74	0.87	0.48	0.74
Social Obligation Management	1.00	0.19	1.00	1.00	0.19	1.00

(9) $\kappa_{\mathrm{tw}} = 1 - \frac{\Sigma(1-\delta(i,j))\cdot P_{oi}j}{\Sigma(1-\delta(i,j))\cdot P_{ei}j}$

where the distance metric δ_{ij} measures disagreement and is a real number normalized in the range between 0 and 1 (P_{oi} and P_{ei} are observed and expected probabilities, respectively). Table 6.2 shows standard and taxonomically weighted kappa scores per ISO 246170-2 dimension, averaged over all annotation pairs, for the DIAMOND corpus.[4]

The agreement scores indicate that human annotators can reliably use a rich, multidimensional annotation scheme like ISO 24617-2 or DIT^{++}. The usability and reliability of an annotation scheme is not just a matter of the size or simplicity of the tag set, but rather of the conceptual clarity of the tags, their definitions and accompanying annotation guidelines.

6.3.2 *Dimension Recognition*

The notion of a dimension, as used in ISO 24617-2 and DIT^{++}, is defined as follows:

(10) *A dimension is a class of dialogue acts concerned with one particular aspect of communication that a dialogue act can address independently from other dimensions* [6].

Geertzen et al. [26] assessed the recognizability of dimensions by human annotators and by automatic means. Three annotators independently annotated dialogues from the DIAMOND and OVIS[5] corpora with dimension tags. Table 6.3 presents agreement scores expressed in terms of Cohen's kappa and tagging

[4]See Geertzen et al. [25].

[5]See http://www.let.rug.nl/vannoord/Ovis/.

Table 6.3 Inter-annotator agreement and tagging accuracy per dimension for the OVIS and DIAMOND corpora

Dimension	Annotator agreement			Accuracy		
	P_o	P_e	κ	P_o	P_e	κ
Task	0.85	0.1	0.83	0.91	0.47	0.81
Auto-Feedback	0.91	0.1	0.90	0.94	0.24	0.92
Allo-Feedback	0.93	0.1	0.92	0.95	0.43	0.91
Turn Management	0.93	0.1	0.92	0.92	0.08	0.92
Time Management	0.99	0.1	0.99	0.99	0.11	0.90
Discourse Structuring	0.99	0.1	0.99	0.87	0.05	0.87
Contact Management	0.99	0.1	0.99	0.91	0.14	0.89
Own Communication Man.	0.99	0.1	0.99	1.00	0.02	1.00
Partner Communication Man.	0.99	0.1	0.99	1.00	0.02	1.00
Social Obligation Man.	0.99	0.1	0.99	0.95	0.09	0.95

Table 6.4 Automatic dimension recognition scores in terms of accuracy (in %), with baseline scores (BL, classifier based on the dimension tag of the previous utterance), for AMI, DIAMOND, and OVIS data sets

Dimension	DIAMOND		AMI		OVIS	
	BL	Accuracy	BL	Accuracy	BL	Accuracy
Task	64.9	70.5	66.8	72.3	60.8	73.5
Auto-Feedback	71.1	85.1	77.9	89.7	66.1	75.9
Allo-Feedback	86.9	96.6	96.7	99.3	52.5	80.1
Turn Management	69.5	90.0	59.0	93.0	89.8	99.2
Time Management	65.6	82.2	69.7	99.4	95.5	99.4
Discourse Structuring	59.0	67.9	98.0	92.5	76.3	89.4
Contact Management	88.0	95.2	99.8	99.8	87.7	98.5
Own Communication Man.	77.4	83.1	89.6	94.1	99.7	99.7
Partner Communication Man.	45.4	62.6	99.7	99.7	99.8	99.8
Social Obligation Management	80.3	92.2	99.6	99.6	96.2	98.4

accuracy (comparing with a gold standard, see [26]). The table shows near perfect agreement between annotators, and moreover that accuracy is very high. Human annotators can apparently recognize the dimensions of the ISO 24617-2 standard almost perfectly.

To assess the machine learnability of dimension recognition, the rule induction algorithm Ripper was applied to data from the AMI, OVIS, and DIAMOND corpora. The features included in the data sets relate to prosody (minimum, maximum, mean, and standard deviation of pitch); energy; voicing; duration; occurrence of words (a bag-of-words vector); and dialogue history: tags of ten previous turns. Table 6.4 presents the scores obtained in tenfold cross-validation experiments. The results indicate that the dimensions of DIT^{++} and ISO 24617-2 are automatically recognizable with fairly high accuracy.

6.3.3 Machine-Learned Dialogue Act Recognition

Petukhova and Bunt (2011) investigated the automatic classification of dialogue acts for unsegmented spoken dialogue. Table 6.5 shows the results of the combined classification of dimension and communicative function, using three different 'local' classifiers that apply to local utterance features. The DER_{sc} error-rate metric is based on the Dialog Act Error Rate (*DER*) defined by Zimmermann et al. [46], which considers a word to be correctly classified if it has been assigned the correct dialogue act type, and it lies in the correct segment. Table 6.6 shows the results for two-step classification (manual segmentation followed by communicative function

Table 6.5 Overview of *F*- and DER_{sc}-scores for joint segmentation and classification in each ISO 24617-2 dimension for Map Task data. Best scores in bold face

Classification task	BL		BayesNet		Ripper	
Dimension	F_1	DER_{sc}	F_1	DER_{sc}	F_1	DER_{sc}
Task	43.8	70.2	**79.7**	41.9	77.7	58.5
Auto-Feedback	64.6	60.6	65.4	55.2	**80.1**	43.9
Allo-Feedback	30.7	91.2	59.3	54.0	**72.7**	51.8
Turn Management	50.3	47.5	70.8	40.9	**81.4**	36.2
Time management	54.2	28.4	72.1	20.3	**83.6**	10.4
Discourse Structuring	33.2	95.1	62.5	44.3	**66.7**	43.5
Contact Management	24.7	93.2	**57.0**	79.5	11.0	93.5
Own Communication Man.	11.2	97.4	**42.9**	64.7	28.6	92.1
Partner Communication Man.	14.3	95.2	61.5	55.2	**66.7**	50.1
Social Obligations Management	08.8	96.2	40.0	71.8	**85.7**	21.4

Table 6.6 Overview of *F*-scores on baseline (BL) and classifiers for two-step segmentation and classification tasks. Best scores in bold face

Classification	BL	NBayes	Ripper	IB1
Task	66.8	71.2	**72.3**	53.6
Auto-Feedback	77.9	86.0	**89.7**	85.9
Allo-Feedback	79.7	**99.3**	99.2	98.8
Turn M.: initial	93.2	92.9	93.2	88.0
Turn M.: final	58.9	85.1	**91.1**	69.6
Time management	69.7	99.2	**99.4**	99.5
Discourse Structuring	69.3	**99.3**	**99.3**	99.1
Contact Management	89.8	99.8	99.8	99.8
Own Communication Management	89.6	90.0	**94.1**	85.6
Partner Communication Management	99.7	99.7	99.7	99.7
Social Obligations Management	99.6	99.6	99.6	99.6

classification), which can be seen to work better for all dimensions except the Task dimension (the most important one).

The fact that dialogue utterances are often multifunctional, having a communicative function in more than one dimension, makes dialogue act recognition a complex task. Splitting up the task may make it more manageable. A widely used strategy is to split a multi-class learning task into several binary learning tasks. Learning multiple classes, however, allows a learning algorithm to exploit interactions among classes. Petukhova and Bunt (2011) split the task in such a way that a classifier needs to learn (1) communicative functions in isolation; (2) semantically related functions together, e.g. all information-seeking functions (all types of questions) or all information-providing functions (all types of answers and informs). In total 64 classifiers were built for dialogue act recognition in AMI data and 43 for Map Task data.

Using local classifiers that produce all possible output predictions ('hypotheses') given a certain input leads to some predictions being false, since a local classifier never revisits a decision that it has made, in contrast with a human interpreter. Decisions should preferably be based not only on local features of the input, but also on broader contextual information. Therefore, Petukhova and Bunt (2011) trained higher-level 'global' classifiers that have, along with features extracted locally from the input data, the partial output predicted so far from all local classifiers. (This technique is also called 'meta-classification' or 'late fusion'.) Five previously predicted class labels were used, taking into account that the average length of a functional segment in the data is 4.4 tokens. This was found to result in a 10–15 % improvement. Some incorrect predictions are still made, since the decision is sometimes based on incorrect previous predictions.

A strategy to optimize the use of output hypotheses is to perform a global search in the output space looking for best predictions. This is not always the best strategy, however, since the highest-ranking predictions are not always correct in a given context. A possible solution is to postpone the decision until some (or all) future predictions have been made for the rest of the current segment. For training, the classifier then uses not only previous predictions as additional features, but also future predictions of local classifiers. This forces the classifier to not immediately select the highest-ranking predictions, but to also consider lower-ranking predictions that could be better in the context.

Table 6.7 gives an overview of the global classification results based on added previous and next predictions of local classifiers. Both classifiers performed very well, outperforming the use of only local classifiers by a broad margin (cf. Table 6.5). It may be noted that the overall performance reported here is substantially better than the results of other approaches that have been reported in the literature. For instance, Reithinger and Klesen [43] report an average tagging accuracy of 74.7 % of applying techniques based on n-gram modelling to Verbmobil data; transformation-based learning applied to the same data achieved an accuracy of 75.1 % [44]. Hidden Markov Models used for dialogue act classification in the Switchboard corpus gave a tagging accuracy of 71 % [45]; and [33]

Table 6.7 Overview of F-scores and DER_{sc} when global classifiers are used for AMI and Map Task data, based on added predictions of local classifiers for five previous and five next tokens. Best scores in bold face

Classification	AMI data				Map Task data			
	BayesNet		Ripper		BayesNet		Ripper	
Dimension	F_1	DER_{sc}	F_1	DER_{sc}	F_1	DER_{sc}	F_1	DER_{sc}
Task	82.6	9.5	**86.1**	8.3	**85.8**	12.2	80.8	9.1
Auto-Feedback	81.9	1.9	**95.1**	0.6	84.4	15.0	**93.0**	7.6
Allo-Feedback	**96.3**	0.6	95.7	0.5	**95.3**	4.6	94.6	6.9
Turn Management:initial	**85.7**	1.5	81.5	1.6	89.5	8.2	**91.0**	8.0
Turn Management:close	90.9	3.8	**91.2**	3.6	**82.9**	17.1	77.2	18.9
Time management	90.4	2.4	**93.4**	1.7	**94.9**	5.5	92.8	6.1
Discourse Structuring	**82.1**	1.7	78.3	1.8	85.7	12.4	**87.4**	8.2
Contact Management	87.9	1.2	**94.3**	0.6	87.4	9.9	**88.3**	7.4
Own Communication Man.	78.4	2.2	**81.6**	2.0	87.2	9.8	**87.4**	7.6
Partner Communication Man.	**71.8**	2.4	70.0	4.6	86.7	11.1	**86.8**	9.8
Social Obligations Man.	98.6	0.4	98.6	0.5	**97.9**	1.1	**97.9**	1.2

report an accuracy of 73.8 % for the application to data from the OVIS corpus of a memory-based approach based on the k-nearest neighbour algorithm.

Altogether, an incremental, token-based approach with global classifiers that exploit the outputs of local classifiers, applied to previous and subsequent tokens, results in excellent dialogue act recognition scores for unsegmented spoken dialogue. This can be seen as strong evidence for the machine learnability of the ISO 24717-2 annotation scheme.

6.3.4 Qualifier Recognition

The recognition of dialogue act qualifiers by human annotators was investigated by Petukhova [36]. The task in these experiments, involving four untrained annotators (undergraduate students), was to assign qualifier values to functional segments in pre-annotated dialogue fragments from the AMI corpus and the TRAINS corpus.[6]

Table 6.8 shows that there are no systematic differences between annotators in assigning values for qualifier tags. They achieved moderate agreement ($0.4 < \kappa < 0.6$) on labelling certainty for the AMI data; the agreement for this category when labelling TRAINS dialogues is substantial ($0.6 < \kappa < 0.8$). The difference can be explained by the fact that AMI dialogues are more difficult to annotate for untrained annotators: AMI meetings are considerably more complex, as they are both multi-party and multi-modal. The best recognized category is

[6]See https://www.cs.rochester.edu/research/speech.

Table 6.8 Cohen's kappa scores for inter-annotator agreement on the assignment of qualifiers per annotator pair for AMI and TRAINS data

Annotator	AMI dialogues			TRAINS dialogues	
pair	Certainty	Conditionality	Sentiment	Certainty	Conditionality
1, 2	0.49	0.79	0.70	0.64	0.88
1, 3	0.48	0.64	0.66	0.70	0.73
1, 4	0.42	0.65	0.25	0.64	0.93
2, 3	0.47	0.85	0.60	0.68	0.64
2, 4	0.35	0.79	0.36	0.71	0.88
3, 4	0.38	0.65	0.30	0.75	0.73

conditionality, for which annotators achieved substantial to near perfect agreement ($\kappa > 0.8$).

Inter-annotator agreement scores for certainty and sentiment were influenced negatively by the fact that one of the values that annotators could choose for these qualifiers was 'neutral'; some annotators assigned this qualifier to every segment that did not clearly express a certainty or a sentiment, while others assigned a certainty or a sentiment qualifier only to those segments which they judged as expressing a particular sentiment or (un)certainty.

6.4 Annotated Corpora

6.4.1 The DBOX Corpus

In the European project DBOX,[7] which aims to develop interactive games based on spoken natural language human-computer dialogues, a corpus has been collected in a Wizard-of-Oz setting. A set of quiz games was designed where the Wizard holds the facts about a famous person's life and the player's task is to guess this person's name by asking questions.

In total 338 dialogues were collected with a total duration of 16 h, comprising about 6000 speaking turns. The collected data has been transcribed and annotated using the ISO 24617-2 annotation scheme. Table 6.9 shows that inter-annotator agreement between two trained annotators ranged between 0.55 and 0.94 in terms of Cohen's kappa for segmentation and between 0.55 and 1.00 for the annotation of dialogue acts in the various dimensions (see [38] for details). For relations between dialogue acts the agreements ranged from 0.66 to 0.88.

[7]Eureka project E! 7152, see https://www.lsv.uni-saarland.de/index.php?id=71.

Table 6.9 Inter-annotator agreement on segmentation and annotation of communicative functions per ISO dimension and on annotation of relations of the ISO relation types

ISO 24617-2 dimension	**Segmentation (κ)**	**Function (κ)**
Task	0.88	0.81
Auto-feedback	0.78	0.79
Allo-Feedback	0.94	0.95
Turn Management	0.71	0.64
Time Management	0.86	0.86
Discourse Structuring	0.88	0.55
Own Communication Management	0.55	0.98
Partner Communication Management	n.a.	n.a.
Social Obligations Management	0.77	1.00
ISO 24617-2 relation type		**Relations**
Functional dependence	0.88	0.68
Feedback dependence	0.88	0.88
Rhetorical relations	0.88	0.68

6.4.2 Youth Parliament Debate Data

As part of the FP 7 European project Metalogue,[8] data have been analysed from three sessions of the UK Youth Parliament (YP). The sessions are video recorded and available on YouTube.[9] In these sessions, the YP members, aged 11–18, debate issues addressing sex education; university tuition fees; and job opportunities for young people.

The annotated corpus consists of 1388 functional segments from 35 speakers. Table 6.10 provides an overview of the relative frequencies of functional tags per ISO-dimension.

Of the dialogue acts in the Task dimension, 41.4 % are *Inform* acts, which are often connected by rhetorical relations. For example:

(11) $D1_{21}$: Let us be clear, sex education covers a wide range of issues affecting young people [*Inform*]
$D1_{22}$: These include safe sex practices, STIs and legal issues surrounding consent and abuse [*Inform Elaboration* $D1_{21}$]

The ISO 24617-2 standard does not prescribe the use of any particular set of rhetorical relations; for the annotation of the DBOX corpus a combination was used of the hierarchy of relations used in the Penn Discourse Treebank (PDTB, [41]) and the taxonomy defined in [27]. Table 6.11 shows the distribution in the corpus of the rhetorical relations associated with *Inform* acts. The corpus is used for designing the Dialogue Manager module of the dialogue system that is built in the Metalogue project.

[8]See http://www.metalogue.eu.

[9]See, for example, http://www.youtube.com/watch?v=g2Fg-LJHPA4. For information about the UK Youth Parliament, see http://www.ukyouthparliament.org.uk/

Table 6.10 Distribution of functional tags across ISO-dimensions in the UK YP corpus

ISO 24617-2 dimension	Frequency (%)
Task	54.9
Auto Feedback	2.9
Allo Feedback	1.0
Turn Management	22.7
Time Management	21.1
Discourse Structuring	10.0
Own Communication Management	7.3
Partner Communication Management	0.0
Social Obligations Management	1.2

Table 6.11 Distribution of rhetorical relations associated with Inform acts in the corpus

Rhetorical relation	Relative frequency	Annotator agreement
Elaboration[a]	28.1	0.67
Evidence[a]	21.4	0.72
Justify[b]	16.1	0.76
Condition[b]	0.7	0.34
Motivation[a]	1.4	0.48
Background[a]	0.3	0.18
Cause[b]	3.4	0.37
Result[b]	2.2	0.26
Reason[c]	10.6	0.33
Conclude[a]	5.7	0.71
Restatement[b]	10.1	0.76

Inter-annotator agreement in terms of Cohen'skappa[a] As defined by Hovy and Maier [27][b] In bothtaxonomies[c] As defined in the PDTB

6.4.3 *The SWBD-ISO Corpus*

Fang and collaborators made an effort to assign ISO 24617-2 annotations to the dialogues in the Switchboard Dialog Act (SWBD-DA) corpus (see Fang et al. [20–22]).[10] This resource contains 1155 5-min conversations, orthographically transcribed in about 1.5 million word tokens. Each utterance in the corpus is segmented in 'slash units', defined as "maximally a sentence; slash units below the sentence level correspond to parts of the narrative which are not sentential but which the annotator interprets as complete" [35]. The corpus comprises 223,606 slash units, which are annotated with a communicative function tag from the SWBD-DAMSL annotation scheme, a variation of the DAMSL scheme defined specifically for this purpose [32]. See example (6.4.3), where 'qy' is the SWBD-DAMSL tag for yes/no questions and 'utt1' indicates the first slash unit within a turn.

[10]The Switchboard corpus is distributed by the Linguistic Data Consortium: https://www.ldc.upenn.edu.

(12) qy A.1 utt1: { D Well, } { F uh, } does the company you work for test for drugs? /

In addition to this marking up of communicative functions, in-line markups are also used to mark 'discourse markers' such as { D Well, }, which often signal a rhetorical relation; filled pauses, like { F uh, }, restarts and repetitions, such as [I think, I think] and some other types of 'disfluencies'.

To assess the possibility of converting SWBD-DA annotations to ISO 24617-2 annotations, first a detailed comparison was made of the two sets of communicative functions, revealing 14 one-to-one correspondences and 26 many-to-one equivalences. These tags can thus be converted automatically to ISO tags, which accounts for 83.97 % of the SWBD-DAMSL tags in the corpus. Six SWBD-DAMSL function tags have a one-to-many correspondence with 26 ISO tags, corresponding to 5.74 % of the Switchboard corpus; about 30 % of these cases can be converted automatically to an ISO tag by taking the tagging of the preceding slash unit into account; for example, an utterance tagged 'aa' (i.e., Accept) following an offer should be assigned the ISO tag *Accept Offer*, while it should be assigned the ISO tag *Accept Request* when following a request. For those cases where such a contextual disambiguation does not help, manual annotation was performed (see Fang et al. [22]).[11]

Altogether, through combined automatic conversion and manual annotation 200.605 utterances (89.71 % of the Switchboard corpus) were assigned ISO 24617-2 communicative function tags. Table 6.12 shows the distribution of function tags in the resulting 'SWBD-ISO' corpus.

6.4.4 *The DialogBank*

In a recent initiative at Tilburg University a publicly available corpus has been created called the **DialogBank**, which consists of dialogues with gold standard annotations in DiAML according to the ISO 24617-2 standard. While recommending the use of XML for representing annotation structures as defined by the DiAML abstract syntax, the standard allows representations in other formats as long as these have the properties of being (1) *complete*, i.e. defining a rendering of any annotation structure defined by the abstract syntax, and (2) *unambiguous*, i.e. every representation encodes only one annotation structure. Representation formats that have these properties can be converted to and from the DiAML-XML format without loss of information. For some of the dialogues in the DialogBank, an alternative tabular representation format was defined that has

[11]The remaining 10.29 % of SWBD-DAMSL tags cannot be converted into ISO tags since they are not really concerned with communicative functions, such as the SWBD-DAMSL tags 'non-verbal', 'uninterpretable', 'quoted material', 'transcription error'.

Table 6.12 Distribution of ISO 24617-2 communicative function tags the SWBD-ISO corpus

ISO 24617-2	Utterances			ISO 24617-2	Utterances		
Comm. functions	#	%	Cum %	Comm. functions	#	%	Cum %
inform	120227	53.767	53.77	instruct	106	0.047	89.44
autoPositive	46382	20.743	74.51	acceptSuggest	99	0.044	89.48
agreement	10934	4.890	79.40	acceptApology	79	0.035	89.52
propositionalQuestion	5896	2.637	82.04	thanking	79	0.035	89.55
confirm	3115	1.393	83.43	offer	71	0.032	89.58
initialGoodbye	2661	1.190	84.62	acceptRequest	65	0.029	89.61
setQuestion	2174	0.972	85.59	signalSpeakingError	56	0.025	89.64
disconfirm	1597	0.714	86.31	promise	41	0.018	89.66
answer	1522	0.681	86.99	correction	29	0.013	89.67
checkQuestion	1471	0.658	87.64	acceptOffer	26	0.012	89.68
completion	813	0.364	88.01	turnTake	18	0.008	89.69
question	680	0.304	88.31	alloPositive	17	0.008	89.70
stalling	580	0.259	88.57	correctMisspeaking	14	0.006	89.70
choiceQuestion	506	0.226	88.80	selfCorrection	8	0.004	89.71
suggest	369	0.165	88.96	acceptThanking	6	0.003	89.71
autoNegative	307	0.137	89.10	declineOffer	3	0.001	89.71
request	278	0.124	89.22	declineRequest	3	0.001	89.71
disagreement	258	0.115	89.34	turnRelease	2	0.001	89.71
apology	112	0.050	89.39	declineSuggest	1	0.000	89.71
non-functional tags					23001	10.29	100.00
Total					223606	100.00	

these properties and that is more convenient for human readers (see Bunt et al. [14]).

The annotations include not only the multidimensional marking up of communicative functions and dimensions, but also of functional dependence relations; feedback dependence relations; rhetorical relations; and qualifiers for certainty, conditionality and sentiment.

The DialogBank currently contains dialogues taken from four English-language corpora: the HCRC Map Task, Switchboard, TRAINS, and DBOX corpora, and four Dutch-language corpora: the OVIS, DIAMOND, Dutch Map Task,[12] and Schiphol[13] corpora. Addition is foreseen of dialogues from the AMI corpus, the YP corpus, and several other corpora.

6.4.4.1 Map Task and DBOX Dialogues

The Map Task and DBOX dialogues in the DialogBank were annotated using the ANVIL tool in which a facility has been created to export annotations in the DiAML-XML reference format of ISO 24617-2 [13]. Example (14) in the Appendix shows the result for a very short dialogue fragment. This format is perfect for machine consumption, but rather inconvenient for human readers, for example for checking the correctness of annotations. The more compact tabular formats shown below are more attractive in that respect.

The DBOX application (quiz game dialogues) called for some small extensions to the ISO annotation scheme, which were made in accordance with the guidelines included in the ISO 24617-2 standard for extending the annotation scheme. Two additional dimensions were introduced: Task Management (also familiar from DAMSL), for dialogue acts where the rules of the game are discussed, and Contact Management, also familiar from DIT^{++}, for dialogue acts where the participants establish, check, or end contact between them.

6.4.4.2 Switchboard Dialogues

The dialogues in the Switchboard corpus were originally represented in a 3-column tabular format where the leftmost column contains an identifier of the slash unit in the third column, and the middle column contains an SWBD-DAMSL function tag.[14] In constructing the SWBD-ISO corpus, all in-line markups of filled pauses were replaced by *Stalling* tags and in-line markups of restarts by *SelfCorrection* tags. The result looks as shown in (13).

---

[12]See http://doc//.ukdataservice.ac.uk/doc/4632/mrdoc/pdf/4632userguide.pdf.

[13]See Prüst et al. [42].

[14]The Switchboard corpus is also available in NXT format [15], without in-line markups.

(13)

sw01-0105-0001-A001-01	setQuestion	A.1 utt1: Jimmy, {D so } how do you get most of your news? /
sw01-0105-0002-B002-01	stalling	B.2 utt1: {D Well, } [I kind of, +
	selfCorrection	{F uh, } I] watch the,
	stalling	{F uh, } national news
	answer	everyday, for one /
sw01-0105-0003-B002-02	answer	B.2 utt2: I also read one or two papers a day /
sw01-0105-0004-B002-03	selfCorrection inform	B.2 utt3: {C and } [I'm a, + I'm pretty much a] news junkie /
sw01-0105-0005-B002-04	answer	B.2 utt4: {C and } I tune in to CNN a lot./
sw01-0105-0006-A003-01	autoPositive	A 3 utt1: {F Oh, } wow /

While convenient for human readers, this format is not optimal for computer processing. The numbering of speaker turns and slash units is redundant (and turns have no special status in the ISO standard), and the rightmost column contains a mixed bag of information types (speaker, turn number, slash unit number within turn, transcribed slash unit, and disfluency and other markups). It could be converted to an XML representation like DiAML-XML by interpreting the first column as the values of the `xml:id` attribute, the second as the values of the `communicativeFunction` attribute, and the third as the values of the `sender` and `target` attributes and the textual rendering of slash units. However, representations like (13) differ from DiAML-annotations in three fundamental respects: (1) slash units do not always correspond to functional segments, which in general form a more fine-grained way of segmenting a dialogue; (2) the use of in-line markups goes against the ISO requirement that annotations should be in stand-off form; and (3) annotations according to ISO 24617-2 contain more information than just communicative functions, in particular also dimensions, qualifiers, and dependence relations, which are semantically indispensable.

These differences are taken into account in the design of a tabular representation format, called 'DiAML-TabSW', that is relatively close to that of (6.4.4.2), and facilitates comparison between the SWBD-DAMSL and ISO annotation schemes. For incorporating annotated Switchboard dialogues into the DialogBank, first, existing annotated dialogues were re-segmented into functional segments, and the functional segments that do not correspond to a slash unit were newly annotated with ISO 24617-2 communicative function tags and dimension tags. Second, a copy was made of the slash unit transcriptions in which all in-line markups were interpreted in terms of communicative functions, rhetorical relations, or qualifiers whenever possible, and removed. Third, the functional segments were represented in stand-off fashion by referring to a file that contains segment definitions in terms of word tokens or time points. Finally, the annotations of functional segments were enriched with functional and feedback dependences, qualifiers, and rhetorical relations.

markables	ID	Dialogue acts	Sp	FS text	Turn transcript
sw01-0105-fs.1	da1	Ta:setQuestion	A	Jimmy, so how do you get most of your news?	Jimmy, {D so } how do you get most of your news? /
			B		{D Well, } [I kind of, + {F uh, } I] watch the, national news every day, for one / I also read one or two papers a day / {C and } [I'm a, + I'm pretty much a] / news junkie {C and } I tune in to CNN a lot /
sw01-0105-fs.2	da2 da3	TiM:stalling TuM:turnTake	B	Well,	
sw01-0105-fs.3	da4	OCM: selfCorrection	B	I kind of, I	
sw01-0105-fs.4	da5	TiM;stalling	B	uh	
sw01-0105-fs.5	da6	Ta:answer (Fu:da1)	B	I watch the national news every day, for one	
sw01-0105-fs.6	da7	TiM:stalling	B	uh	
sw01-0105-fs.7	da8	Ta:answer (da2) {Expansion: foregr da7}	B	I also read one or two papers a day	
sw01-0105-fs.8	da9	TuM:turnKeep	B	and	
sw01-0105-fs.9	da10	OCM: selfCorrection	B	I'm a, I'm pretty much a	
sw01-0105-fs.10	da11	Ta:inform	B	I'm pretty much a news junkie	
sw01-0105-fs.11	da12	TuM:turnKeep	B	and	
sw01-0105-fs.12	da13	Ta:answer (Fu:da1) {Expansion: foregr da7, da9}	B	I tune in to CNN a lot	
sw01-0105-fs.13	da14	AuF:autoPositive (Fe: da6 ,da8, da13)	A	Oh, wow.	Oh, wow

Fig. 6.1 ISO 24617-2 annotation of dialogue fragment in example (6.4.4.2), represented in DiAML-TabSW format. (Ta = Task, TiM = Time Management, TuM = Turn Management, OCM = Own Communication Management, AuF = Auto-Feedback)

Figure 6.1 shows the resulting representation. The first four columns represent the annotations proper: (1) functional segment identifiers; (2) dialogue act identifiers; (3) dialogue acts; and (4) sender, with much of the information concentrated in the third column: dimension, communicative function, dependences (as in "Ta: answer (da2)"), qualifiers and rhetorical relations. The fifth and sixths, s, containing functional segment texts and turn transcripts, column have been added for the convenience of human readers, and have no formal status.

6.4.4.3 Other Annotated Dialogues and Their Representation

The dialogues in the DIAMOND corpus were originally annotated with the DIT^{++} annotation scheme, for which the DitAT annotation tool was developed [23]; this tool produces representations in a multi-column tabular format with a separate column for each dimension. For inclusion of ISO 24617-2 versions of these annotations in the DialogBank, a new multi-column tabular format was defined, the 'DiAML-MultiTab' format, with one column identifying functional segments in stand-off fashion, as in the DiAML-TabSW format, one column indicating the speaker, and one column per dimension for representing communicative functions, qualifiers, dependence relations, and rhetorical relations. Figure 6.2 illustrates this format, which was proven to be convertible without loss of information to DiAML-XML and vice versa [14]. In the example, those columns have been suppressed that correspond to dimensions in which no communicative functions were marked up for this fragment.

The DiAML-MultiTab format was used also for representing re-annotated dialogues from the OVIS and TRAINS corpora, and newly annotated Schiphol dialogues.

markables	sp	fs text	turn transcript	Task	Auto-Feedback	Turn Man.	Time Man.	Discourse Struct.	SocialObl. Man.
			hello, can I help you						
TR1-fs.1	s	hello							da1:Init. Greeting
TR1-fs.2	s	can I help you						da2:Offer	
			uhm, yes hello,maybe, I'd like to take a tanker...						
TR1-fs.3	u	uhm				da3: Turn Take	da4: Stalling		
TR1-fs.4	u	yes hello			da5:Pos. (Fe:da1)				
TR1-fs.5	u	yes maybe						da6: Accept Offer [uncertain] (Fu:da2)	
TR1-fs.6	u	I'like to take...		da7: Inform					

Fig. 6.2 ISO 24617-2 annotation of TRAINS dialogue fragment represented in DiAML-MultiTab format

6.5 Conclusions and Perspectives

The ISO 24617-2 standard for dialogue annotation has as its main features a rich taxonomy of clearly defined communicative functions, including many functions from previously developed annotation schemes such as DAMSL, DIT^{++}, and ICSI-MRDA; the distinction of nine dimensions, inherited from the DIT^{++} schema; functional and feedback dependence relations that account for semantic dependences between dialogue acts; the use of qualifiers for expressing (un−)certainty, conditionality and sentiment; and rhetorical relations among dialogue acts. In this chapter, experiences and experiments were discussed that investigate how these features play out in human and automatic dialogue annotation.

New and emerging corpora were discussed that contain dialogues, annotated according to the ISO 24617-2 standard, notably the DBOX, YP, and DialogBank corpora. Such resources offer a promising basis for the study of human communication as well as for the design and training of modules in dialogue systems, such as recognizers of communicative functions in human interactive behaviour, and dialogue managers in speech-based or multimodal dialogue systems.

Appendix

This appendix shows the ISO 24617-2 annotation of the first two utterances of a Map Task dialogue in the DialogBank corpus, as produced with the ANVIL tool and exported in DiAML format. In a TEI-compliant way,[15] the first part identifies the two dialogue participants (“p1” and “p2”), followed by a second part that identifies the word tokens in the audio-video input stream, and a third part that identifies the functional segments in terms of the word tokens. The last part represents the dialogue act annotations in the DIAML format of the ISO standard.

(14) G: right
G: go south and you’ll pass some cliffs on your right
F: okay

```
<?xml version="1.0" encoding="UTF-8"?>
<TEI xmlns="http://www.tei-c.org/ns/1.0">
 <profileDescr xmlns="">
  <particDescr xml:id="p1">
   <p>the 1. participant</p>
  </particDescr>
  <particDescr xml:id="p2">
   <p>the 2. participant</p>
```

[15]Text Encoding Initiative: www.tei.org.

```
  </particDescr>
 </profileDescr>
 <text>
  <body />
  <div>
   <head>The dialogue turns, segmented into words
   (TEI-compliant)</head>
   <u>
   <w xml:id="w1">right</w>
    <w xml:id="w2">go</w>
    <w xml:id="w3">south</w>
    <w xml:id="w4">and</w>
    <w xml:id="w5">you'll</w>
    <w xml:id="w6">pass</w>
    <w xml:id="w7">some</w>
    <w xml:id="w8">cliffs</w>
    <w xml:id="w9">on</w>
    <w xml:id="w10">your</w>
    <w xml:id="w11">right</w>
    <w xml:id="w12">okay</w>
    ...
    </u>
   </div>
  <div>
   <head>Identification of functional segments</head>
   <spanGrp xml:id="ves1" type="functionalVerbalSegment">
    <span xml:id="ts1" type="textStretch" from="w1" to="w1" />
   </spanGrp>
   <fs type="functionalSegment" xml:id="fs1">
    <f name="verbalComponent" fVal="#ves1" />
   </fs>
   <spanGrp xml:id="ves2" type="functionalVerbalSegment">
    <span xml:id="ts2" type="textStretch" from="w2" to="w11" />
   </spanGrp>
   <fs type="functionalSegment" xml:id="fs2">
    <f name="verbalComponent" fVal="#ves2" />
   </fs>
   <spanGrp xml:id="ves3" type="functionalVerbalSegment">
    <span xml:id="ts3" type="textStretch" from="w12" to="w12" />
   </spanGrp>
   <fs type="functionalSegment" xml:id="fs3">
    <f name="verbalComponent" fVal="#ves3" />
   </fs>
  </div>
  <diaml xmlns="http://www.iso.org/diaml">
```

```
  <dialogueAct xml:id="da1"
  target="#fs1" sender="#p1" addressee="#p2"
  dimension="turnManagement" communicativeFunction="turnTake" />
  <dialogueAct xml:id="da2"
  target="#fs1" sender="#p1" addressee="#p2"
 dimension="discourseStructuring" communicativeFunction="opening" />
  <dialogueAct xml:id="da3"
  target="#fs2" sender="#p1" addressee="#p2"
  dimension="task" communicativeFunction="instruct" />
  <dialogueAct xml:id="da4"
  target="#fs3" sender="#p2" addressee="#p1"
  dimension="autoFeedback" communicativeFunction="autoPositive"
  feedbackDependence="#fs2" />
  </diaml>
 </text>
</TEI>
```

References

1. Alexandersson, J., Buschbeck-Wolf, B., Fujinami, T., Kipp, M., Koch, S., Maier, E., et al. (1998). *Dialogue acts in VERBMOBIL-2 (second edition).* Verbmobil Report 226. Saarbrücken: DFKI.
2. Allen, J., & Core, M. (1997). *DAMSL: Dialogue act markup in several layers (Draft 2.1).* Technical Report. Rochester, NY: University of Rochester.
3. Allwood, J. (1992). On dialogue cohesion. Gothenburg University, Department of Linguistics.
4. Bunt, H. (1994). Context and dialogue control. *Think Quarterly, 3*(1), 19–31.
5. Bunt, H. (2000). Dialogue pragmatics and context specification. In H. Bunt & W. Black (Eds.), *Abduction, belief and context in dialogue. Studies in computational pragmatics* (pp. 81–150). Amsterdam: John Benjamins.
6. Bunt, H. (2006). Dimensions in dialogue annotation. In *Proceedings 5th International Conference on Language Resources and Evaluation (LREC 2006)*, Genova, Paris. ELRA.
7. Bunt, H. (2009). The DIT^{++} taxonomy for functional for dialogue markup. In D. Heylen, C. Pelachaud, R. Catizone, & D. Traum (Eds.), *Proceedings of EDAML-AAMAS Workshop "Towards a Standard Markup Language for Embodied Dialogue Acts"*, Budapest (pp. 36–48).
8. Bunt, H. (2011). Multifunctionality in dialogue. *Computer, Speech and Language, 25*, 222–245.
9. Bunt, H. (2015). On the principles of semantic annotation. In *Proceedings 11th Joint ACL-ISO Workshop on Interoperable Semantic Annotation (ISA-11)*, London (pp. 1–13).
10. Bunt, H., Alexandersson, J., Carletta, J., Choe, J.-W., Fang, A., Hasida, K., et al. (2010). Towards and ISO standard for dialogue act annotation. In *Proceedings 7th International Conference on Language Resources and Evaluation (LREC 2010)*, Malta, Paris. ELDA.
11. Bunt, H., Alexandersson, J., Choe, J.-W., Fang, A., Hasida, K., Petukhova, V., et al. (2012). ISO 24617-2: A semantically-based standard for dialogue annotation. In *Proceedings of 8th International Conference on Language Resources and Evaluation (LREC 2012)*, Istanbul. Paris: ELDA.

12. Bunt, H., Fang, A., Cao, J., Liu, X., & Petukhova, V. (2013). Issues in the addition of ISO standard annotations to the Switchboard corpus. In *Proceedings 9th Joint ISO - ACL SIGSEM Workshop on Interoperable Semantic Annotation (ISA-9)*, Potsdam (pp. 59–70).
13. Bunt, H., Kipp, M., & Petukhova, V. (2012). Using DiAML and ANVIL for multimodal dialogue annotation. In *Proceedings 8th International Conference on Language Resources and Evaluation (LREC 2012)*, Istanbul. Paris: ELRA.
14. Bunt, H., Petukhova, V., Malchanau, A., & Wijnhoven, K. (2016). The DialogBank. In *Proceedings 10th International Conference on Language Resources and Evaluation (LREC 2016)*, Portoroz, Slovenia. Paris: ELRA.
15. Calhoun, S., Carletta, J., Brenier, J., Mayo, N., Jurafsky, D., Steedman, M., et al. (2010). The NXT-format Switchboard corpus: A rich resource for investigating the syntax, semantics, pragmatics and prosody of dialogue. *Language Resources and Evaluation, 44*(4), 387–419.
16. Carletta, J., Isard, S., Kowtko, J., & Doherty-Sneddon, G. (1996). *HCRC dialogue structure coding manual*. Technical Report HCRC/TR-82, University of Edinburgh.
17. Cohen, J. (1960). A coefficient of agreement for nominal scales. *Education and Psychological Measurement, 20*, 37–46.
18. Cohen, J. (1968). Weighted kappa: Nominal scale agreement with provision for scaled disagreement or partial credit. *Psychological Bulletin, 70*, 213–261.
19. Dhillon, R., Bhagat, S., Carvey, H., & Schriberg, E. (2004). *Meeting recorder project: Dialogue labelling guide*. ICSI Technical Report TR-04-002. University of California at Berkeley.
20. Fang, A., Cao, J., Bunt, H., & Liu, X. (2011). Relating the semantics of dialogue acts to linguistic properties: A machine learning perspective through lexical cues. In *Proceedings IEEE-ICSC 2011 Workshop on Semantic Annotation for Computational Linguistic Resources*, Stanford, CA.
21. Fang, A., Cao, J., Bunt, H., & Liu, X. (2012). The annotation of the Switchboard corpus with the new ISO standard for dialogue act analysis. In *Proceedings 8th Joint ISO - ACL SIGSEM Workshop on Interoperable Semantic Annotation (ISA-8)*, Pisa (pp. 13–18).
22. Fang, A., Cao, J., Bunt, H., & Liu, X. (2012). Applicability verification of a new ISO standard for dialogue act annotation with the Switchboard corpus. In *Proceedings of EACL 2012 Workshop on Innovative Hybrid Approaches to the Processing of Textual Data*, Avignon.
23. Geertzen, J. (2007). DitAT: A flexible tool to support web-based dialogue annotation. In *Proceedings 7th International Conference on Computational Semantics (IWCS-7)*, Tilburg (pp. 320–323).
24. Geertzen, J., & Bunt, H. (2006). Measuring annotator agreement in a complex, hierarchical dialogue act schema. In *Proceedings SIGDIAL 2006*, Sydney.
25. Geertzen, J., Girard, Y., & Morante, R. (2004). The DIAMOND project. In *Proceedings of 8th Workshop on the Semantics and Pragmatics of Dialogue (CATALOG 2004)*, Barcelona.
26. Geertzen, J., Petukhova, V., & Bunt, H. (2008). Evaluating dialogue act tagging with naive and expert annotators. In *Proceedings 6th International Conference on Language Resources and Evaluation (LREC 2008)*, Marrakech. Paris: ELDA.
27. Hovy, E., & Maier, E. (1995). *Parsimonious or profligate: How many and which discourse structure relations?* ISI Research Report. Marina del Rey: Information Sciences Institute, University of Southern California.
28. Ide, N., & Romary, L. (2004). International standard for a linguistic annotation framework. *Natural Language Engineering, 10*, 211–225.
29. ISO (2011). *ISO 24612: Language Resource Management - Linguistic Annotation Framework (LAF)*. Geneva: ISO.
30. ISO (2012). ISO 24617-2: Language Resource Management - Semantic Annotation Framework (SemAF) - Part 2: Dialogue Acts. Geneva: ISO.
31. ISO (2016). ISO 24617-6: Language Resource Management - Semantic Annotation Framework (SemAF) - Part 6: Principles of Semantic Annotation. Geneva: ISO.
32. Jurafsky, D., Shriberg, E., & Biasca, D. (1997). *Switchboard SWBD-DAMSL shallow-discourse-function annotation: Coders manual, Draft 1.3*. University of Colorado.

33. Lendvai, P., van den Bosch, A., Krahmer, E., & Canisius, S. (2004). Memory-based robust interpretation of recognised speech. In *Proceedings 9th International Conference on Speech and Computer (SPECOM'04)*, St. Petersburg (pp. 415–422).
34. Lesch, S., Kleinbauer, T., & Alexandersson, J. (2005). A new metric for the evaluation of dialog act classification. In *Proceedings 9th Workshop on the Semantics and Pragmatics of Dialogue (DIALOR)*, Nancy.
35. Meteer, M., & Taylor, A. (1995). *Dysflency annotation stylebook for the Switchboard corpus*. Washington: Linguistic Data Consortium.
36. Petukhova, V. (2011). Multidimensional dialogue modelling. Ph.D. dissertation. Tilburg University.
37. Petukhova, V., & Bunt, H. (2011). Incremental dialogue act understanding. In *Proceedings Ninth International Conference on Computational Semantics* (IWCS 2011), Oxford (pp. 235–244).
38. Petukhova, V., Gropp, M., Klakow, D., Eigner, G., Topf, M., Srb, S., et al. (2014). The DBOX corpus collection of spoken human-human and human-machine dialogues. In *Proceedings 9th International Conference on Language Resources and Evaluation (LREC 2014)*, Reykjavik, Iceland.
39. Petukhova, V., Prévot, L., & Bunt, H. (2011). Multi-level discourse relations between dialogue units. In *Proceedings 6th Joint ACL-ISO Workshop on Interoperable Semantic Annotation (ISA-6)*, Oxford (pp. 18–28).
40. Popescu-Belis, A. (2005). *Dialogue acts: One or more dimensions?* ISSCO Working Paper 62. Geneva: ISSCO.
41. Prasad, R., Dinesh, N., Lee, A., Miltsakaki, E., Robaldo, L., Joshi, A., et al. (2008). The Penn Discourse TreeBank 2.0. In *Proceedings 6th International Conference on Language Resources and Systems (LREC 2008)*, Marrakech.
42. Prüst, H., Minnen, G., & Beun, R.-J. (1984). *Transcriptie dialooogesperiment juni/juli 1984*. IPO Rapport 481. Institute for Perception Research, Eindhoven University of Technology.
43. Reithinger, N., & Klesen, M. (1997). Dialogue act classification using language models. In *Proceedings of Eurospeech-97* (pp. 2235–2238).
44. Samuel, K., Carberry, S., & Vijay-Shanker, K. (1998). Dialogue act tagging with transformation-based learning. In *Proceedings ACL 1998*, Montreal (pp. 1150–1156).
45. Stolcke, A., Res, K., Coccaro, K., Shriberg, E., Bates, R., Jurafsky, D., et al. (2000). Dialogue act modeling for automatic tagging and recognition of conversational speech. *Computational Linguistics, 26*(3), 339–373.
46. Traum, D. (2000). 20 questions on dialogue act taxonomies. *Journal of Semantics, 17*(1), 7–30.
47. Zimmermann, M., Lui, Y., Shriberg, E., & Stolcke, A. (2005). Toward joint segmentation and classification of dialogue acts in multiparty meetings. In *Proceedings of the Multimodal Interaction and Related Machine Learning Algorithms Workshop (MLMI-05)* (pp. 187–193). Berlin: Springer.

Chapter 7
Six-Layered Model for Multimodal Interaction Systems

Kouichi Katsurada, Tsuneo Nitta, Masahiro Araki, and Kazuyuki Ashimura

Abstract We have proposed a six-layered model for multimodal interaction (MMI) systems as an Information Technology Standards Commission of Japan (ITSCJ) standard. It specifies an architecture of an MMI system composed of six layers: application layer, task control layer, a-modal dialogue control, a-modal ⇔ multimodal conversion, modality-dependent layer, and input–output devices. The standard defines the role of each layer in an MMI system, its granularity, and the events transferred between the layers. The EMMA format is employed as the container of the input results. In this chapter, we introduce the outline of the proposed model and show its practical implementation as a Web-based MMI system.

7.1 Background

Multimodal interaction (MMI) is expected to be a future human–machine interface to communicate with smart phones, car navigation systems, information appliances, robots, and so on. Many researchers have developed advanced MMI systems, and

---

K. Katsurada (✉)
Department of Information Science, Tokyo University of Science, Noda-shi, Chiba, Japan
e-mail: katsurada@rs.tus.ac.jp

T. Nitta
Green Computing Systems Research Organization, Waseda University, Shinjuku-ku, Tokyo, Japan

M. Araki
Department of Information Science, Kyoto Institute of Technology, Kyoto, Japan

K. Ashimura
Keio Research Institute at SFC, Keio University, Fujisawa-shi, Kanagawa, Japan

D.A. Dahl (ed.), *Multimodal Interaction with W3C Standards*,
DOI 10.1007/978-3-319-42816-1_7

demonstrated them at some exhibitions and research conferences [1]. However, they are not regularly used in daily life because researchers do not usually publish the detailed specifications of their MMI systems such as module configuration and message transfer. Therefore, if a developer would like to reuse parts of an existing MMI system, it is difficult because its modules are a kind of black box. Hence, it is necessary to specify some standards that help rapid implementation of the modules that are connected in an MMI system in a complicated way and reduce troubles in building the whole system.

The W3C MMI working group [2] has proposed some standards to cope with the problem. The group launched in 2002 and published five specifications and some notes on system architecture: some markup languages to represent input interpretation, emotion, and so on. For system architecture, they published a Russian Doll model based on the model-view-controller (MVC) paradigm [3]. Because the Russian Doll model enables modules that are designed with arbitrary granularity, it has a great flexibility in designing the modules in an MMI system. The group also proposed an XML-based language, EMMA (Extensible Multimodal Annotation markup language), for annotating the interpretation of user inputs to be used in a higher level of dialogue management [4].

In contrast, our Information Technology Standards Commission of Japan (ITSCJ) standard for MMI architecture defines more restricted modules adapted for their roles in an MMI system [5, 6]. The modules in the architecture are positioned in one of six layers: application layer, task control layer, a-modal (modality independent) dialogue control, a-modal $\Leftrightarrow$ multimodal conversion, modality-dependent layer, and input–output devices. In the architecture, the modules in each layer have different responsibilities in an MMI system. Therefore, the events transferred between layers differ according to the layers. The advantage of this architecture is that it is easy to understand the range of tasks when replacing or newly developing the modules [7].

In the following sections, we present the outline of the ITSCJ standard for the MMI architecture and the events transferred between the layers. We also introduce an example of an MMI system conforming to this architecture.

7.2 Six-Layered Model for Multimodal Interaction Systems

In our proposed MMI architecture, an MMI system is composed of the modules classified into six layers and an external user/device model. Figure 7.1 shows the overview of an MMI system. The events transferred between layers are different in each layer. In this section, we outline the roles of each layer, discuss event transfer between layers, and show examples of an event transfer sequence that appears in an MMI system in some situations.

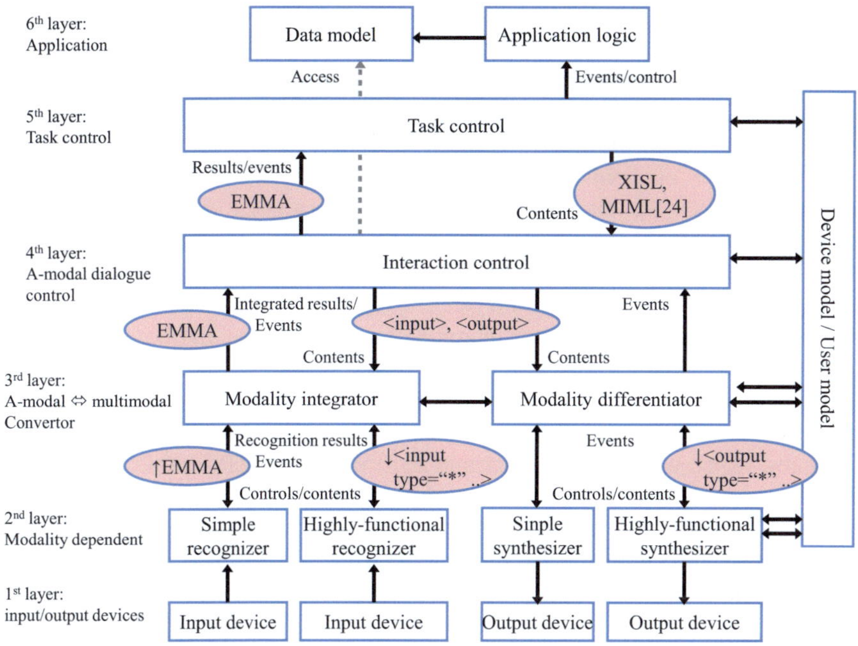

Fig. 7.1 Overview of the six-layered model of MMI systems [24]

7.2.1 Six Layers and User/Device Model

The six layers are composed of the input–output devices layer (the first layer), a modality-dependent layer (the second layer), an a-modal ⇔ multimodal conversion layer (the third layer), an a-modal dialogue control layer (the fourth layer), a task control layer (the fifth layer), and an application layer (the sixth layer). The granularity changes from fine (that is, close to the device) to coarse (that is, far from the device) with increasing layer level. The system also has a user/device model component outside these layers. The component can be used for managing the status of the user and the devices. This section presents the outlines of each layer and the user/device model and their roles in an MMI system.

7.2.1.1 Input–Output Device Layer (First Layer)

This layer controls the input–output devices. The modules in this layer are the wrapper programs between the devices and the MMI system. Because the data format for accessing devices depends on their types, OS, and modalities, we do not define a concrete interface between this layer and the second layer.

7.2.1.2 Modality-Dependent Layer (Second Layer)

The second layer controls each modality such as speech inputs/outputs, animation of the anthropomorphic agent, and inputs/outputs using a touch screen. The modality-dependent information is received from the third layer. It interprets and processes these elements defined in markup languages as follows: `<rect>` element in SVG [8], `<prompt>` element in VoiceXML [9], and `<listen>` element in SSML [10] or SALT [11]. The recognition modules reply with the EMMA format results [4] to the third layer, whereas the synthesis modules return some response events to the third layer. The EMMA format is employed because it is best suited to represent input interpretations whose granularities are different among layers.

Some modules in this layer are implemented as simple recognizers/synthesizers that work as wrapper programs for the modality recognizers/synthesizers. The other modules, which would not be discriminated from the simple modules in the third layer, are highly functional recognizers/synthesizers that control multiple simple submodules (such as an animated agent with speech synthesis). This type of complex module is implemented by combining modules that perform the following three functions.

1. Distributor

   A function to distribute an output to multiple modules and send the inputs from multiple modules to a module/device.

2. Filter

   Input format conversion and information extraction function.

3. Trigger

   A function to call conditional operation according to input values.

Figure 7.2 illustrates a sample configuration of the second layer module. In this example, a speech recognition result is delivered to two filters through a distributor; a copy is sent to a GUI output module through the trigger, and the other is given to the modality integrator in the third layer.

7.2.1.3 A-Modal ⇔ Multimodal Conversion Layer (Third Layer)

The modality-dependent result obtained in the second layer should be converted into an a-modal semantic result that is handled in the dialogue management layer. This layer is responsible for converting the modality-dependent information into modality-independent information, and vice versa. It also integrates inputs, differentiates outputs, and controls the synchronization of multiple inputs and outputs. For example, interpretation of sequential/simultaneous inputs and synchronization of sequential/simultaneous outputs are executed in this layer. This layer is

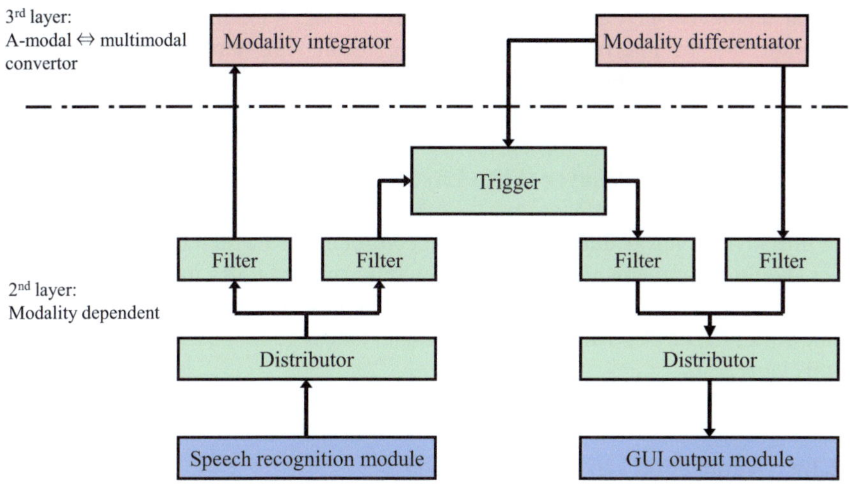

Fig. 7.2 A sample configuration of the second layer module

composed of two modules: the modality integration module and the modality differentiation module.

The modality integration module accepts some restrictions on a user's input (such as data type) from the fourth layer and sends the converted information (such as speech recognition grammar) to the second layer. The information from the fourth layer may include some modality-dependent information. This module accepts the input result from the second layer in the EMMA format and sends it to the fourth layer after eliminating the modality-dependent information. The inputs from multiple modules in the second layer are integrated in this layer based on the rules defined in some markup languages such as the `<operation>` element in XISL [12].

The modality differentiation module accepts the output contents from the fourth layer and sends them to the second layer after converting into the modality-dependent formats (such as sentences processed in the speech synthesis module). The contents sent from the fourth layer may include modality-dependent information. The details of differentiation rules such as synchronization of timing should be defined by markup languages such as SMIL [13] or XISL.

7.2.1.4 A-Modal Dialogue Control Layer (Fourth Layer)

Management of form filling and dialogue state transition are executed in this layer. The fourth layer should be implemented in a single module because it receives a-modal dialogue control information (possibly including modality-dependent information) described in markup languages (such as `<form>` element in VoiceXML2.1 with FIA, or `<form>` element of XHTML [14]) from the fifth layer, interprets it, and executes it. This layer is responsible for managing task internal dialogues such as prompting for form filling, handling a barge-in from a user, and system interruption control. The results are sent to the fifth layer according to the description in the markup language.

7.2.1.5 Task Control Layer (Fifth Layer)

The fifth layer controls dialogue tasks and communicates with the sixth layer. This layer should be implemented as a single module and generates interaction patterns that are sent to the fourth layer. A controller description written in the Rails framework [15] or SCXML [16] is a candidate language for describing the activity of this layer.

7.2.1.6 Application Layer (Sixth Layer)

The data model and application logic are implemented in this layer. The definition of an API is required to communicate with the fifth layer. We do not define any specification for this layer.

7.2.1.7 User Model and Device Model

This component provides an API to contact the user/device model variables that are defined in some external ontology. It provides the user's and device's information to the layers second to fifth. For the lower layers, set/get functions to change/obtain the user/device status are prepared. For the higher layers, publish/subscribe functions to notify the changes of status are provided.

7.2.2 Events Transferred Between Layers

7.2.2.1 The Interface Between Second (Modality Dependent) and Third (A-Modal ⇔ Multimodal Conversion) Layers

The role of the second layer is to unify multiple input–output devices by combining the distributor, filter, and trigger functions. It provides a unified interface to the third layer. Its configuration is almost the same as the Russian Doll model proposed in the W3C MMI architecture [3]. The components accessed from the third layer should be designed with consideration for extensibility, versatility, and efficiency of description in the third layer. Most of the input–output events transferred between the second and third layers are asynchronous events that are integrated and synchronized in the third layer. Therefore, the interface between the second and third layers should be a set of asynchronous event pairs.

Table 7.1 Transferred events between the second and third layers

Input events	Output events	Common events
StartSessionRequest	StartSessionRequest	NewContextRequest
UpdateParameterRequest	UpdateParameterRequest	PrepareRequest
StartInputRequest	SetOutputContentRequest	ClearContextRequest
StopInputRequest	StartOutputRequest	ExtentionNotification
EndSessionRequest	SuspendOutputRequest	StatusRequest
ErrorNotification	ResumeOutputRequest	
InputNotification	StopOutputRequest	
HelpInputNotification	EndSessionRequest	
NoinputNotification	WaitingInputNotification	
NomatchNotification	InformCurrentInputNotification	
	ErrorNotification	
	EndOutputNotification	

Corresponding "Response" events are prepared for all "Request" events

The events defined in this layer are based on those defined in Sect. 6.2 (Standard Life Cycle Events) in the W3C MMI architecture [3]. However, some events related to input–output modalities are modified for compatibility with the upper layers. The events transferred between the second and third layers are shown in Table 7.1.

7.2.2.2 The Interface Between Third (A-Modal ⇔ Multimodal Conversion) and Fourth (A-Modal Dialogue Control) Layers and Inside Third Layer

The events transferred between the third and fourth layers include a unit of possible inputs, an output execution, and their results. The same kinds of events are transferred between the input integration module and the output differentiation module inside the third layer. The internal third layer events are used to send some detailed information that should not be intermediated by the dialogue control layer. The events transferred between the third and fourth layers and inside the third layer are listed in Table 7.2.

7.2.2.3 The Interface Between the Fourth (A-Modal Dialogue Control) and Fifth (Task Control) Layers

The fifth layer sends the events that contain the form to be executed (filled) in the fourth layer when the task-level operation determines the action. The published event can contain the list of possible field values as its optional argument. These values are calculated based on the information from the current state of the dialogue, the status of the sixth layer or the user model.

Table 7.2 Transferred events between the third and fourth layers and inside the third layer

From fourth layer to input integrator	From fourth layer to output differentiator	Internal third layer
StartSessionRequest	StartSessionRequest	StartInputRequest
InitializeRequest	InitializeRequest	StopInputRequest
UpdateParameterRequest	UpdateParameterRequest	WaitingInputRequest
SetInputContentRequest	SetOutputContentRequest	InformCurrentInputNotification
StartInputRequest	StartOutputRequest	StartOutputRequest
StopInputRequest	RepromptRequest	RepromptRequest
EndSessionRequest	RetryRequest	RetryRequest
StatusRequest	SuspendOutputRequest	SuspendOutputRequest
ErrorNotification	ResumeOutputRequest	ResumeOutputRequest
ExtensionNotification	StopOutputRequest	StopOutputRequest
	StatusRequest	ErrorNotification
	EndSessionRequest	EndOurputNotification
	ErrorNotification	ExtensionNotification
	ExtensionNotification	
From input integrator to fourth layer	**From output differentiator to fourth layer**	
AcceptNormalInput Notification	EndOutputNotification	
NoinputNotification	ErrorNotification	
NomatchNotification	ExtensionNotification	
HelpInputNotification		
ErrorNotification		
ExtensionNotification		

Corresponding "Response" events are prepared for all "Request" events

From the fourth layer, the events to notify the status of a form are sent to the fifth layer. The events in the EMMA format are published when the form is completely filled or a field value is changed during the form filling. The fifth layer decides which form to execute using application logic after updating the user model according to the results received from the fourth layer.

Table 7.3 shows the list of events transferred between the fourth and fifth layers.

7.2.2.4 The Interface Between Each Layer and Device/User Model Component

Each layer has a different interface with the device/user model component because the layers need different information held in the component. The information required in each layer is shown below.

- The device model sends the start events to the second layer when the MMI system is booted up because this layer must activate the recognition/synthesis

Table 7.3 Transferred events between the fourth and fifth layers

Fifth to fourth events	Fourth to fifth events
StartForm	FilledForm
StopForm	ChangedField
Common events	
NewContextRequest	NewContextResponse
PrepareRequest	PrepareResponse
ClearContextRequest	ClearContextResponse
ExtentionNotification	StatusResponse
StatusRequest	

modules. While the MMI system is running, the second layer sends the change of module status and user model variables to the device model and the user model, respectively.

- The third layer obtains the information on the change of the second layer's status through the device model.
- The fourth layer is independent of modality. Therefore, it communicates with the user model only.
- The fifth layer is also independent of modality. Therefore, it communicates with the user model only. Moreover, the fifth layer does not update the user model.

Table 7.4 enumerates the transferred events between each layer and the device/ user model component.

7.2.3 Some Examples of Event Transfer

This section presents two examples of event transfer among layers. The first example, shown in Fig. 7.3, illustrates the session start and end procedures. At first, the higher layers send *StartSessionRequest* events to lower layers and receive corresponding responses. The initialization process is then executed. If some parameter must be changed in a lower layer, the *UpdateParameterRequest* events are sent to the lower layers. The *EndSessionRequest* events are sent to the lower layers when the session ends.

The second example shows how the user's inputs are processed in an MMI system. As shown in Fig. 7.4, some prompts are output before starting the input standby. If the system does not accept a barge-in, the *StartInputRequest* events that start input standby are sent to the lower layers after accepting the *EndOutputNotification* event as shown in the upper part of Fig. 7.4. If the system accepts the barge-in, the *StartInputRequest* events are sent to the lower layers before accepting the *EndOutputNotification* event as illustrated in the lower part of Fig. 7.4. In this case, the *StopOutputRequest* events are sent to the output modules if the system accepts an input.

Table 7.4 Transferred events between the device/user model component and each layer

The second layer	The third layer
Device model to second layer	Device model to third layer
StartSession	SetAvailableModality
EndSession	SuspendModality
Second layer to device model	Third layer to device model
SessionClosed	SetDeviceModelAttribute
	GetDeviceModelAttribute
	SubscribeDeviceModelAttribute
	UnsubscribeDeviceModelAttribute
Second layer to user model	Third layer to user model
RecognizedInteractionError	SetUserModelAttribute
	GetUserModelAttribute
	SubscribeUserModelAttribute
	UnsubscribeUserModelAttribute
The fourth layer	**The fifth layer**
Fourth layer to user model	Fifth layer to user model
SetUserModelAttribute	SubscribeUserModelAttribute
GetUserModelAttribute	UnsubscribeUserModelAttribute
SubscribeUserModelAttribute	
UnsubscribeUserModelAttribute	

7.3 Practical Implementation in the Web-Based MMI System

7.3.1 Outline of the Web-Based MMI System

In this section, we introduce a Web-based MMI system [17] constructed based on the proposed architecture. This system uses some modules developed in the Galatea project [18] which aims to construct an anthropomorphic agent-based MMI system. In the Web-based MMI system, we used the speech recognizer Julius [19], the speech synthesizer Galatea talk [20], the facial image generator Galatea FSM [21], and the dialogue scenario description language XISL and its interpreter [22], which are provided in the Galatea toolkit [23]. Figure 7.5 illustrates the configuration of the Web-based MMI system. The first layer is implemented on a Web browser and a server side program. The second layer is composed of some modules provided in the Galatea toolkit and their wrapper programs to adapt to the six-layered model. Modality integration and differentiation in the third layer are achieved by interpreting some elements described in XISL that specify multimodal inputs and outputs. The fourth and fifth layers are combined into an XISL interpreter in this system.

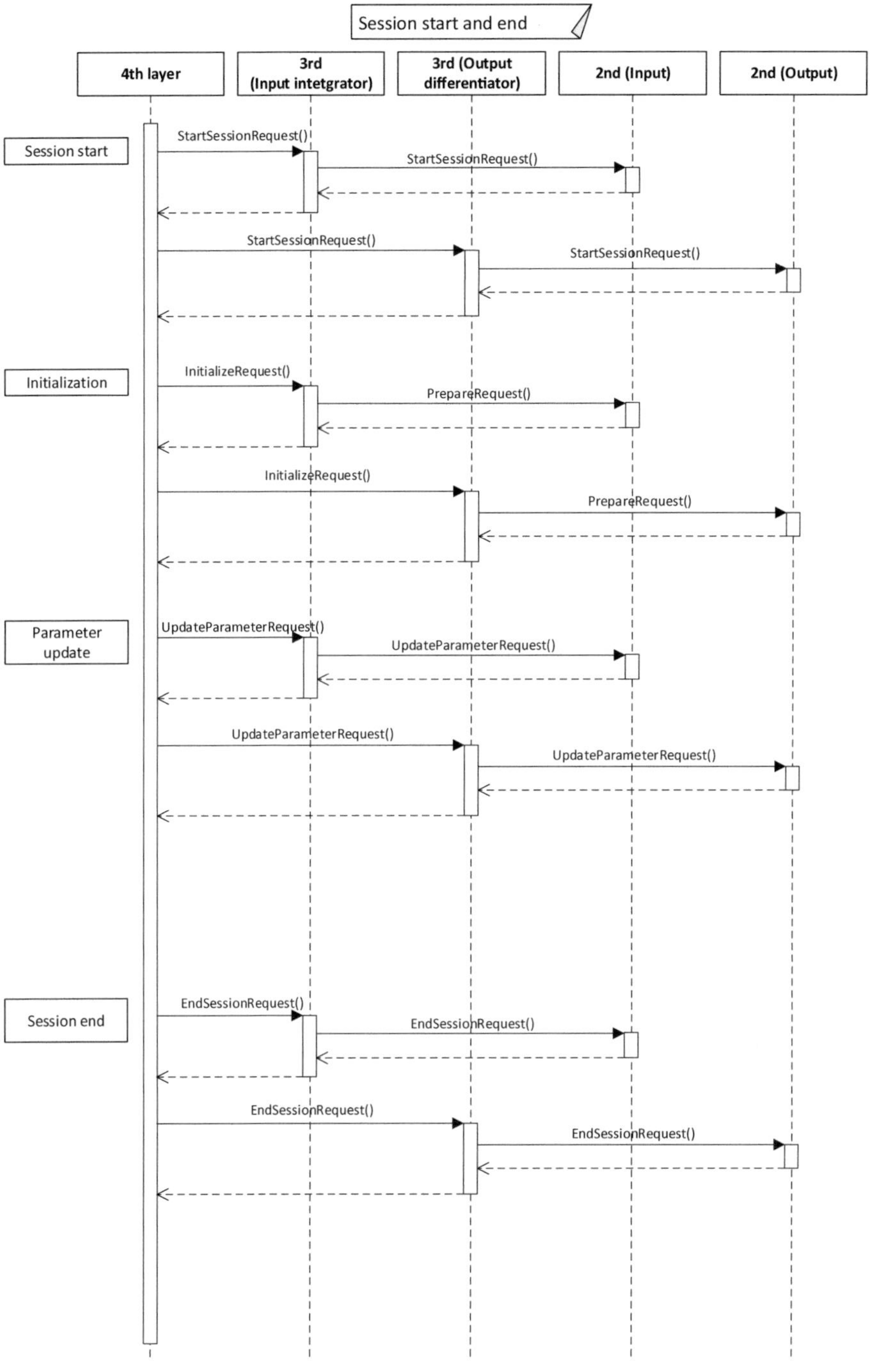

Fig. 7.3 Example of event transfer: session start and end

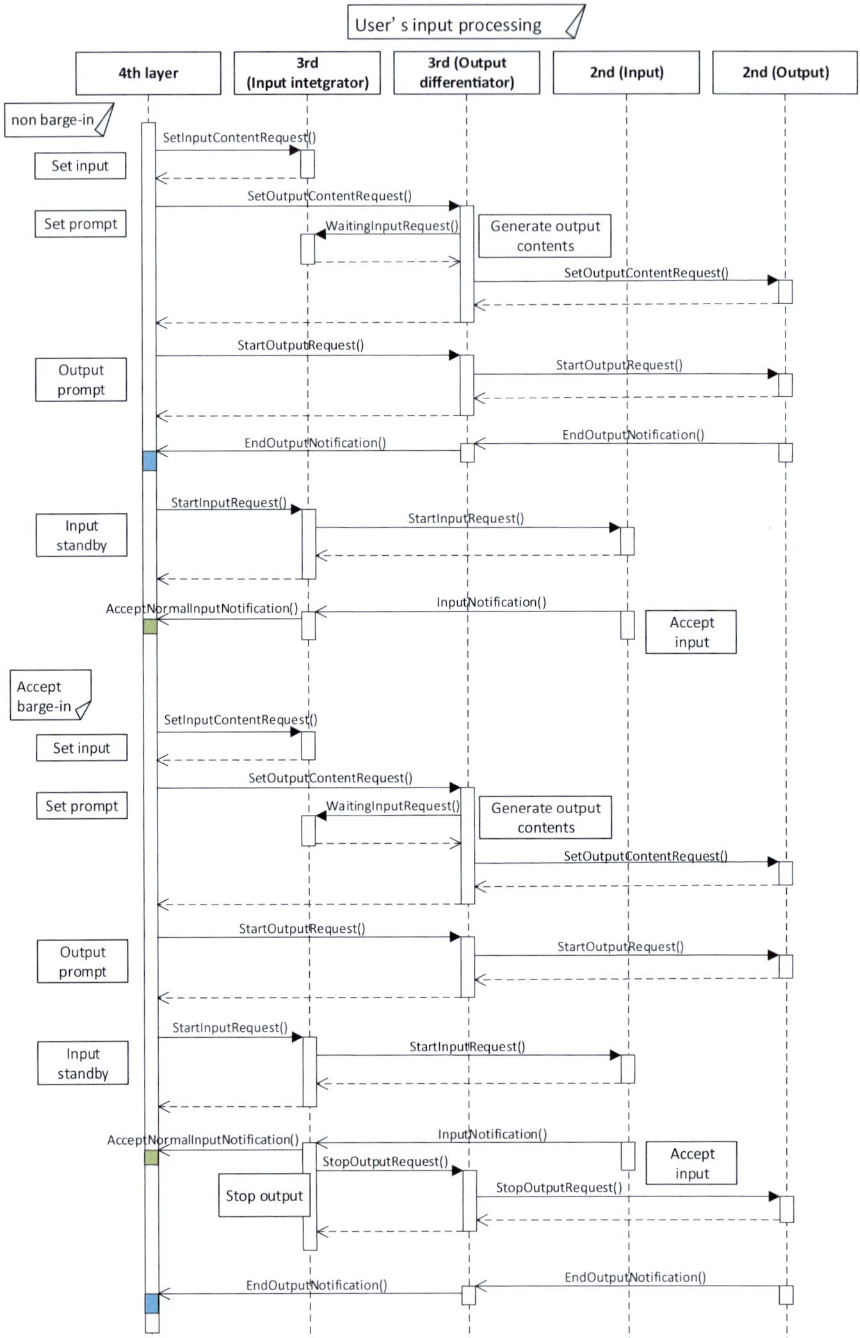

Fig. 7.4 Example of event transfer: processing a user's input

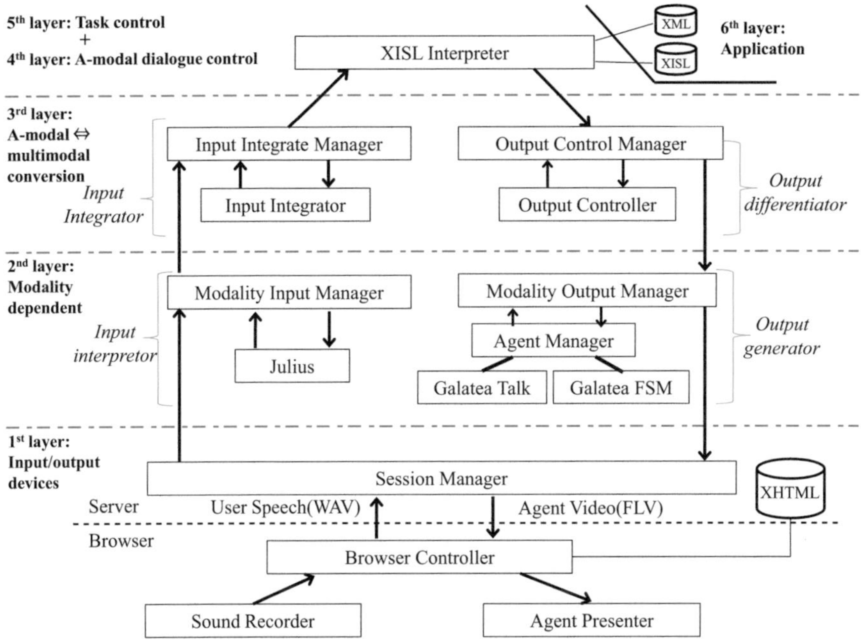

Fig. 7.5 Configuration of the web-based MMI system

7.3.2 Implementation of Each Layer

7.3.2.1 Implementation of the First Layer

The first layer in our Web-based MMI system is composed of two parts: a browser side module and a server side module. The browser side module is a user interface. It records the speech from the user and outputs the sounds and the anthropomorphic agent. It is implemented using JavaScript, Adobe Flash, and the Java Applet for execution on a Web-browser. The server side module is a wrapper program between the second layer and the Web browser. It is implemented as a servlet using the Java language.

7.3.2.2 Implementation of the Second Layer

The modality input manager accepts pointing and speech inputs. The speech inputs are recognized using Julius. The recognition results and pointing information are sent to the third layer by the events in the EMMA format. The modality output manager accepts the speech synthesis and the anthropomorphic agent information from the third layer. It invokes Galatea talk and Galatea FSM through the agent manager and accepts the synthesized speech and facial animation. These contents are sent to the first layer and output to the Web browser.

7.3.2.3 Implementation of the Third Layer

The input integration manager and the output control manager are the wrapper modules for event handling. They also manage the input integrator and output controller. This system structure makes it easy to replace the input integrator or output controller when different algorithms are appropriate for integration and differentiation.

In this MMI system, the input integrator accepts the `<operation>` element in an XISL description. This element includes a set of inputs to be accepted in a given situation. After interpreting it, the integrator sends modality-dependent information (such as speech grammar) to the second layer. The accepted inputs sent from the second layer are stored in the integrator and integrated in this module according to the description of the `<operation>` element. A sample of the `<operation>` element is shown in Fig. 7.6. This `<operation>` element accepts a speech input or a click input alternatively.

The output controller accepts the `<prompt>` and `<action>` elements in an XISL description that contains the output contents. The controller controls the output timing (such as sequential, parallel) described in the XISL and sends it to the second layer. A sample of the `<prompt>` element is shown in Fig. 7.7. This `<prompt>` element outputs two sentences of speech “Hello, this is MMI online shop.” and “Please select an item.” sequentially with some expressions of a female anthropomorphic agent.

```
<operation comb="alt">
  <input type="speech" event="recognize" match="menu" return="item" />
  <input type="touch" event="click" match="menu" return="item" />
</operation>
```

Fig. 7.6 Example of the `<operation>` element

```
<prompt>
  <output type="agent" event="speech">
    <character>female01</character>
    <expression>HAPPY</expression>
    <text> Hello, this is MMI online shop.</text>
  </output>
  <output type="agent" event="speech">
    <character>female01</character>
    <expression>NEUTRAL</expression>
    <text>Please select an item.</text>
  </output>
</prompt>
```

Fig. 7.7 Example of the `<prompt>` element

7.3.2.4 Implementation of the Fourth and Fifth Layers

The fourth and fifth layers are integrated into a module in this Web-based MMI system. This module controls the dialogue state and tasks. XISL is used for this purpose. The module is implemented as an extension of the existing XISL interpreter. Figure 7.8 describes a sample XISL document.

7.3.2.5 An Application Constructed Using the Web-Based MMI System

We developed an MMI application using the Web-based MMI system. Its screen shot is illustrated in Fig. 7.9. An anthropomorphic agent produced from a real facial image is displayed on the left side of the browser. The right side constitutes an

```
<?xml version="1.0" encoding="UTF-8"?>
<xisl xmlns="http://www.mmi.cs.tut.ac.jp/xisl" version="1.1">
  <body>
    <dialog id="main">
      <exchange>
        <var name="item" />
        <prompt>
          <output type="agent" event="speech">
            <character>female01</character>
            <expression>HAPPY</expression>
            <text>Hello, this is MMI online shop.</text>
          </output>
          <output type="agent" event="speech">
            <character>female01</character>
            <expression>NEUTRAL</expression>
            <text>Please select an item.</text>
          </output>
        </prompt>
        <operation comb="alt">
          <input type="speech" event="recognize" match="menu" return="item" />
          <input type="touch" event="click" match="menu" return="item" />
        </operation>
        <action>
          <output type="agent" event="speech">
            <character>female01</character>
            <expression>HAPPY</expression>
            <text>You ordered <value name="item" />.</text>
          </output>
          <goto  next=”how_many.xisl”  namelist=”item”/>
        </action>
      </exchange>
    </dialog>
  </body>
</xisl>
```

Fig. 7.8 Example XISL document

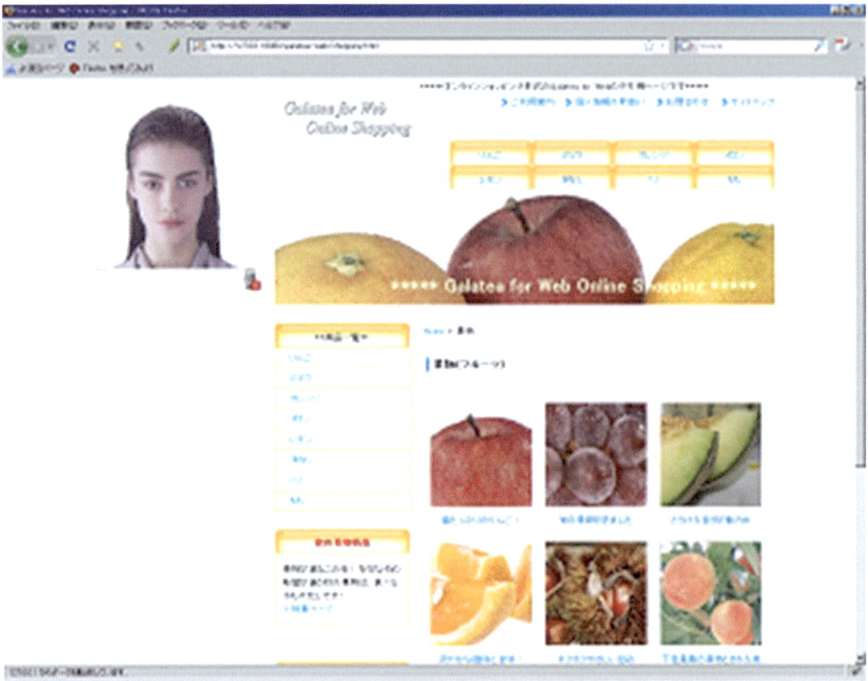

Fig. 7.9 Screen shot of the web-based MMI system

online shopping site. A user can interact with this system using speech or/and pointing. The output from the MMI system is given in speech with an anthropomorphic agent and a Web page.

7.4 Conclusions

In this chapter we presented a six-layered model for MMI systems published as an ITSCJ standard. This standard clarifies the role of each module by placing it on one of the six layers. The standard also provides the events transferred between layers that are designed to send necessary and sufficient information to the other layers. Because the modules that conform to this standard have well-defined role and interface, it is easy for a developer to replace a module with another one. This is a desirable feature for facilitating the development of MMI systems by reusing existing systems.

In the future, we would like to provide a reference system and a toolkit to develop MMI systems based on the proposed architecture.

References

1. Turk, M. (2014). Multimodal interaction: A review. *Pattern Recognition Letters, 36*, 189–195.
2. W3C Multimodal Interaction Working Group. https://www.w3.org/2002/mmi/.
3. Multimodal Architecture and Interfaces. https://www.w3.org/TR/2012/REC-mmi-arch-20121025/.
4. EMMA1.0. https://www.w3.org/TR/2009/REC-emma-20090210/.
5. ITSCJ standard for hierarchical MMI architecture. https://www.itscj.ipsj.or.jp/ipsj-ts/ts0012/mmi-arch.html (in Japanese).
6. Six-layered MMI architecture (position paper). https://www.w3.org/2007/08/mmi-arch/papers/position.pdf.
7. Nitta, T., Katsurada, K., Araki, M., Nishimoto, T., Amakasu T., & Kawamoto, S. (2007). Proposal of a hierarchical architecture for multimodal interactive systems. IPSJ SIG Technical Reports, Tokyo, Japan 2007-SLP-68 (pp. 7–12) (in Japanese).
8. SVG. https://www.w3.org/TR/SVG/.
9. VoiceXML2.1. https://www.w3.org/TR/voicexml21/.
10. SSML1.1. https://www.w3.org/TR/speech-synthesis11/.
11. Wang, K. (2002). SALT: A spoken language interface for web-based multimodal dialog systems. In *Proceedings of ICSLP '02, Denver, USA*, pp. 2241–2244.
12. Katsurada, K., Nakamura, Y., Yamada, H., & Nitta, T. (2003). XISL: A language for describing multimodal interaction scenarios. In *Proceedings of ICMI '03, Vancouver, Canada*, pp. 281–284.
13. SMIL3.0. https://www.w3.org/TR/2008/REC-SMIL3-20081201/.
14. XHTML2.0. https://www.w3.org/TR/xhtml2/.
15. Araki, M., & Mizukami, Y. (2011). Development of a data-driven framework for multimodal interactive systems. In *Proceedings of the Paralinguistic Information and Its Integration in Spoken Dialogue Systems Workshop, Granada, Spain* pp. 91–101.
16. SCXML. https://www.w3.org/TR/scxml/.
17. Katsurada, K., Kirihata, T., Kudo, M., Takada, J., & Nitta, T. (2008). A browser-based multimodal interaction system. In *Proceedings of ICMI '08, Chania, Greece* pp. 195–196.
18. Kawamoto, S., Shimodaira, H., Nitta, T., Nishimoto, T., Nakamura, S., Itou, K., Morishima, S., Yotsukura, T., Kai, A., Lee, A., Yamashita, Y., Kobayashi, T., Tokuda, K., Hirose, K., Minematsu, N., Yamada, A., Den, Y., Utsuro, T., & Sagayama, S. (2004). Galatea: Open-source software for developing anthropomorphic spoken dialog agents. In H. Prendinger & M. Ishizuka (Eds.), *Life-like characters* (pp. 187–211). Heidelberg: Springer.
19. Kawahara, T., Kobayashi, T., Takeda, K., Minematsu, N., Itou, K., Yamamoto, M., et al. (1998). Sharable software repository for Japanese large vocabulary continuous speech recognition. In *Proceedings of ICSLP '98, Sydney, Australia* pp. 3257–3260.
20. Yoshimura, T., Tokuda, K., Kobayashi, T., Masuko, T., & Kitamura, T. (1999). Simultaneous modeling of spectrum, pitch and duration in hmm-based speech synthesis. In *Proceedings of EUROSPEECH '99, Budapest, Hungary* pp. 2347–2350.
21. Yotsukura, T., & Morishima, S. (2002). An open source development tool for anthropomorphic dialog agent-face image synthesis and lip synchronization. In *Proceedings of IEEE MMSP2002, St. Thomas, USA*, pp. 272–275.
22. Katsurada, K., Ootani, Y., Nakamura, Y., Kobayashi, S., Yamada, H., & Nitta, T. (2002). A modality independent MMI system architecture. In *Proceedings of ICSLP '02, Denver, USA*, pp. 2549–2552.
23. Katsurada, K., Lee, A., Kawahara, T., Yotsukura, T., Morishima, S., Nishimoto, T., et al. (2009). Development of a toolkit for spoken dialog systems with an anthropomorphic agent: Galatea. In *Proceedings of APSIPA09, Sapporo, Japan* pp. 148–153.
24. Araki, M., & Tachibana, K. (2006). Multimodal dialog description language for rapid system development. In *Proceedings of SigDIAL06, Sydney, Australia*, pp. 109–116.

Chapter 8
WebRTC: Handling Media on the Web

Daniel C. Burnett

Abstract WebRTC is a growing set of JavaScript APIs for HTML5 that make it easy to capture media from cameras and microphones, screens, windows, and applications, and even from other HTML elements and then stream that media, live, directly to other web browsers. Different from the traditional client–server model of the web, this client-to-client communication enables more efficient communications, particularly in combination with the built-in firewall traversal capabilities. Also built in are critical signal processing capabilities and top-quality codecs. Ultimately, the most significant aspect of WebRTC is how it enables communications to be embedded into existing web applications, deepening engagement, and enhancing community.

8.1 Introduction

WebRTC is a growing set of JavaScript Application Programming Interfaces (APIs) for the web's Hypertext Markup Language (HTML5) [1] that make it easy to capture media from cameras and microphones, screens, windows, and applications, and even from other HTML elements and then stream that media, live, directly to other web browsers. This is actually quite new for the web, since the traditional model on the web is client–server, where end users are represented by web browser clients that connect to cloud-based servers. For the first time, it is now possible to establish a media (and data) connection directly between two client web browsers. This can have significant advantages of scale when media can be sent peer-to-peer, but WebRTC also includes ICE, Interactive Connectivity Establishment [2], a protocol designed to figure out the best way to get media from one client to another despite Network Address Translation (NAT) devices such as routers, gateways, and firewalls in the way. This is all provided for free.

D.C. Burnett (✉)
StandardsPlay, Lilburn, GA, USA
e-mail: dan@burnettconsultingservices.com

D.A. Dahl (ed.), *Multimodal Interaction with W3C Standards*,
DOI 10.1007/978-3-319-42816-1_8

Traditionally, writing a real-world communications application was made more difficult by the need to build echo cancellation, packet loss concealment, and other basic streamed audio processing capabilities. Now this is built-in as well, allowing for even the simplest of JavaScript applications to make use of these advanced, yet critical, capabilities. Additionally, the royalty-free, real-time streaming audio and video codecs that are now provided in the browsers are high-quality, state-of-the-art codecs, which means that audio and video quality right out of the box is limited more by the cameras, microphones, and speakers, than by the technology or even bandwidth (within reason).

Most importantly, WebRTC allows for communications to be embedded in existing web applications rather than being separate. This allows for all of the web context to be pulled along into the call or chat, improving customer interactions while keeping customers within the enterprise's application scope. This, along with the data channel for sending arbitrary data in addition to the streaming media, is particularly helpful for online gaming as well.

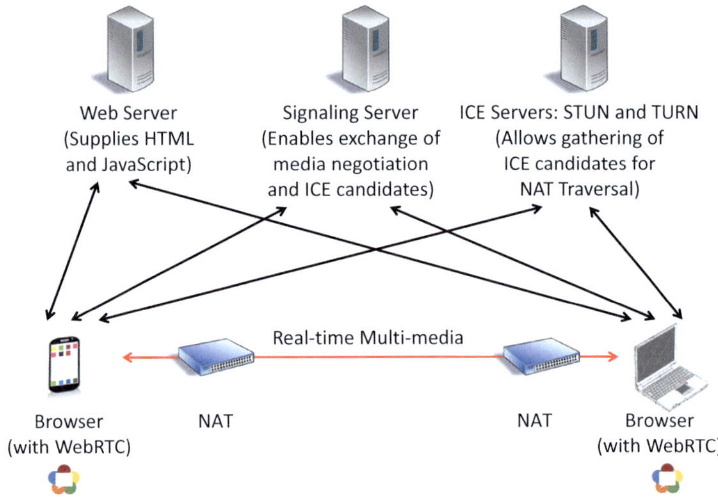

8.2 The Mechanics

Before talking about the technologies in WebRTC, it is important to understand that WebRTC work is happening at two different levels: the JavaScript API level and the Internet protocol level.

The public APIs for WebRTC are JavaScript extensions to HTML 5 and are defined in the WebRTC Working Group [3] at the World Wide Web Consortium (W3C) [4]. As a part of HTML, these APIs are implemented by web browsers such as Google Chrome and Mozilla Firefox and are available to the world of web application developers. To make these APIs work, the web browsers need to implement lower-level protocols designed for transporting media over the Internet. Originally designed for use with SIP-based [5] Voice over IP calls, protocols such as RTP [6], RTCP [6], DTLS [7], SCTP [8], STUN [9], TURN [10], and ICE [2] are what is actually used to establish connections and send and receive media. These protocols are defined in the RTCWEB and other working groups within the Internet Engineering Task Force (IETF) [11]. Thankfully, many of the same organizations and individuals are involved in both W3C and IETF, simplifying the coordination between the efforts at the two different levels.

Officially, WebRTC is only two new technologies: the ability to capture media (audio and video) without a plugin and the ability to transport that media directly from one web browser to another. In practice, there are many new technologies involved. Media capture only requires that browsers be able to access cameras, microphones, and other media sources attached to the device the browser is running on. Media transmission, however, has required that browsers support a whole host of new technologies:

- Codecs

  To be sent electronically, media must be encoded in some way. There are several reasons for this: digitization, redundancy, compression, and security. Let's use audio as an example. Audio in the real world, as registered by human ears, is a compression wave in the air. A microphone is merely a flexible membrane that converts wave-like movements of air into wave-like changes in voltage and/or current in an electronic circuit. Note that both of these are smoothly varying, continuous movements over time. Digitization is the process of sampling these continuous movements at regular time intervals and converting them into numeric values within some range. Thus, an 8 kHz, 8 bit pulse-code-modulated (PCM) digitization is merely a sampling of the microphone voltage or current 8000 times per second (8 kHz) and a conversion of the sampled value into a number between 0 and 255 (8 bits), giving us a stream of 8 bit values at a rate of one every 125 μs. This is an encoding but is considered trivial, since a media stream in this encoding is essentially ready to be converted back into voltage on a circuit for playback over a speaker. However, let's say it's a stereo audio stream, meaning that our stream of bits actually represents two separate audio channels. In that case the bits from the digitized left and right channels have to be combined into this stereo encoding and then split apart again (or decoded) into the individual streams in order to be played out on separate left and right speakers. The software or hardware that pairs together an encoder and a decoder is a called a coder–decoder, or a codec for short.

  Codecs don't just combine together affiliated media as in the stereo example above. Codecs also sometimes have redundant information such as repeated bits, checksums, or error-correcting bits to deal with the fact that corruption or loss of bits in transmission can occur. Also, codecs nowadays usually have significant compression included in order to send the maximum amount of data using the smallest number of bits, allowing for better-quality audio (and video) even in conditions with low bandwidth. Finally, modern codecs either incorporate encryption directly or are intended to be encrypted in order to reduce the potential for abuse by either malicious interception or passive monitoring.

Browsers supporting WebRTC now include support for today's most advanced audio and video codecs. G.711 [12], the equivalent of basic digitization, is supported for backwards compatibility, but WebRTC browsers are also required to support Opus [13], the most powerful and flexible audio codec available today. Opus automatically adjusts what it sends based on the currently available network performance. On the video side, browsers are officially required to support both VP8 [14] and H.264 [15]. Both are approximately equivalent in performance but vary significantly in known intellectual property challenges, licensing terms, and configurability.

- Echo cancellation and packet loss concealment

  Real-time communications stacks have to deal with the problems of both audio and unreliable Internet-based transmission. One of the former problems is echo, where the playback of audio gets fed back into the input microphone. Although arbitrary feedback cannot be cancelled, a first level of feedback elimination is crucial to making played audio be intelligible. An Internet transmission problem is the fact that packets of audio data can be lost. There are a variety of Internet protocols whose sole purpose is to conceal the loss of packets. These, and other signal processing technologies, are now built in to web browsers. This is wonderful for web developers. With no changes to application code, over the past few years the quality of the audio and video in WebRTC browsers has improved purely due to fixes and enhancements in the browser signal processing code.

- Congestion control

  TCP [16] has, built into it, some automatic controls to deal with network congestion. RTP, the protocol used to send real-time audio and video, does not. However, there are protocols to add congestion control to RTP media flows. Between support for those and other decisions made for WebRTC, the browsers now have substantial code to make intelligent decisions regarding congestion, from adjusting the parameters of codecs on the fly to make them adapt properly as network bandwidth and throughput degrade, to switching to other codecs, to prioritizing packets of some media flows over others, to suspending high-bandwidth flows such as video as needed.

- NAT traversal

  The Internet Protocol (IP) [17] was created in order to connect between different local area networks, which is why it is an inter- (between) network, or internet, protocol. At the time, there were a variety of other protocols used for local area networks, Ethernet [18] and Token Ring [19] being the most common. Fairly quickly researchers and network administrators realized that it was convenient to just give every machine a unique IP (IPv4, actually) address, a 4-byte string such as 15.2.3.4, and to then use the Internet Protocol as a "meta-protocol" in order to hide the details of the differences in the more transmission-specific protocols such as Ethernet, Token Ring, and now Wi-Fi [20]. As we all know by now, the set of 4-byte addresses available for use has run out. Many years ago Internet Service Providers (ISPs), as well as organizational network administrators, began using only a small number of public Internet addresses that were then mapped by a Network Address Translator (NAT) to a larger number of private internal IP addresses, typically beginning with either 10. or 192.168. (addresses reserved as private). These NATs (called routers or gateways today) often added firewall and policy functions, making them crucial for more than just network address translation (also abbreviated "NAT").

These NATs are a problem for peer-to-peer traffic. Why? Because when computer A wishes to send media to computer B, it's not clear which IP address to use. Sending media to the private IP address of B definitely won't work, and sending it to the public address of B's NAT won't work unless the NAT knows the private address to which it should forward the media packets. This is where ICE comes in. Interactive Connectivity Establishment (ICE) makes use of STUN (Secure Tunneling around NATs) and TURN (Tunneling using Relays around NATs) to set all of this up. Basically, each computer asks a STUN server it knows about to tell it what public IP address the STUN server sees for it. This also causes the NAT to watch for return packets that will be sent back to the private address, setting up a "mapping" that will be used when media flows later. Both computers then tell each other what IP addresses they can be reached at, and both start sending test packets to see which addresses work. Once they both find addresses that work, they can send packets.

This firewall "hole punching" is something that game players first worked out in order to get peer-to-peer connections to work over the Internet. Now it is built into web browsers and used automatically by WebRTC.

The tremendous advantage of these technologies that are now built into WebRTC browsers is that they greatly simplify the process of creating functional web communications. The simpler they can be made to work, the more power is given to the millions of web developers in the world.

8.3 Developing with WebRTC

There are three main new pieces of code required in a full WebRTC application: media capture, signaling, and media transmission. Although a full explanation of the WebRTC APIs is beyond the scope of this chapter, a brief highlight of the core APIs is necessary in order to understand what can be done with it.

8.3.1 Promises

First, a brief summary of Promises is in order.

A new feature of ECMAScript (JavaScript) 6 is Promises [21], a way to represent the result of an asynchronous function call. A Promise is an object that will eventually resolve with a value or reject with a "reason" (an error value). Two nice properties of Promises are that handlers can easily be set to execute upon the resolve or reject conditions and that Promise objects can be chained together in a way similar to JQuery [22] objects.

An example might help to clarify. The `setTimeout()` function in JavaScript executes a callback after a specified number of milliseconds:

```
setTimeout(callback, millisecondDelay);
```

This can be turned into a Promise as follows:

```
function pleaseWait(duration) {
  var p = new Promise(function(resolve, reject) {
    setTimeout(function() {
      resolve("I waited");
    }, duration);
  };
  return p;
}
```

Now it can be used as follows:

```
pleaseWait(5)
  .then(function(x) {
    console.log("promise fulfulled: " + x);
  })
  .catch(function(y) {
    console.log("promise rejected: " + y);
  });
```

Note that the then() method runs the given handler if/when the Promise resolves, while the catch() method runs its handler if/when the Promise fails. In this particular example we resolve the Promise (the first bit of code) with the value "I waited" after the given number of milliseconds has expired. So in this example, "promise fulfilled: I waited" would be sent to the console.

Let's say that we had a function that logged a value to the console and then returned a Promise. We could then chain our method calls as follows:

```
pleaseWait(5)
  .then(logSomethingAndReturnPromise)
  .then(function() {
    return pleaseWait(10);
  })
  .then(function() {
    console.log("all done");
  })
  .catch(function(y) {
    console.log("failed somewhere: " + y);
  });
```

In this case our code would wait 5 ms, log something, wait 10 ms, and log "all done." If for some reason the code failed *at any point in the chain*, it would log "failed somewhere" to the console, because an uncaught failure (Promise rejection) will "fall down" through the chain until it reaches a catch method.

There are a variety of helpful resources and tutorials on the web [23–26] to learn how to work with Promises and, particularly, with the tricky edge cases they can have.

8.3.2 Media Capture

Defined in the Media Capture and Streams specification [27], the `getUserMedia()` call is used to get access to cameras and microphones on the computer or mobile device. The most basic call looks like this:

```
<script>
     ...
       // find video element in my app
       var myHTMLVideoElement =
         document.getElementById("myVideo");
       // get microphone and camera
       navigator.mediaDevices.getUserMedia(
         "audio": true; "video": true)
         .then(function(s) {
            myAudioAndVideoStream = s;
       // play audio and video in video element
       myHTMLVideoElement.srcObject = s;
     };
  ...
  </script>
  ...
<video id="myVideo" autoplay="autoplay"
        muted="true"/>
```

The `getUserMedia()` Promise resolves with a `MediaStream`, which is a collection of zero or more `MediaStreamTrack` objects. Each `MediaStreamTrack` object represents one flow of media of one type. In the example above, the returned MediaStream would contain two tracks, one for the audio and one for the video. Of course, tracks can be added to or removed from a MediaStream, so it could contain any number of tracks of any type, but each individual track only contains audio from a single source or video from a single source.

In the example above, the new '`srcObject`' property on the video element takes a MediaStream as a value and plays it as if it were a URI on the web that had been assigned to the "`src`" property of the video element.

An important additional function of the `getUserMedia()` call is to obtain permission from the end user to access the requested sources. If the access is for a secure origin, as in localhost or a site accessed with HTTPS [28], the browser will ask once and, if permission is granted, save the permission for that origin (site). If the access is for an insecure origin, such as when using HTTP [29–34] to access a site, Google Chrome will not even allow permission to be given, while Mozilla Firefox will ask every time for permission.

8.3.3 Signaling

One of the most confusing aspects of WebRTC when first getting started is the fact that signaling is not standardized. The term "signaling" is typically used in communication systems to mean the information used to set up a call, as opposed to the audio and/or video of the call itself which is referred to as "media." In traditional telephony systems SS7, or Signaling System 7 [35], is the most common in recent history. In Voice over IP (VoIP) systems, where telephone calls are made over the Internet, the Session Initiation Protocol (SIP) is used for call setup (signaling), while the Real-Time Protocol (RTP) is used for the media. Note that both SS7 and SIP are *standard* signaling protocols. WebRTC, by design, does not standardize a signaling protocol. WebRTC does not define how setup information is to be exchanged between two browsers that wish to communicate, although it does define the process by which the two browsers must agree on the details of what they send each other (called the "negotiation").

WebRTC developers will often refer to "the signalling channel," by which they mean whatever method the web developer has chosen to convey setup information from one browser to the other. This choice is completely up to the developer but will typically require relaying through some server on the web since there is no way (other than the peer connection that will be set up after the signaling channel) for two browsers to communicate directly with one another. Common ways to do this relaying are via Web Sockets [36] or HTTP polling to a web server, via a web service such as PubNub [37] or FireBase [38], or via a traditional signaling server such as a SIP registrar/proxy server if using SIP as the signaling protocol. All of these are possible through either custom JavaScript code or libraries such as JsSip [39] and sipML5 [40].

A key consequence/principle to keep in mind here is that notions of identity, authentication, authorization, location, and gatewaying are all included in signaling. WebRTC assumes that all decisions about which browser to contact, how to contact that browser, whether contact with that browser is permitted, etc., have already been made in the process of setting up "the signaling channel."

8.3.4 Media Transmission

The WebRTC specification [41] defines a set of JavaScript APIs that cause things to happen at a protocol level. The key things a web developer needs to do are

- Create a peer connection
- Indicate that media should be sent
- Negotiate the media connection

Creating a peer connection can be done by executing something like

```
var pc = new PeerConnection();
```

A track can then be scheduled to be sent over the peer connection by calling

```
pc.addTrack(myVideoTrack);
```

Negotiation of the media connection is a two-part process on each end. In order to send media, each side needs an SDP (media description) offer and an SDP answer. One of these must be for itself (the "local" end), and the other must be for the "remote" end. Each side can only create one description and must get the other one, the remote one, from the other browser. This negotiation process is referred to as "SDP offer/answer."

The offer/answer protocol used in WebRTC was originally designed for use with SIP. The idea is that one side (usually the one placing a call) would generate and send a media description, including a list of acceptable codecs and their parameters, for each media flow, in the format of SDP (Session Description Protocol). This description is known as the offer. The other side would review this offer and construct a subset of it to send back, called the answer. In this way both sides have agreed not only on what can be sent and received, but also the formats that are acceptable. The JavaScript Session Establishment Protocol (JSEP) [42] defines a version of this SDP offer/answer protocol for use in WebRTC. In the WebRTC version of offer/answer, SDP is still used as the text format for offered codecs and their parameters, and the browsers generate this SDP for the developer, but the WebRTC application is responsible for exchanging the offers and answers using the signaling channel.

In JavaScript, the way all of this happens is that the calling side executes something like

```
pc.createOffer()
    .then(function(localdescription) {
      pc.setLocalDescription(localdescription);
      //...also send localdescription to the
      // other browser over the signaling channel
     })
```

The receiving side (the other browser), upon receiving the remote offer over the signaling channel, executes something like

```
pc.setRemoteDescription(remotedescription)
    .then(function(remotedescription) {
      pc.createAnswer()
        .then(function(localdescription) {
          pc.setLocalDescription(localdescription);
          //...also send localdescription to the
          // other browser over signaling channel
        });
     });
```

Note that the setting of the local description is the same process on both browsers. The only difference is that in one case it's an offer and in the other it's an answer.

For more information about SDP, there is a nice tutorial at [43].

8.4 Support

WebRTC work officially began in 2011 with the launch of webrtc.org and the creation of the standards working groups at W3C and IETF. Since then there have been numerous implementation milestones, even though the standards are not yet done. Most importantly, several years ago Chrome and Firefox demonstrated basic call interoperability, something they have maintained since. At the time of this writing Apple has not yet announced support for WebRTC, although they have hired for WebRTC implementer positions. Microsoft Internet Explorer and Edge both support the `getUserMedia()` call. Edge plans support for Object RTC (ORTC) [44], a separate (and originally competing) effort from WebRTC that provides lower-level controls. Just last year the ORTC and WebRTC groups got together and decided that all official standards work for both after WebRTC 1.0 will happen in the WebRTC Working Group, meaning that the efforts are effectively merging after WebRTC 1.0. Although the groups also agreed that WebRTC 1.0

applications will be expected to work in the next version of WebRTC (even though work on the next version has not yet begun), it is not clear at this time what Microsoft plans to support of WebRTC 1.0.

What is a developer to do? With the standards still in flux, with varying levels of implementation support so far in the different browsers, and with unknowns around Microsoft and Apple, it can be challenging to know what can be done today. Several sites have sprung up that are useful. First, webrtc.org [45] has not only good info but also libwebrtc, Google's C++ library that underlies the Chrome implementation. Second, iswebrtcreadyyet.com [46] maintains a dashboard showing which features are implemented by which browsers. Third, webrtchacks.com [47] has a wealth of blog posts describing various features and how to access them on the different browsers, whether they are prefixed or implemented as the standard says.

For developers just getting started with WebRTC, the first WebRTC book [48] and the WebRTC School online training [49] may be helpful. Also, the "`adapter.js`" library [50] attempts to provide a standards-compliant shim layer that will work on current and past versions of Chrome and Firefox. This library file is maintained by both Google and Mozilla. Using this library initially is a great way to get started playing with WebRTC, and then reviewing the contents of the adapter file is a great way to learn what the differences are between Chrome and Firefox.

Although this entire chapter talks only about web browser use of WebRTC, the APIs are also available in the C++ library mentioned above. Google has committed to having the C++ APIs match as closely as possible the JavaScript APIs defined in the standards. A number of companies have built more complete system libraries on top of this one, including the Temasys WebRTC plugin [51].

On Android devices today, both Chrome and Opera support WebRTC. Ericsson's Bowser browser for iOS also has WebRTC support. Anything else on mobile has to be done using the C++ library above or something that integrates it.

8.5 Tools and Services

There are quite a few tools and services available for developing with WebRTC, ranging from alternate APIs built on top of WebRTC, to signaling libraries, to services for signaling and for TURN (media relay).

Some APIs built on top of WebRTC are those by TokBox [52], APIDaze [53], APIzee [54], Twilio [55], and Tropo [56] (now part of Cisco).

The largest group of signaling libraries is for SIP signaling, which can be quite helpful when building an application that is required to interoperate with an existing SIP infrastructure. These libraries include JsSIP, sipML5, and SIP.js [57]. Asterisk [58], OverSIP [59], and Kamailio [60] are three different SIP servers that have support for clients using those SIP libraries.

Third-party signaling services, or services that can easily be used as such, include PubNub and FireBase. Both Twilio and Xirsys [61] provide pay-as-you-go TURN servers.

8.6 Uses of WebRTC Today

WebRTC use is mainstream, yet, paradoxically, still not widespread. Some of the more well-known uses of WebRTC are in Google Hangouts, Amazon Mayday, and Facebook Messenger. Virtually every telecommunications carrier now has a web front-end to their network that uses WebRTC, and every unified communications platform does as well. However, what is most interesting are the uses of WebRTC to add communications into existing applications. Most of these are proprietary and restricted to members, but that's the point. For example, a site that already connects together a support group for alcoholics could add the ability to do virtual group sessions if that would help its members. Pipe.com [62] makes use of a feature this paper has not really covered yet, and that is the data channel. In addition to sending and receiving streaming real-time media, WebRTC can also establish data channels peer-to-peer that can carry arbitrary data. Originally envisioned as a way for game developers to send position data, events such as shooting or being shot, etc., the data channel is definitely unlimited in its uses. Pipe.com uses it to provide peer-to-peer file sharing that is incredibly easy to use.

8.7 Multimodal Use

Given the subject of this book, it is appropriate to say a few words about how WebRTC can be used to build or enhance multimodal web applications. Obviously the acquisition of local cameras and microphones allows for video input that can be used for gesture recognition and/or sign language, augmented reality, and object or situation recognition, and audio input that can be modified and/or mixed locally, or sent to a remote speech recognition engine (or anywhere else, for that matter).

The ability to send media directly between browsers, when the network permits, is obviously useful for multimodal applications that need either remote processing of media (such as advanced gesture or speech recognition) or media from one endpoint that needs to be conveyed to another. An example might be video of an intersection where facial recognition or crowd formation analysis is being done. So many devices already contain an IP stack, particularly in the Internet of Things world, that adding a WebRTC stack can be relatively inexpensive.

References

1. Hickson, I., Berjon, R., Faulkner, S., Leithead, T., Navara, E. D., O'Connor, E., et al. (2014). HTML5: A vocabulary and associated APIs for HTML and XHTML. https://www.w3.org/TR/html5/. Accessed 21 Apr 2016.
2. Rosenberg, J. (2010). Interactie connectivity establishment (ICE): A protocol for network address translator (NAT) traversal for offer/answer protocols, RFC5245. https://tools.ietf.org/rfc/rfc5245.txt. Accessed 21 Apr 2016.
3. World Wide Web Consortium (2016). Web real-time communications working group. https://www.w3.org/2011/04/webrtc/. Accessed 21 Apr 2016.
4. World Wide Web Consortium (2016). https://www.w3.org/. Accessed 21 Apr 2016.
5. Rosenberg, J., Shulzrinne, H., Camarillo, G., Johnston, A., Peterson, J., Sparks, R., et al. (2002). SIP: Session initiation protocol, RFC3261. https://www.ietf.org/rfc/rfc3261.txt. Accessed 21 Apr 2016.
6. Schulzrinne, H., Casner, S., Frederick, R., & Jacobson, V. (2003). RTP: A transport protocol for real-time applications, RFC3550. https://tools.ietf.org/rfc/rfc3550.txt. Accessed 21 Apr 2016.
7. Rescorla, E., & Modadugu, N. (2006). Datagram transport layer security, RFC4347. https://tools.ietf.org/rfc/rfc4347.txt. Accessed 21 Apr 2016.
8. Stewart, R. (2007). Stream control transmission protocol, RFC4960. https://tools.ietf.org/rfc/rfc4960.txt. Accessed 21 Apr 2016.
9. Rosenberg, J., Mahy, R., Matthews, P., & Wing, D. (2008). Session traversal utilities for NAT (STUN), RFC5389. https://tools.ietf.org/rfc/rfc5389.txt. Accessed 21 Apr 2016.
10. Mahy, R., Matthews, P., & Rosenberg, J. (2010). Traversal using relays around NAT (TURN): Relay extensions to session traversal utilities for NAT (STUN), RFC5766. https://tools.ietf.org/rfc/rfc5766.txt. Accessed 21 Apr 2016.
11. Internet Engineering Task Force (2016). https://www.ietf.org/. Accessed 21 Apr 2016.
12. International Telecommunication Union (1988). Pulse Code Modulation (PCM) of voice frequencies, ITU-T G.711. https://www.itu.int/rec/T-REC-G.711-198811-I/en. Accessed 21 Apr 2016.
13. Valin, J. M., Vos, K., & Terriberry, T. (2012). Definition of the opus audio codec. https://tools.ietf.org/rfc/rfc6716.txt. Accessed 21 Apr 2016.
14. Westin, P., Lundin, H., Glover, M., Uberti, J., & Galligan, F. (2016). RTP payload format for VP8 video, RFC7741. https://tools.ietf.org/html/rfc7741. Accessed 21 Apr 2016.
15. International Telecommunications Union (2016). H.264: Advanced video coding for generic audiovisual services, ITU-T H.264. http://www.itu.int/rec/T-REC-H.264. Accessed 21 Apr 2016.
16. Information Sciences Institute, University of Southern California (1981). Transmission control protocol, RFC793. https://www.ietf.org/rfc/rfc793.txt. Accessed 21 Apr 2016.
17. Information Sciences Institute, University of Southern California (1981). Internet protocol, RFC791. https://www.ietf.org/rfc/rfc0791.txt. Accessed 21 Apr 2016.
18. Instititute of Electrical and Electronics Engineers (2012). ETHERNET, IEEE 802.3. http://standards.ieee.org/getieee802/download/802.3-2012.zip. Accessed 21 Apr 2016.
19. Institute of Electrical and Electronics Engineers (1989). IEEE standard for local area networks: Token ring access method and physical layer specifications, IEEE 802.5-1989. https://standards.ieee.org/findstds/standard/802.5-1989.html. Accessed 21 Apr 2016.
20. Institute of Electrical and Electronics Engineers (2012). IEEE 802.11: Wireless LANs. http://standards.ieee.org/about/get/802/802.11.html. Accessed 21 Apr 2016.
21. Cavalier, B., & Denicola, D. (2014). Promises/A+. https://promisesaplus.com/. Accessed 21 Apr 2016.
22. jQuery Foundation (2016). jQuery. https://jquery.com/. Accessed 21 Apr 2016.
23. Archibald, J. (2014). JavaScript promises: There and back again. http://www.html5rocks.com/en/tutorials/es6/promises/. Accessed 21 Apr 2016.

24. Erdi, B. (2015). JavaScript promises: A tutorial with examples. http://www.toptal.com/javascript/javascript-promises. Accessed 21 Apr 2016.
25. Franklin, J. (2015). Embracing promises in JavaScript. http://javascriptplayground.com/blog/2015/02/promises/. Accessed 21 Apr 2016.
26. Mozilla Foundation (2016). Promise. https://developer.mozilla.org/en-US/docs/Web/JavaScript/Reference/Global_Objects/Promise. Accessed 21 Apr 2016.
27. Burnett, D. C., Bergkvist, A., Jennings, C., & Narayan, A. (2016). Media capture and streams. https://www.w3.org/TR/mediacapture-streams/. Accessed 21 Apr 2016.
28. Rescorla, E. (2000). HTTP over TLS, RFC2818. https://tools.ietf.org/rfc/rfc2818.txt. Accessed 21 Apr 2016.
29. Fielding, R., & Reschke, J. (2014). Hypertext transfer protocol (HTTP/1.1): Message syntax and routing, RFC7230. https://tools.ietf.org/rfc/rfc7230.txt. Accessed 21 Apr 2016.
30. Fielding, R., & Reschke, J. (2014). Hypertext transfer protocol (HTTP/1.1): Semantics and content, RFC7231. https://tools.ietf.org/rfc/rfc7231.txt. Accessed 21 Apr 2016.
31. Fielding, R., & Reschke, J. (2014). Hypertext transfer protocol (HTTP/1.1): Conditional requests, RFC7232. https://tools.ietf.org/rfc/rfc7232.txt. Accessed 21 Apr 2016.
32. Fielding, R., & Reschke, J. (2014). Hypertext transfer protocol (HTTP/1.1): Range requests, RFC7233. https://tools.ietf.org/rfc/rfc7233.txt. Accessed 21 Apr 2016.
33. Fielding, R., & Reschke, J. (2014). Hypertext transfer protocol (HTTP/1.1): Caching, RFC7234. https://tools.ietf.org/rfc/rfc7234.txt. Accessed 21 Apr 2016.
34. Fielding, R., & Reschke, J. (2014). Hypertext transfer protocol (HTTP/1.1): Authentication, RFC7235. https://tools.ietf.org/rfc/rfc7235.txt. Accessed 21 Apr 2016.
35. International Telecommunications Union (1993). Q.700 introduction to CCITT signalling system No. 7. http://www.itu.int/rec/T-REC-Q.700-199303-I/e. Accessed 21 Apr 2016.
36. Fette, I., & Melnikov, A. (2011). The WebSocket Protocol, RFC6455. https://tools.ietf.org/rfc/rfc6455.txt. Accessed 21 Apr 2016.
37. PubNub (2016). PubNub. https://www.pubnub.com/. Accessed 21 Apr 2016.
38. Firebase (2016). Firebase. https://www.firebase.com/. Accessed 21 Apr 2016.
39. Versatica (2015). JsSIP: The JavaScript SIP library. http://www.jssip.net/. Accessed 21 Apr 2016.
40. Doubango Telecom (2016). sipML5 API. https://www.doubango.org/sipml5/. Accessed 21 Apr 2016.
41. Bergkvist, A., Burnett, D. C., Jennings, C., Narayan, A., & Aboba, B. (2016). WebRTC 1.0: Real-time communication between browsers. https://www.w3.org/TR/webrtc/. Accessed 21 Apr 2016.
42. Uberti, J., Jennings, C., & Rescorla, E. (2016). Javascript session establishment protocol. https://tools.ietf.org/id/draft-ietf-rtcweb-jsep-14.txt. Accessed 21 Apr 2016.
43. Banerjee, K. (2005). SDP introduction. http://siptutorial.net/SDP/index.html. Accessed 21 Apr 2016.
44. World Wide Web Consortium (2016). ORTC (Object Real-Time Communications) Community Group. https://www.w3.org/community/ortc/. Accessed 21 Apr 2016.
45. The WebRTC project authors (2011). WebRTC. https://webrtc.org/. Accessed 21 Apr 2016.
46. &yet, 710 George Washington Way Ste A, Richland, WA 99352 (2016). Is WebRTC ready yet? Browser support scorecard. http://iswebrtcreadyyet.com/. Accessed 21 Apr 2016.
47. Hart, C., Ávila, V. P., Levent-Levi, T., & Hancke, P. (2016). webrtcHacks. https://webrtchacks.com/. Accessed 21 Apr 2016.
48. Johnston, A. B., & Burnett, D. C. (2014). *WebRTC: APIs and RTCWEB protocols of the HTML5 real-time web*. St. Louis: Digital Codex.
49. The SIP School (2016). WebRTC school. https://www.webrtcschool.com/. Accessed 21 Apr 2016.
50. The WebRTC project authors and GitHub, Inc. (2016). adapter.js: WebRTC adapter. https://webrtc.github.com/adapter/adapter-latest.js. Accessed 21 Apr 2016.

51. Temasys (2016). WebRTC plugins. http://confluence.temasys.com.sg/display/TWPP. Accessed 21 Apr 2016.
52. TokBox (2015). OpenTok platform. https://tokbox.com/developer/. Accessed 21 Apr 2016.
53. Widget4Call (2016). Apidaze. http://www.apidaze.io/. Accessed 21 Apr 2016.
54. Apizee (2016). ApiRTC. https://apirtc.com/. Accessed 21 Apr 2016.
55. Twilio (2016). Twilio. https://www.twilio.com/. Accessed 21 Apr 2016.
56. Cisco (2016). Tropo. https://www.tropo.com/. Accessed 21 Apr 2016.
57. onsip (2016). SIP.js. http://sipjs.com/. Accessed 21 Apr 2016.
58. Digium, Inc. (2016). Asterisk. http://www.asterisk.org/. Accessed 21 Apr 2016.
59. Versatica (2016). OverSIP. http://www.oversip.net/. Accessed 21 Apr 2016.
60. Kamailio SIP Server Project (2015). Kamailio: The open source SIP server. http://www.kamailio.org/w/. Accessed 21 Apr 2016.
61. XirSys (2014). XirSys. https://xirsys.com/. Accessed 21 Apr 2016.
62. Pipe Dream Technologies GmbH (2016). The pipe service. https://pipe.com/. Accessed 21 Apr 2016.

Part II
Implementations

Chapter 9
Developing Portable Context-Aware Multimodal Applications for Connected Devices Using the W3C Multimodal Architecture

Raj Tumuluri and Nagesh Kharidi

Abstract Cue-me™ is one of the reference implementations of the W3C's multimodal interaction (MMI) architecture and is a context-aware multimodal authoring and run-time platform that securely houses various modality components and facilitates cross-platform development of multimodal applications. It features several multimodal elements such as Face Recognition, Speech Recognition (ASR) and Synthesis (TTS), Digital annotations/gestures (Ink), Motion Sensing and EEG-headset based interactions that were developed using W3C MMI Architecture and Markup Languages. The MMI architecture described elsewhere in this volume facilitates single-authoring of multimodal applications and shields the developers from the nuances of the implementation of individual modality components or their distribution.

9.1 Introduction

Given the proliferation of mobile devices/operating systems, developing and maintaining applications is a challenge for authors of applications. Ensuring usability, which largely depends on the availability of modes-of-interaction, while being mobile is even a bigger challenge, with all the complexities of dealing with various modality components such as Speech Recognizers, Synthesizers, Sensory interfaces, etc. Built as a reference implementation of W3C's Multimodal Architecture [1], Cue-me Multimodal Platform ("Cue-me") provides a multiplatform framework for the development of applications across mobile devices. Applications developed using the Cue-me architecture can run on desktops, laptops, tablets and a host of smartphone, wearable, and robotic devices.

R. Tumuluri (✉) • N. Kharidi
Openstream Inc, Somerset, NJ, USA
e-mail: raj@openstream.com

D.A. Dahl (ed.), *Multimodal Interaction with W3C Standards*,
DOI 10.1007/978-3-319-42816-1_9

9.2 Architecture Overview

The Cue-me architecture is based on the W3C MMI Framework [2]. The goal of the design is to provide a general and flexible framework providing interoperability among remote and local components on the device. The framework is based on the MVC Design Pattern widely used in languages such as Java and C++. The design pattern proposes three main parts: *a Data Model* that represents the underlying logical structure of the data and associated integrity constraints, one or more *Views* which correspond to the objects that the user directly interacts with, and *a Controller* which sits between the data model and the views. The separation between data and user interface provides considerable flexibility in how the data is presented and how the user interacts with that data. While the MVC paradigm has been traditionally applied to graphical user interfaces, it lends itself to the broader context of multimodal interaction where the user is able to use a combination of visual, aural, and tactile modalities.

9.2.1 Run-Time Framework

At the core, the Cue-me run-time framework consists of the Interaction Manager, interacting with several components in the framework. The Core components are HTML-GUI and Voice. Other components can be added based on the application interaction requirements. This approach provides the following flexibilities (Fig. 9.1):

- Granular Application Footprint Control for different devices and applications
- Platform extensibility using the component architecture
- Component level update/control

Figure 9.1 depicts the Cue-me platform with component interaction between the GUI, Voice, and other modalities. The application itself (Application Server and Database) can be local or remote.

9.2.2 Core Run-time Framework Components

The core run-time framework components consist of:

- GUI Modality (HTML)
- Voice Modality (providing speech interaction capability).

The core components provide the framework for building a multimodal application across platforms. The Modality Components can support local and remote interaction based on the platform choice and application configuration.

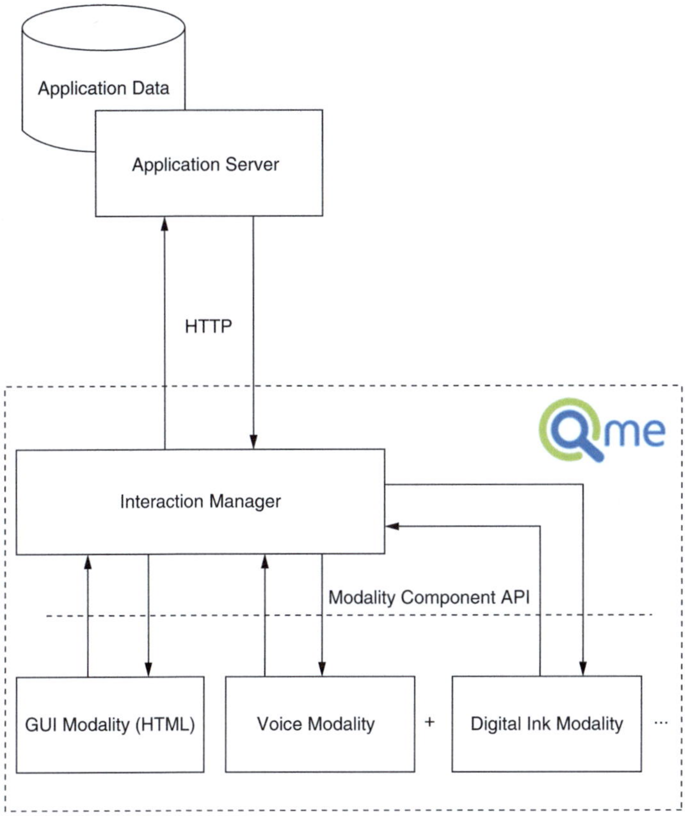

Fig. 9.1 Cue-me MMI architectural overview

For example, voice processing can be done remotely on a server or locally based on the choice of platform and the type of voice processing required by the application.

9.2.3 Other Components for Multimodal Applications

Mobility Applications require platform level component support. Cue-me achieves application portability by implementing mobile functionality using the component framework via the Interaction Manager. Some examples of other components are

- Ink Component using InkML [3]
- Signature Component
- DB Sync Component

  . . ..

Because the Cue-me architecture uses an open framework for component integration (W3C MMI framework), components can be built by third parties and plugged in.

Hence, Cue-me is an *extensible* platform—components specific to the target platform can be plugged into the framework architecture for interaction via the Interaction Manager.

9.2.4 Anatomy of a Multimodal Application

A multimodal application developer must provide XML and text application files for each of the following components:

- GUI: The visual application delivers HTML or XHTML and the various supporting images and files including JavaScript and CSS to the web browser.

  It is very important that the GUI application is tailored to look and feel and run on a wide variety of mobile devices and wearables. The visual application should therefore be tested on as many different devices as possible before it is deployed.

  Two web sites which perform an analysis on the performance and quality of the visual mobile application are dev.mobi (http://ready.mobi) and the W3C mobile web initiative (http://validator.w3.org/mobile/).
- Voice: The voice application consists of a number of grammar and TTS files delivered to a speech engine for processing. The formats of the grammar and TTS files depend on what is supported by the speech engine. Grammar file formats include JSGF, ABNF, and SRGS. TTS file formats are either raw text or SSML. A developer should know what file formats the speech engine supports.

Other XML file formats may be used according to various requirements such as support for a legacy IVR application ported to run on Cue-me. This application may run one or more VoiceXML and/or CCXML files.

- Interaction Manager: The Cue-me Interaction Manager runs a variant of State Chart XML (SCXML) [4, 5], known as X-SCXML on the mobile phone or PDA. X-SCXML which implements SCXML, (with the exception of parallel states for want of resource optimization on mobile devices), can be used to integrate the interaction between the visual, voice, ink, and other modalities.

9.2.5 Basic Application Development Steps

Here are the basic application development steps, as shown by Fig. 9.3, below:

1. Create a Dynamic web project within a development environment such as Eclipse. Eclipse should be version 3.4.2 or higher and at a minimum the Web Standard Tools (WST) must be installed.

2. Create the visual web application, or import a legacy application into the project as required. The web application should be adapted for browsers running on mobile devices with limited memory and processing resources.
3. Identify the various states in the application. Each state may be associated with a separate web page or a web page may have more than one state, if the page can be associated with more than one set of active grammars. Generally, a unique set of active grammars represents a separate state.
4. Create the voice grammars, or modify for Cue-me if already available. There should be one or more grammars representing all controls and fields for each application web page. Usually there are grammars which remain active across several or all web pages.
5. Each grammar should also have a corresponding text file for displaying for the user what the user can say when the grammar is active.
6. The X-SCXML document is developed to integrate the voice and visual user interactions. When complete, a cookie is added to the application's home web page specifying the URL location of the X-SCXML document:

```
<%
Cookie
cookie = new Cookie ("IM-Loc", "http://www.ex.com/im/im.scxml");
cookie.setMaxAge (10);
response.addCookie (cookie);
%>
```

   - A web page can set the Interaction Manager to another state using a cookie to specify the new state. For example, a JSP may have:

```
<%
Cookie cookie = new Cookie ("IM-State", "trade");
cookie.setMaxAge (10); // Do NOT set Max Age more than 10-20 s
response.addCookie (cookie);
%>
```

   - The X-SCXML may direct the GUI component to execute a JavaScript function (e.g., in response to a voiceResult event). This function must of course be defined in the active web page.

7. The application and the voice proxy (one of the Cue-me deliverables) must both be deployed to a web server. After configuring the voice and proxy, the application is tested on a mobile phone running the Cue-me multimodal platform.
   - The web application and voice proxy may be deployed on different application servers (Figs. 9.2 and 9.3).

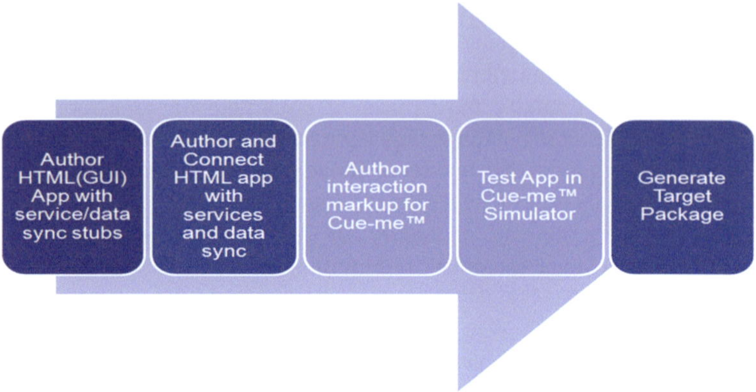

Fig. 9.2 Overall application development flow

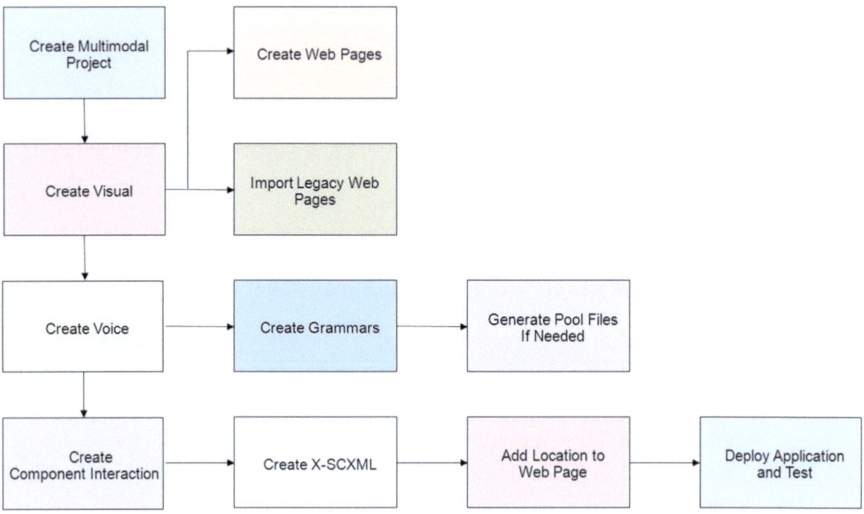

Fig. 9.3 Basic multimodal application development steps

9.2.6 *Application Components Overview*

A basic multimodal web application should have markup and software for processing by the following components:

- Application Server and Database (for business logic)
- Interaction Manager (for multimodal interaction integration)
- Voice Client Modality
- GUI Modality
- Access to a Cue-me Multimodal Server to provide voice interactions among other multimodal features.

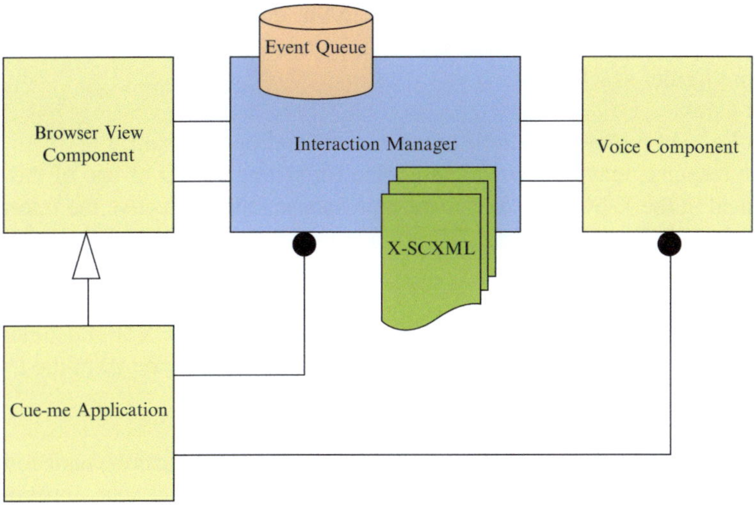

Fig. 9.4 Cue-me components object model

Figure 9.4 shows the object model of the Cue-me components on the client.

The components which reside on the client device are the Interaction Manager, the GUI (or HTML) modality, and the Voice client. GUI modality interactions, as represented by HTML, are deployed by an Application Server such as Apache Tomcat. The Cue-me Multimodal Server runs on a remote server and provides support for voice interactions.

When the voice recognition and TTS are rendered by a Cue-me Multimodal Server, the voice modality spans both client and server with a voice component on the client communicating with the server. However, voice recognition and TTS may both be rendered on the device, or recognition may be rendered by the server and TTS rendered on the device, or vice versa.

Both the GUI and Voice components on the device communicate with an Interaction Manager, which performs the modality integration by processing an X-SCXML document. X-SCXML is covered in a later section in this document.

9.2.6.1 Application Server

The Application Server deploys the application for the GUI modality represented by HTML and the X-SCXML for processing by the Interaction Manager. For an Application Server such as IBM WebSphere, the application is comprised of JSP's, Servlets, and static HTML pages.

Typically the application requires one or more databases for storing user information.

The application server returns some flavor of HTML or XHTML to the GUI component of Cue-me in response to an HTTP GET or POST request. This response may contain the URL location of the X-SCXML document required by Cue-me's Interaction Manager Component. The URL location is returned to the client in the response header as a Cookie:

Set-Cookie:

IM-Loc=http://www.example.com/ChineseFood/im/im.scxml;

The response may also contain a cookie for setting the ID of one of the states contained in the X-SCXML document. In response to this directive the Interaction Manager will move to this state. The cookie is specified as follows:

Set-Cookie:

IM-State=home;

The application server may specify the location of the X-SCXML document and set the state at one time by appending a "#" followed by the state ID to the IM-Loc cookie:

Set-Cookie:

IM-Loc=http://www.example.com/ChineseFood/im/im.xscxml#changeorder;

If the application sets a cookie for IM-Loc with the initial state appended, as shown as above, and also sets a cookie for IM-State, the cookie for IM-State is ignored. This is because the contents of the IM-Loc cookie always have priority. Best practice is to set a separate IM-Loc and IM-State cookies and not append the state to the IM-Loc URL.

The application server may return an X-SCXML document in response to the HTTP request. Support for processing X-SCXML should be specified in the HTTP Accept header as follows:

Accept: application/xscxml+xml, text/html, text/plain, image/*

9.2.6.2 Interaction Manager

The Interaction Manager (IM) processes one or more X-SCXML documents for a single multimodal application, where X-SCXML is an XML language derived from SCXML. The X-SCXML directs the IM to retrieve resources, add event listeners to specified events, and to process the events generated by each multimodal component.

While X-SCXML is similar to SCXML, a number of small changes were made to facilitate processing on the client. This "SCXML-like" language is processed by the IM as follows (examples included):

(1) Set each modality component's base URI for documents retrieved from the server:

```
<base id="x-voice" url="http://A.B.C.D/sb/voice"/>
```

or, preferably:

```
<send event="base" url="$APP_BASE/voice"/>
```

As shown above, the Cue-me Platform supports an $APP_BASE macro containing the application's host address and application context. According to the example above,

$APP_BASE = http://A.B.C.D/sb

$HOST is the other supported macro and contains the value of the application's host address. According to the example above,

$HOST = http://A.B.C.D

(2) Send initial (e.g., initialize or prepare) commands to the modality components:

```
<send event="prepare" to="x-voice" url="http://.../voice"/>
```

(3) On entry to a state send one or more commands to the modality components:

```
<send event="addGrammar" to="x-voice" url="login.jsgf"/>
<send event="addHelp" to="x-html" url="login.txt"/>
```

(4) Add an event listener to listen for a click event and in response send a command to the voice modality to response with TTS (e.g., "you said hello"):

```
<go on="click" node="hello_id">
<send event="playUrl" to="x-voice" url="voice/sayHello.txt"/>
</go>
```

(5) Go to the menu state if the event data contains the "order" string:

```
<go on="location" to="menu" if="event.name=='order'"/>
```

The Interaction Manager is composed of the following components.

9.2.7 *XML Language Parser*

The parser is a SAX parser which parses the X-SCXML into a state machine object. While it parses the X-SCXML it also generates events for initial processing and adds event handlers to the modality components.

9.2.8 *State Machine*

The state machine maintains the current state and retrieves the actions for an event which triggers an event handler. It also returns the actions contained by the *onentry* and *onexit* tags for each state, and the final tag for the document.

9.2.9 *Event Queue*

All events are placed on an internal event queue to be processed one-by-one by the Interaction Manager.

9.2.10 Data Model

The current application may have a data model specified by the X-SCXML document. The data model is a set of EcmaScript variables, which may be assigned in the X-SCXML with the <assign> tag and accessed by the <send> tag's "*expr*" attribute. The <script> tag may also be used to declare, assign, and access variables and functions.

9.2.11 X-SCXML Markup Language

The X-SCXML Markup language defines the actions and event listeners for a set of states enabling multimodal interactions between the user of a mobile device and a web application running on the device.

X-SCXML is a simplified version of State Chart XML (SCXML) [4, 5], an XML language that represents the execution environment of a state machine based on Harel state charts.

A state machine as defined by X-SCXML is a set of:

- States: A state embodies information about the current situation in the system.
- Events: An event is an input message to the state machine.
- Transitions: A transition is either a change from one state to another, usually triggered by an event, or it is a set of actions to be performed when triggered by an event.
- Data model variables.
- Actions: An action is an activity to be performed by the state machine at a given point in time.
- Primitives: to express guard conditions on transitions.

X-SCXML is processed by an Interaction Manager (IM) residing on a limited resource mobile client. The IM's role on the client is generally to enable multimodal interaction between a web application and the user, as defined by the W3C Multimodal Interaction Working Group's architecture.

According to this architecture:

- A state is the set of currently active grammars, files, web pages, etc., in the multimodal system.
- A life-cycle event is a message sent between a modality component (Speech, Visual, SMS, Ink, etc.) and the IM.
- An action is an activity the IM requests the modality component to perform.
- A transition moves the system either to another multimodal state or triggers a set of actions to be performed by one or more modality component. For the latter case, a transition defines an event handler.

X-SCXML is a mobile profile of SCXML (https://www.w3.org/TR/scxml/), in that it currently excludes the SCXML <parallel> tag which can capture concurrent behavior within a state machine, as it is not a requirement for most mobile applications.

Here is an example X-SCXML document representing a "Food Order" application. Its outermost state is identified as "home." The "home" state has one "order" state. For any voice result event received by the IM while within the nested states, the view component calls a JavaScript function "handleCommand" with a parameter string contained in the "event.value" property.

The data model declares an EcmaScript variable, "_data.welcome." Upon entering the "home" state, the voice component is instructed to play, "Welcome to Chinese Food Order!."

```
<?xml version="1.0" encoding="UTF-8"?>
<xscxml initialstate="home" version="1.0">
  <initial>
<send event="base" to="x-html" url="$APP_BASE/gram/"/>
            <send event="base" to="x-voice" url="$APP_BASE/gram/"/>
    <send event="prepare" to="x-voice" url="$HOST/mmivoice"/>
  </initial>
  <datamodel>
    <data id="welcome" expr="'Welcome to Chinese Food Order!'">
  </datamodal>
  <state id="home">
            <onentry>
     <send event="addGrammar" to="x-voice" url="chineseord.gram"/>
     <send event="addHelp" to="x-html" url="chineseord.txt"/>
     <send event="playText" to="x-html" expr="_data.welcome"/>
            </onentry>
            <go on="voiceResult" from="x-voice">
              <send event="execute" data="event.value" to="x-html"
          target="handleCommand" />
            </go>
            <state id="order">
      <onentry>
        <send event="addGrammar" to="x-voice" url="foods.gram"/>
   <send event="addHelp" to="x-html" url="foods.txt"/>
      </onentry>
      <onexit>
        <send event="removeGrammar" to="x-voice" url="foods.gram"/>
   <send event="removeHelp" to="x-html" url="foods.txt"/>
      </onexit>
</state>
  </state>
  <final>
    <send event="unPrepare" to="x-voice" url="$HOST/mmivoice"/>
  </final>
</xscxml>
```

9.2.12 Cue-me Multimodal Server

The Multimodal Server is required for a voice component configured for remote speech recognition and synthesis. The Multimodal Server handles communication and data transfers between the Voice Component running on the client.

The Multimodal Server processes speech input against one or more provided grammars, returns results, and generates TTS audio to be rendered by the device. This is the remote speech processing model and no application resources need to be stored on the Server.

9.2.13 Voice Component

The Voice Component records the speech input and plays the output (TTS) as directed by the Interaction Manager. It will send the recorded input to either a remote or local speech engine for processing, depending on how the mobile device is configured.

Events and data from the Multimodal Server are forwarded to the Interaction Manager as events for processing.

9.2.14 GUI Modality

The GUI modality is the mobile phone's Web Browser, except that it also adds event listeners as directed by the IM, sends events to the IM, and handles events from the IM. Event listeners added to the GUI modality add listeners for DOM events to the current HTML page.

The GUI modality can listen for the any of the HTML DOM Events supported by HTML 5.

9.3 X-SCXML Processing

9.3.1 Introduction

The X-SCXML document is processed on a mobile device after being downloaded from a remote server. While it is being parsed all initial actions as well as on-entry actions for the initial state are placed on the internal event queue for processing by the IM.

For each state the parser encounters it also saves all go transitions to a state object identified by the state's identifier. If the state is the initial state it adds an "add event listener" event to the event queue for each go transition.

Once the state objects are constructed for all the states the state machine is created to store the states as well as the final section of the X-SCXML document. The state machine sets the current state to the state identified by the “initialstate” attribute on the <xscxml> tag.

The X-SCXML is comprised of the following:

9.3.2 Data Model

The <datamodel> tag declares the data model for the X-SCXML application. One or more <data> tags below <datamodel> declare a set of EcmaScript variables for storing and accessing information during X-SCXML document processing.

9.3.3 Initial and Final

The <initial> tag contains a set of actions, each denoted by the <send> tag, which one or more modality components are to perform with the initialization of the X-SCXML document.

The <final> tag complements the <initial> tag, containing the set of <send> actions which the modality components are to perform upon exiting the application.

9.3.4 States

At any time a multimodal application may be in one of a set of interaction states. Each active state has its own set of the files (e.g., grammars, web page, etc.) and event handlers. Some or all of these files and handlers may be replaced when there is a transition to another state.

States may be nested. There are several advantages of having nested states: (1) the outermost state can be the global state for the application. That is, event handlers declared in the outermost state apply to all the nested states below. (2) The X-SCXML can model applications which are more complex. (3) X-SCXML is more compliant with Harel State charts and SCXML. Transformations between SCXML and X-SCXML are thereby simpler.

9.3.4.1 State Transitions

A state transition is declared with the <go> tag and is usually triggered by an event from a modality component. The next state is declared by the <go> tag’s “to” attribute.

Here is an example state transition:

```
<go on="voiceResult" from="x-voice" to="foodorder" node="myhome"/>
```

9.3.4.2 Event Handlers

An event handler is declared with the <go> tag and is usually triggered by an event from a modality component. For example, a "voiceResult" event from the voice modality might trigger an action to update a field within a web page with the data returned with the "voiceResult" event.

A <go> tag which is an event handler won't have the "to" attribute and has one or more <send> tags as children. If there are conditionals specified by <if>, <elseif>, and <else> tags, only <send> actions below the conditionals which evaluate to true upon receipt of the event are performed by the modality components.

Here is an example event handler:

```
<go on="voiceResult" from="x-voice">
        <send event="execute" data="event.data" to="x-html"
              target="handleCommand" />
    </go>
```

9.3.4.3 On Entry and On Exit Processing

A state may contain one <onentry> and one <onexit> tag. When the state machine transitions to a new state the actions contained by the <onentry> tag below the new state are sent by the IM to the modality components.

The <onexit> tag is the complementary to <onentry>; the actions contained below <onexit> for the old state are performed when the state machine transitions to a new state.

9.3.5 Event Types

Following the W3C MMI Life-cycle API, the modality components support various API calls as appropriate.

9.3.5.1 Voice Client Event Types

The Voice Client event types are independent of the location of the speech rendering with the exception of the "prepare" and "unPrepare" events. The latter events are required by the Voice Proxy for setting up and tearing down resources on the remote Voice Server.

9.3.5.2 Visual (GUI) Modality Event Types

Event types emitted by visual modality

Event	Description	Detail
click	User clicks on a DOM node	MouseEventDetail
getFieldResponse	Returns content requested with the getField event	Target is the id of an HTML tag containing text
newPage	A new web page has been received	Title of the Web Page URL of the Web page
newContextRequest	A new X-SCXML document is available (discovered in HTML header or cookie)	URL of the X-SCXML document
nextState	Move to a new X-SCXML state	New state ID[a]

[a]The Interaction Manager ignores the nextState event if currently in the state specified by the event's new state ID.

9.3.6 Example getField, getFieldResponse, and playText

A typical use case for the getField and getFieldResponse event combination is to get the contents of a text area or paragraph and forward to the voice client to be played as TTS. For example, the X-SCXML can be programmed to request a paragraph identified by "art_text_id" when requested by the user. When the getFieldResponse event is received, the contents of the associated event.property can be included as data with the playText event sent to the voice component. The X-SCXML snippet is shown below:

```
<go on="voiceResult" from="x-voice">
  <if node="readId"/>
<send event="getField" to="x-html" target="art_text_id" />
  </if>
</go>
<go on="getFieldResponse" from="x-html">
  <send event="playText" data="event.value" to="x-voice" />
</go>
```

9.3.7 Example Noinput and Message

A possible use case for the noinput and message event combination is to display a message for the user when the voice client emits a noinput event. The message lets the user know that the voice client is waiting for speech input. The X-SCXML snippet is shown below:

```
<go on="noinput" from="x-voice">
<send event="message" to="x-html" data="Please begin \
        talking. Press help if you need help as to what to say." />
</go>
```

9.3.7.1 The Event Object

An event object is included with every event put on the internal queue by a modality component. This object is most useful in capturing the voice result when a user's speech input has matched against a grammar. The event object has two properties:

- Name—the node name or the name associated with a name/value pair of a semantic interpretation result.
- Value—the raw result of the value associated with a name/value pair of a semantic interpretation result.

The event object properties include confidence levels and other properties conforming to the W3C EMMA language [6–8] for annotating recognized user input.

9.3.7.2 URL Macros

The IM will process the following two macros when encountered in an X-SCXML document:

- $HOST: This macro represents the server host of the current X-SCXML document's URL. It consists of protocol + host name + [optional] port number. For example, $HOST would contain "http://example.com:5650" if the X-SCXML was retrieved from "http://example.com:5650/im/example.xml."
- $APP_BASE: This macro represents the host plus the application context root retrieved from the current X-SCXML document's URL. For the above example, $APP_BASE would contain "http://example.com:5650/im."

As already explained *IM is the linchpin for both modality and context*, and forms the control plane driving/adapting the behavior of the application for the given context/device or situation.

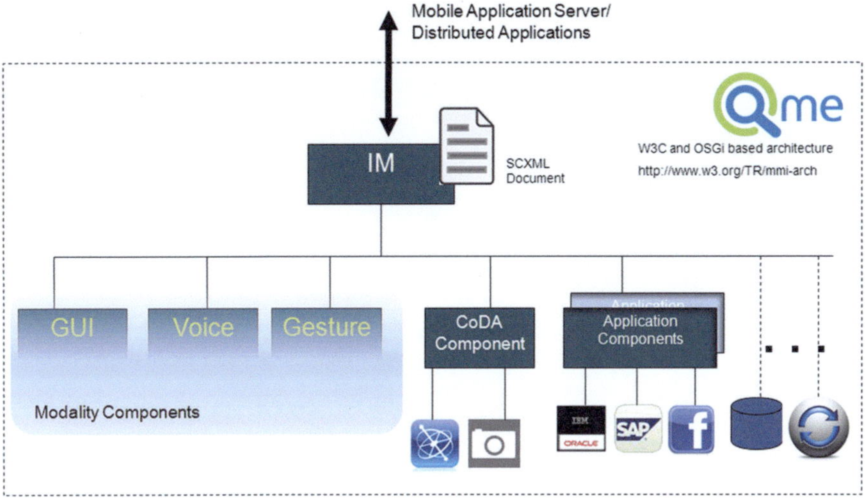

Fig. 9.5 Anatomy of a Cue-me™ application

The ContextDeliveryArchitecture *CoDA component (aka Delivery Context Component)* delivers device context information like location, camera, acceleration, etc. The application uses this context information and drives behavior via state transitions in the IM.

Several Enterprise Utility and Infrastructure components like *Database*, *Synchronization*, etc., are available and co-ordinated through the Interaction Manager (IM) for Enterprise application use.

For the purposes of on-device integration of both Enterprise and Cloud Application services, the *component architecture is extensible* to include application level components connecting to external services and creating services based mashups (Fig. 9.5).

Upon application launch, the *Interaction Manager* starts the Interaction and initializes one or more components based on the Application's SCXML document. The SCXML document itself may be present locally, or obtained from a URL resource. As part of the HTML Component initialization, HTML resources are either obtained from the Cue-me™ sandbox or via URL request to the appropriate server. Speech, Gesture, and other components are initialized and the Application is ready to execute.

The typical execution model is depicted in Fig. 9.6.

9.4 Application Development

Resources for the application are created and packaged using Open Web IDEs or an Eclipse IDE with Cue-me™ Studio Plugins.

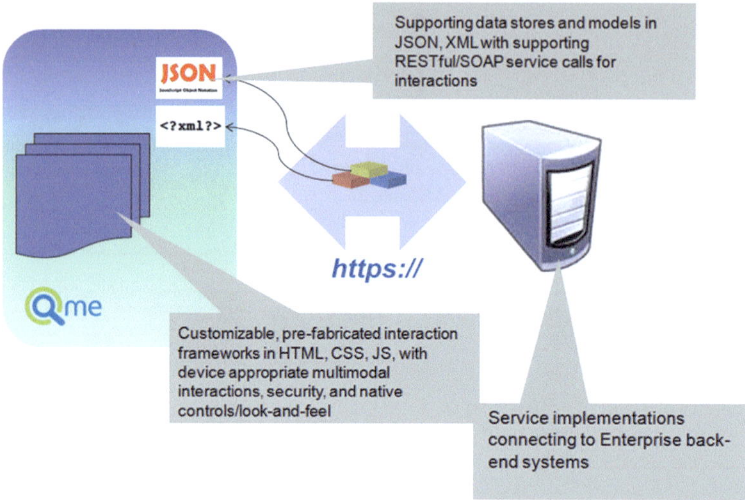

Fig. 9.6 MMI (cue-me) application execution model

9.4.1 HTML Editor with Interaction Context (SCXML) Palette

Interaction Context defines the interaction modes and device peripherals that a Cue-me app can interact with. Interaction Context is device-specific because devices capabilities and peripherals differ.

Cue-me Studio provides an HTML Editor with SCXML palette. The SCXML palette allows users to bind the SCXML components to the elements of HTML document. For example, a Text input field can have a Gesture and Scanner inputs in addition to the keyboard input.

The following picture shows the HTML Editor with the SCXML Palette on its right (Fig. 9.7).

9.4.2 Multimodal Interactions (SCXML)

SCXML Editor is an XML editor with enhanced code assist for multimodal components and events. The editor can be used to edit a multimodal application's Interaction Manager (IM) document that was generated.

9.4.2.1 Components

Components like "x-html, x-voice, etc.," handle the action events from IM and perform requested actions. Components can also raise events after performing the

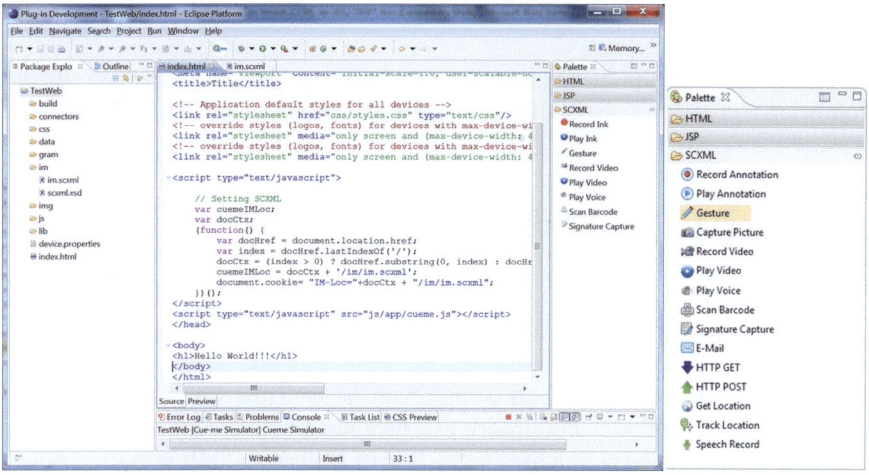

Fig. 9.7 HTML editor with SCXML palette

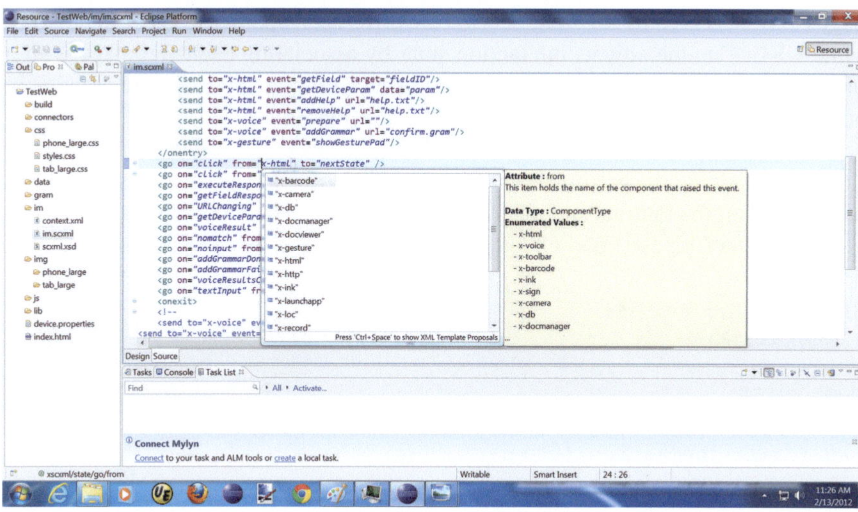

Fig. 9.8 SCXML editor

requested actions. In the following picture, code assist after (ctrl+space bar), "to=" displays the list of components (Fig. 9.8).

9.4.2.2 Action Events

Action Events are the events that can be sent to Components like "x-html." In the following picture, code assist after (ctrl+space bar), "event=" displays the list of events that can be sent to components.

9.4.2.3 Raised Events

Raised Events are the events that can be raised on Actions like "go." In the following picture, code assist displays the list of raised events (Fig. 9.9a, b).

9.5 Example: The Chinese Food Order Application

This section will put all the previous sections together by describing an example multimodal application and how all the pieces, HTML, grammars, and X-SCXML, fit together.

The following two images show the main page of the Chinese Food Order application. In most cases developers will be adding multimodal interaction to a legacy web application so it makes sense to start with the HTML pages or JSPs.

All pages will generally have a set of global controls for navigating the application while other fields and buttons can be said to be specific to each of the pages.

It is important that each web page has a unique title. Because multiple web pages may be served by the same servlet, the different responses (pages) can refer to same URL. Cue-me adjusts interaction state by associating each web page with a unique combination of URL and title. Requiring a unique title is a good practice in any case and <title> is a required XHTML tag (Fig. 9.10).

The next step is to add the voice interaction to this page. Adding voice interaction means adding the grammar and help files which will perform the same actions as clicking on the buttons and entering text into the fields with a keypad or stylus.

Here is the Chinese Food Order grammar in SRGS or ABNF format:

```
#ABNF 1.0 iso-8859-1;
language en-US;

mode voice;
root $chineseorder;
tag-format <semantics/1.0>;

public $chineseorder = [I would like | I'd like] [to (order|get)]
[please] [give me] ((change [my | the] order {$.changeOrder=true})
       | (what is (my | the) order {$.what=true})
       | (submit [my | the] [order] {$.done=true})
| ([I'm] (done | finished | ready) [with] [my | the] [order] {$.done=true})
       | ([and] ([an] egg roll {$.appEggRoll=true}
     | [some] pork dumplings {$.appPkDump=true}
     | [some] crispy spring rolls {$.appCriSpr=true}
     | [with] [a|an|some] fried rice {$.riceFried=true}
     | roast pork noodle soup {$.spRoastPk=true}
     | hot and sour soup {$.spHotSour=true}
     | chicken with sweet corn soup {$.spChicken=true}
     | mixed vegetables in oyster sauce {$.entree="Mixed Vegetables"}
```

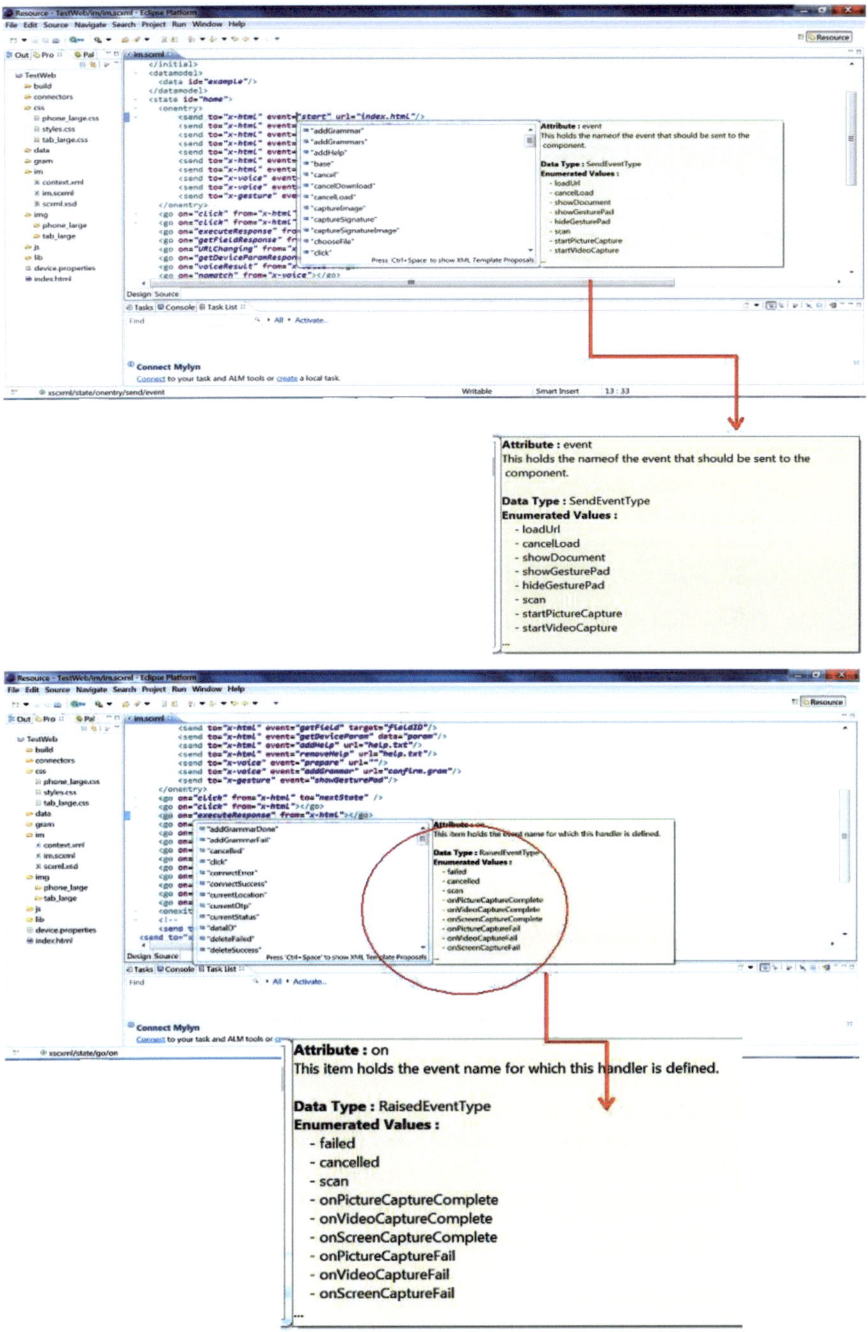

Fig. 9.9 (**a**) Action events, (**b**) Raised events

Fig. 9.10 Chinese food order demo

```
| pineapple sweet and sour pork {$.entree="Sweet and Sour Pork"}
  | moo shoo pork {$.entree="Moo Shu Pork"}
  | ginger chicken {$.entree="Ginger Chicken"}
  | beef with vegetables {$.entree="Beef"}
  | deep fried shrimp {$.entree="Shrimp"}))<1->);
```

9.5.1 Chinese Food Order Global Grammar

Grammars for Cue-me should always be written in the above format using semantic interpretation to set the node and value. The node and value set with semantic interpretation correspond to the name and value properties of the event object associated with the voiceResult event processed by X-SCXML. For example, "$. entree="Shrimp" sets the node to "entree" and the value to "Shrimp."

Associated with each grammar should be a help file which tells the user what he or she can say on the page. This is a text file stored on a server along with the grammar file. Here is an example help file for the above Chinese Food Order grammar:

You may specify most or all of your order at one time, if you already know what you want.

For example, you could say:

> I would like moo shoo pork, with
> an eggroll, hot and sour soup,
> and some fried rice.

To change your order you can say:
    I'd like to change my order.

When done with your order you can say:
    Submit my order.

To hear your order you can say:

    What is my order?

9.5.2 *Chinese Food Order Grammar Help Text*

9.5.2.1 Interaction Integration with X-SCXML

The Interaction Manager integrates the visual and voice interactions as directed by the X-SCXML document associated with the Chinese Food Order application. Each modality component has an identifier: "x-html" identifies the GUI and "x-voice" identifies the voice. The X-SCXML document's initial state is "home."

In the initial section of the X-SCXML, each component's base URL is set, the voice component is directed to prepare the voice modality for this client session, add the global Chinese Food Order grammar, and get the help text to be displayed by the GUI.

```
<?xml version="1.0" encoding="UTF-8"?>
<xscxml initialstate="home">
  <initial>
   <send event="base" to="x-html" url="$APP_BASE/gram/"/>
        <send event="base" to="x-voice" url="$APP_BASE/gram/"/>
   <send event="prepare" to="x-voice" url="$HOST/.../mmivoice"/>
  </initial>
```

9.5.3 *Initial Section*

The data model declares EcmaScript variables for use during the X-SCXML document processing. The Chinese Food Order application defines one variable, "welcome" which is accessed as "_data.welcome" within the application.

```
<datamodel>
   <data id="welcome" expr="'Welcome to Chinese Food Order!'"/>
  </datamodel>
```

9.5.4 Data Model Section

The Chinese Food X-SCXML has a home state and a nested state below the home state so to update the single page dynamically as if the user was engaged in a dialog.

On entry to the home state for the page the grammar and help text events for the page are added. The voice component is also sent the welcome message to play contained in the "_data.welcome" variable.

The X-SCXML <go> tag specifies an event handler. For example, when a voice result event is received by the Interaction Manager, the Interaction Manager will send a JavaScript execute event to the GUI component. The JavaScript function *submit_form* will be called if the event name is "submitButton," and in other words when the user clicked on the button with ID attribute set to "submitButton." If the global attribute is set to "true," this event handler is active for all the states in the grammar. This is the global event handler associated with the global Chinese Food Order grammar.

Here are the order form and nested dialog state sections:

```
<state id="orderform" initialstate="dialog">
          <onentry>
           <send event="addGrammar" to="x-voice" url="chineseorder.gram"/>
           <send event="addHelp" to="x-html" url="chineseorder.txt"/>
           <send event="playText" to="x-voice" expr="_data.order"/>
          </onentry>
          <state id="dialog">
    <go on="voiceResult" from="x-voice" to="dialog"
        cond="event.name!='changeOrder'&amp;&amp;event.name!='done'&amp;&amp;event.name!='what'">
           <send event="setField" data="event.value" to="x-html" target="event.name"/>
    </go>
   </state>
           <go on="click" from="x-html" node="submitButton" to="orderform">
         <send event="playText" to="x-voice" data="Thanks for your order!"/>
                 <send event="execute" to="x-html" target="submit_form"/>
           </go>

  <go on="voiceResult" from="x-voice" node="done" to="orderform">
        <send event="playText" to="x-voice" data="Thanks for your order!"/>
          <send event="execute" to="x-html" target="submit_form"/>
   </go>

  <go on="voiceResult" from="x-voice" to="changeorder" node="changeOrder"/>
  <go on="voiceResult" from="x-voice" node="what" global="true">
         <send event="execute" to="x-html" target="getOrder"/>
  </go>

        <go on="executeResponse" from="x-html" node="getOrder" global="true">
         <send event="playText" to="x-voice" expr="event.value"/>
        </go>
        <onexit>
  <send event="removeGrammar" to="x-voice" url="chineseorder.gram"/>
                 <send event="removeHelp" to="x-html" url="chineseorder.txt"/>
                 </onexit>
        </state> ...
```

9.5.5 Order Form State

On exit from the order form state the order form grammar and help files are removed from the speech engine.

The Interaction Manager can direct the GUI component to update a field with the "setField" event. For example, when the voice result has an event node set to anything besides "changeOrder," "done," or "what", the GUI is directed to set the field with ID set to the value contained in the event object's name property, with a value contained in the value property.

```
<go on="voiceResult" from="x-voice" to="dialog"
      cond="event.name!='changeOrder'&amp;&amp;event.name!='done'&amp;&amp;event.name!='what'">
         <send event="setField" data="event.value" to="x-html" target="event.name"/>
</go>
```

9.5.6 Set Field with ID Contained in the Event Object's Name Property

In the final section of the X-SCXML document, the voice proxy is told to unprepare the current session.

```
<final>
 <send event="unPrepare" to="x-voice" url="/MMoic...Proxy/mmivoice"/>
</final>
```

9.5.7 Final Section

9.5.7.1 Add a Cookie to the Home Page

When development of the X-SCXML document is complete, a cookie must be added to the home page specifying its URL location. For example, for the Chinese Food Order application, the cookie is added to the market's summary JSP as follows:

```
<!DOCTYPE HTML PUBLIC "-//W3C//DTD HTML 4.0//EN" "http://www.w3.org/TR/REC-html40/loose.dtd">
<html>
<%
Cookie cookie =
  new Cookie ("IM-Loc", "http://example.xml/Chinese/im/ch.xscxml");
cookie.setMaxAge(10); // Do NOT set Max Age more than 10-20 s
cookie.setVersion(1);
response.addCookie(cookie);
 %>
```

9.5.7.2 "IM-Loc" Cookie for the X-SCXML URL

The "IM-Loc" cookie should not have a max age set to more than 20 s. Otherwise, it may remain active after the user leaves the Chinese Food Order application.

9.6 Extending the MMI Architecture to Connected Devices

9.6.1 Development of a Prototype Multimodal Robot Using MMI

Let's call our robot EasyHealthAssistant EHA. EHA uses Text-To-Speech (TTS), Speech recognition, Face Recognition, Motion sensing and Mindwave (EEG) interactions to effectively engage patients and care-givers. The robot features bluetooth connectivity to third party health devices and wearables such as a Blood Pressure Monitor or electronic fitness trackers. EHA keeps care-givers informed by providing real time updates concerning the patient's adherence, vital readings, and any new developments in the condition of the patient. The robot facilitates video conference calls leveraging WebRTC with the physician or the care-giver on command (speech/touch) (Fig. 9.11).

9.6.2 Design of EHA

EHA is a static robot which has a revolving head and three passive Infrared Motion (PIR) sensors. The camera, display, microphone, and speakers from a phone are located on the revolving head. When motion is detected, the head turns in the corresponding direction and faces the user (Fig. 9.12).

One of the core functions of EHA is the dispensing of medication to the correct user at the prescribed time. While, the servos, motors, PIR sensors, and Wi-Fi connectivity are controlled using an Arduino microcontroller, the robot itself is controlled by EHA program implemented using the W3C Multimodal architecture, leveraging Openstream's Cue-me multimodal platform.

The control-logic (Interaction Manager) is implemented in SCXML, while the modality components are implemented in other languages with events flowing between Interaction Manager and modality components including the Arduino controller (Fig. 9.13).

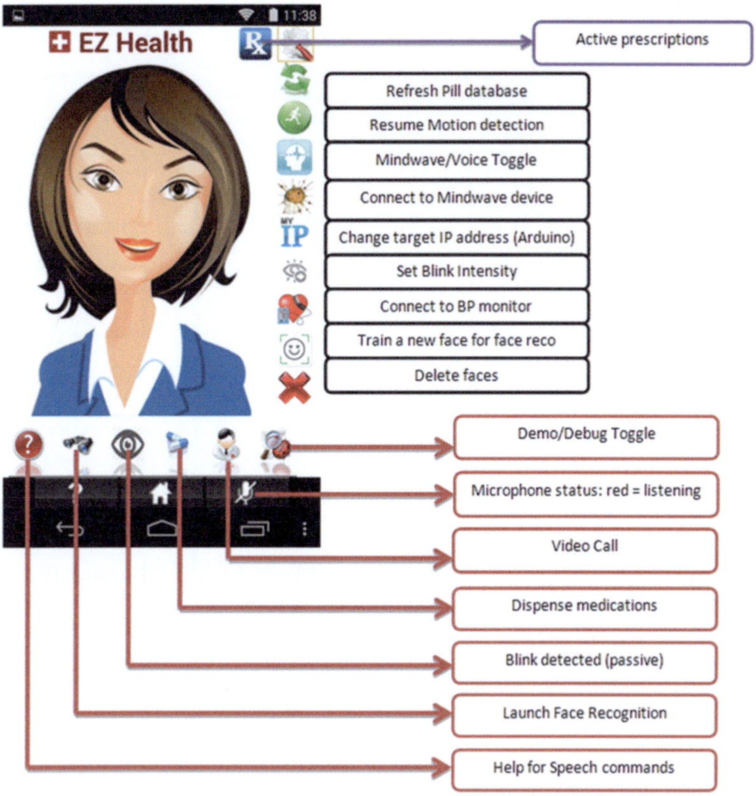

Fig. 9.11 Feature-overview

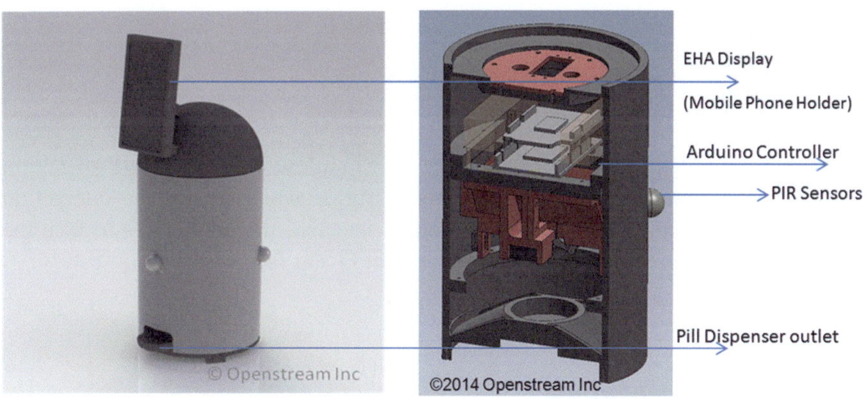

Fig. 9.12 Functional prototype (fully built and sectional views)

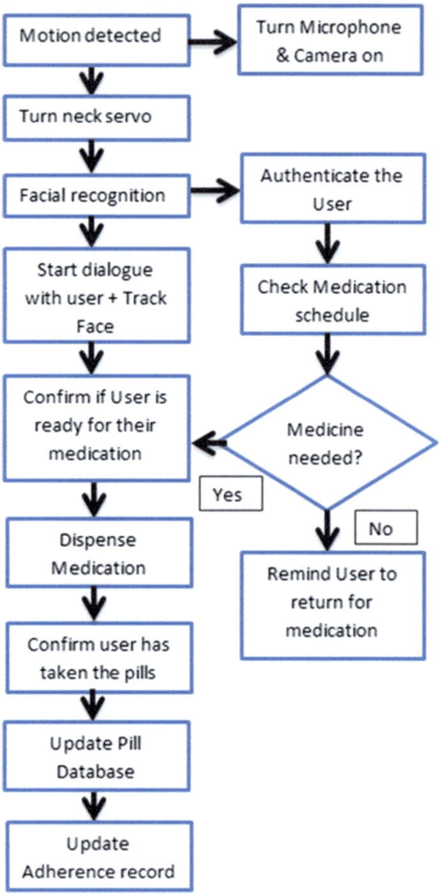

Fig. 9.13 Sample flow and high-level events

9.6.3 Motion Detection

There are three PIR sensors located 120° apart to allow for a full range of motion sensing. The "motion detected" event is the trigger which begins the interaction between the user and EHA. When motion is detected, the robot turns to face the direction of motion and launches face recognition and face tracking.

9.6.4 Face Recognition

The face recognition engine serves the purpose of authentication and personalization for the user. The authentication of the user gives EHA context for any upcoming interaction, such as reminding the user when it is time to take medication

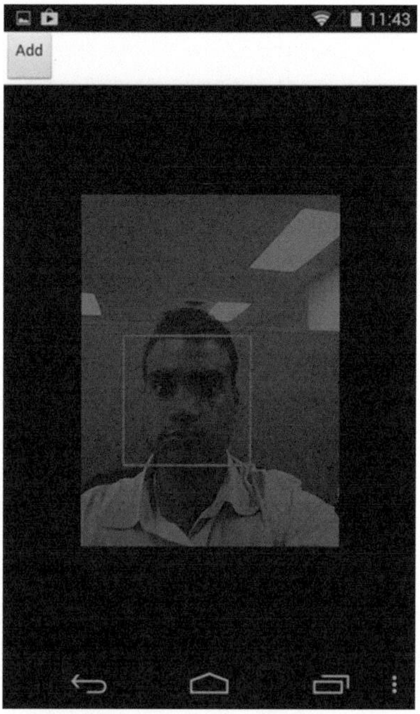

Fig. 9.14 Face-recognition (enhanced camera-component)

or if the care-giver has changed the regimen, while it improves the quality of the interaction by referring to the user with her first name (Fig. 9.14).

An EMMA 2.0-based representation of the face recognition result is given below. The result includes an ID and name corresponding to the recognized match. More than one match can be returned, in which case, the confidence score can be used to arrive at the most likely match.

```
<emma:emma version="2.0"
      xmlns:emma="http://www.w3.org/2003/04/emma"
      xmlns:xsi="http://www.w3.org/2001/XMLSchema-instance"
      xsi:schemaLocation="http://www.w3.org/2003/04/emma
      http://www.w3.org/TR/2009/REC-emma-20090210/emma.xsd"
      xmlns="http://www.example.com/example">
  <emma:one-of    id="r1"    emma:start="1087995961542"    emma:
  end="1087995963542"
      emma:medium="visual"    emma:mode="video"    "emma:device-
      type:"camera"  emma:source="http://example.com/camera/LFDJ-
      43U">
    <emma:interpretation id="int1" emma:confidence="0.85">
      <matchId>43879</matchId>
      <matchName>John Doe</matchName>
    </emma:interpretation>
```

```
    <emma:interpretation id="int2" emma:confidence="0.27">
      <matchId>90328</matchId>
      <matchName>Mark Johnson</matchName>
    </emma:interpretation>
  </emma:one-of>
</emma:emma>
```

9.6.5 Face Tracking

Face tracking allows EHA to respond to the users movements in a more human-like manner. Once motion is detected and the head turns in the corresponding direction, the camera locates the user and sends commands to the neck servo to face the user squarely. As the user moves around, EHA will continue to track the user's face and adjust its head-position accordingly.

An EMMA 2.0-based representation of the face tracking result is given below. The tracking information is provided in terms of the top-left and bottom-right co-ordinates of the "box" that contains the face in the picture captured by the video camera. It is assumed that there is a single face in the picture being interpreted. The incremental results feature of EMMA 2.0 is used here so that face tracking data can be continuously generated as the user's movements are tracked.

```
<emma:emma version="2.0"
  xmlns:emma="http://www.w3.org/2003/04/emma"
  xmlns:xsi="http://www.w3.org/2001/XMLSchema-instance"
  xsi:schemaLocation="http://www.w3.org/2003/04/emma
  http://www.w3.org/TR/2009/REC-emma-20090210/emma.xsd"
  xmlns="http://www.example.com/example">
  <emma:one-of id="r1" emma:start="1087995961542" emma:
  end="1087995963542"
  emma:medium="visual" emma:mode="video" emma:device-type:"camera"
  emma:source="http://example.com/camera/LFDJ-43U">
    <emma:interpretation id="int1" emma:confidence="0.7"
      emma:tokens="{'top-left':{'x':323.219, 'y':643.980}, 'bottom-
      right':{'x':732.543, 'y':132.376}} ">
      emma:token-type="json"
      emma:stream-id="s1"
      emma:stream-seq-num="0"
      emma:stream-status="begin"
      emma:stream-token-span="0-1"
      emma:stream-full-result="true"
    </emma:interpretation>
```

```
    <emma:interpretation id="int2" emma:confidence="0.43"
      emma:tokens="{'top-left':{'x':196.088, 'y':778.692}, 'bottom-
      right':{'x':478.011, 'y':389.289}} ">
      emma:token-type="json"
      emma:stream-id="s1"
      emma:stream-seq-num="0"
      emma:stream-status="begin"
      emma:stream-token-span="0-1"
      emma:stream-full-result="true"
    </emma:interpretation>
  </emma:one-of>
</emma:emma>
```

9.6.6 TTS and Speech Recognition

EHA is equipped with TTS and local speech recognition to make the interaction with the user more natural. The speech recognizer uses a lead word to reduce false positive recognition and end of speech recognition to simulate a normal conversation (Fig. 9.15).

9.6.7 EEG Headset

There are two modes for EHA; a passive mode, where the user provides commands and the robots executes them, and an active mode where EHA initiates and drives the interaction. Once paired with the NeuroSky Mindwave headset, EHA asks binary questions in a logical progression and a redundant manner. The user has to voluntarily blink for an affirmative response. This method of interaction is capable of communicating with individuals that are incapable of speech or are disabled in a way that prevents from using normal interaction (Fig. 9.16).

9.6.8 Bluetooth Integration

EHA leverages Cue-me™'s Bluetooth Low Energy (BLE) component and can read directly from Bluetooth Health-care Device Profile (HDP) and other compatible health-monitoring devices and medical instrumentation obviating the need for user to enter the readings manually off the instrument-displays.

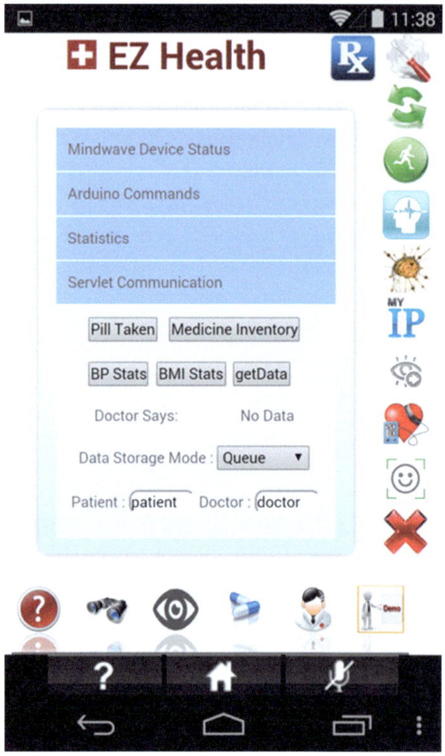

Fig. 9.15 Monitoring vitals

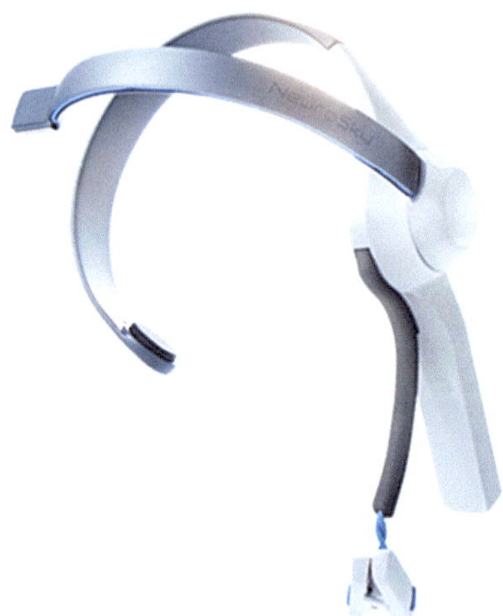

Fig. 9.16 Mindwave headset

9.6.9 Ink and Speech Annotation

EHA facilitates capture of images using camera and high-lighting the picture with Ink and Voice annotations. The annotations are exchanged as W3C Synchronized Multimedia Integration Language (SMIL) documents for maintaining the isochronous nature of the multimodal annotations as shown below:

Sample SMIL document:

```
<smil>
  <head>
    <layout>
      <root-layout width="240" height="299" />
    </layout>
  </head>
  <body>
  <par>
          <audio src="audio_10_0.wav" begin="500" dur="6486"
          region="main" />
          <audio src="audio_10_1.wav" begin="15029" dur="10696"
          region="main" />
          <img src="image_10.jpg" region="main" />
          <ref src="inkml_10.xml" type="application/inkml+xml"
          region="main" />
  </par>
  </body>
</smil>
```

The corresponding InkML document referenced in the SMIL document is as given below (truncated for brevity) :

```
<?xml version="1.0" encoding="utf-8" standalone="yes"?>
<inkml:ink xmlns:inkml="http://www.w3.org/2003/InkML">
  <inkml:definitions>
    <inkml:context xml:id="ct_0" >
      <inkml:inkSource xml:id="inksrc_0" >
        <inkml:traceFormat>
              <inkml:channel name="X" type="integer" max="480"
              units="dev"/>
              <inkml:channel name="Y" type="integer" max="598"
              units="dev"/>
        </inkml:traceFormat>
        <inkml:channelProperties>
            <inkml:channelProperty channel="X" name="resolution"
            value="0" units="1/dev"/>
```

```
            <inkml:channelProperty channel="Y" name="resolution"
            value="0" units="1/dev"/>
        </inkml:channelProperties>
    </inkml:inkSource>
  </inkml:context>
  <inkml:brush xml:id="br_0">
    <inkml:brushProperty name="color" value="#FF0000" />
    <inkml:brushProperty name="width" value="3" units="dev" />
    <inkml:brushProperty name="style" value="1" />
  </inkml:brush>
  <inkml:brush xml:id="br_1">
    <inkml:brushProperty name="color" value="#FF0000" />
    <inkml:brushProperty name="width" value="3" units="dev" />
    <inkml:brushProperty name="style" value="1" />
  </inkml:brush>
  <inkml:brush xml:id="br_2">
    <inkml:brushProperty name="color" value="#FF0000" />
    <inkml:brushProperty name="width" value="3" units="dev" />
    <inkml:brushProperty name="style" value="1" />
  </inkml:brush>
  <inkml:brush xml:id="br_3">
    <inkml:brushProperty name="color" value="#FF0000" />
    <inkml:brushProperty name="width" value="3" units="dev" />
    <inkml:brushProperty name="style" value="1" />
  </inkml:brush>
  <inkml:brush xml:id="br_4">
    <inkml:brushProperty name="color" value="#FF0000" />
    <inkml:brushProperty name="width" value="3" units="dev" />
    <inkml:brushProperty name="style" value="1" />
  </inkml:brush>
  <inkml:brush xml:id="br_5">
    <inkml:brushProperty name="color" value="#FF0000" />
    <inkml:brushProperty name="width" value="3" units="dev" />
    <inkml:brushProperty name="style" value="1" />
  </inkml:brush>
  <inkml:brush xml:id="br_6">
    <inkml:brushProperty name="color" value="#FF0000" />
    <inkml:brushProperty name="width" value="3" units="dev" />
    <inkml:brushProperty name="style" value="1" />
  </inkml:brush>
  <inkml:brush xml:id="br_7">
    <inkml:brushProperty name="color" value="#FF0000" />
    <inkml:brushProperty name="width" value="3" units="dev" />
    <inkml:brushProperty name="style" value="1" />
  </inkml:brush>
```

```
    <inkml:brush xml:id="br_8">
      <inkml:brushProperty name="color" value="#FF0000" />
      <inkml:brushProperty name="width" value="3" units="dev" />
      <inkml:brushProperty name="style" value="1" />
    </inkml:brush>
    <inkml:brush xml:id="br_9">
      <inkml:brushProperty name="color" value="#FF0000" />
      <inkml:brushProperty name="width" value="3" units="dev" />
      <inkml:brushProperty name="style" value="1" />
    </inkml:brush>
    <inkml:brush xml:id="br_10">
      <inkml:brushProperty name="color" value="#FF0000" />
      <inkml:brushProperty name="width" value="3" units="dev" />
      <inkml:brushProperty name="style" value="1" />
    </inkml:brush>
  </inkml:definitions>
                <inkml:trace contextRef="#ct_0" brushRef="#br_0"
duration="1284" timeOffset="1001" >281 48,281 48,277 48,277 48,273
48,273 48,269 50,269 50,264 51,264 51,261 51,261 51,255 53,255 53,247
55,247 55,237 57,237 57,229 59,229 59,225 60,225 60,221 62,221 62,212
64,212 64,208 65,208 65,201 69,201 69,197 70,197 70,196 71,196 71,189
75,189 75,187 77,187 77,181 83,181 83,177 87,177 87,175 92,175 92,173
96,173 96,171 103,171 103,171 106,171 106,169 114,169 114,171 117,171
117,172 124,172 124,175 128,175 128,176 133,176 133,180 139,180
139,184 144,184 144,189 149,189 149,193 152,193 152,199 155,199
155,203 157,203 157,208 160,208 160,216 162,216 162,221 163,221
163,233 164,233 164,248 165,248 165,257 164,257 164,263 164,263
164,268 164,268 164,277 163,277 163,283 162,283 162,292 161,292
161,296 160,296 160,301 159,301 159,311 157,311 157,316 157,316
157,327 154,327 154,332 153,332 153,343 150,343 150,352 148,352
148,356 146,356 146,359 145,359 145,363 143,363 143,369 139,369
139,372 137,372 137,375 135,375 135,377 132,377 132,393 109,393
109,395 107,395 107,395 103,395 103,395 101,395 101,393 97,393 97,393
96,393 96,392 90,392 90,392 86,392 86,391 84,391 84,391 83,391 83,389
79,389 79,387 76,387 76,384 74,384 74,381 72,381 72,379 70,379 70,376
68,376 68,373 66,373 66,371 64,371 64,367 63,367 63,364 61,364 61,360
59,360 59,357 57,357 57,353 55,353 55,348 54,348 54,345 53,345 53,340
51,340 51,337 50,337 50,332 47,332 47,325 45,325 45,321 44,321 44,319
43,319 43,315 43,315 43,308 43,308 43,304 43,304 43,296 44,296 44,291
45,291 45,281 47,281 47,276 48,276 48,276 48,276 48</inkml:trace>
  <inkml:trace contextRef="#ct_0" brushRef="#br_1" duration="955"
timeOffset="3537" >545 69,545 69,541 70,541 70,537 70,537 70,533
70,533 70,529 71,529 71,525 72,525 72,521 74,521 74,517 74,517 74,512
76,512 76,507 78,507 78,504 79,504 79,501 81,501 81,499 83,499 83,493
86,493 86,492 87,492 87,489 89,489 89,485 93,485 93,481 96,481 96,480
```

98,480 98,455 130,455 130,452 138,452 138,451 141,451 141,451 144,451 144,451 148,451 148,451 152,451 152,455 162,455 162,456 165,456 165,460 169,460 169,464 174,464 174,468 177,468 177,477 185,477 185,484 189,484 189,488 192,488 192,491 193,491 193,496 194,496 194,505 196,505 196,509 197,509 197,520 197,520 197,525 197,525 197,535 195,535 195,555 190,555 190,565 186,565 186,569 184,569 184,573 181,573 181,577 179,577 179,584 173,584 173,591 169,591 169,593 165,593 165,597 161,597 161,599 157,599 157,601 151,601 151,604 143,604 143,607 136,607 136,607 132,607 132,608 124,608 124,609 121,609 121,609 118,609 118,609 115,609 115,608 110,608 110,607 107,607 107,605 104,605 104,604 102,604 102,601 97,601 97,600 95,600 95,600 93,600 93,597 88,597 88,596 87,596 87,592 83,592 83,591 81,591 81,588 79,588 79,585 78,585 78,579 76,579 76,573 75,573 75,571 75,571 75,567 75,567 75,561 74,561 74,556 73,556 73,556 73,556 73</inkml:trace>

<inkml:trace contextRef="#ct_0" brushRef="#br_2" duration="647" timeOffset="8272" >59 220,59 220,60 222,60 222,64 229,64 229,67 233,67 233,68 237,68 237,71 242,71 242,73 249,73 249,75 253,75 253,76 256,76 256,79 262,79 262,80 266,80 266,80 267,80 267</inkml:trace>

<inkml:trace contextRef="#ct_0" brushRef="#br_3" duration="110" timeOffset="8941" >76 241,76 241,79 239,79 239,81 237,81 237,85 234,85 234,88 232,88 232,93 229,93 229,93 229,93 229</inkml:trace>

<inkml:trace contextRef="#ct_0" brushRef="#br_4" duration="262" timeOffset="9314" >91 214,91 214,91 217,91 217,92 218,92 218,95 226,95 226,97 232,97 232,99 235,99 235,107 250,107 250,111 251,111 251,111 251,111 251</inkml:trace>

<inkml:trace contextRef="#ct_0" brushRef="#br_5" duration="354" timeOffset="9854" >121 245,121 245,123 243,123 243,125 242,125 242,127 239,127 239,132 234,132 234,133 230,133 230,136 225,136 225,135 220,135 220,133 218,133 218,132 215,132 215,124 213,124 213,117 217,117 217,113 222,113 222,112 225,112 225,112 234,112 234,113 238,113 238,121 240,121 240,125 241,125 241,128 240,128 240,132 240,132 240,139 238,139 238,141 236,141 236,144 234,144 234,147 232,147 232,147 230,147 230,147 230,147 230</inkml:trace>

<inkml:trace contextRef="#ct_0" brushRef="#br_6" duration="361" timeOffset="10382" >151 201,151 201,155 203,155 203,156 205,156 205,159 207,159 207,160 210,160 210,163 214,163 214,165 218,165 218,167 221,167 221,167 225,167 225,164 222,164 222,164 220,164 220,164 218,164 218,164 216,164 216,163 214,163 214,163 211,163 211,163 209,163 209,161 202,161 202,163 200,163 200,164 198,164 198,165 197,165 197,168 195,168 195,168 195,168 195</inkml:trace>

<inkml:trace contextRef="#ct_0" brushRef="#br_7" duration="346" timeOffset="10962" >200 210,200 210,203 208,203 208,204 206,204 206,204 204,204 204,205 202,205 202,205 200,205 200,205 194,205 194,204 189,204 189,203 187,203 187,200 185,200 185,191 189,191

189,188 192,188 192,185 197,185 197,184 201,184 201,184 204,184 204,184 209,184 209,185 211,185 211,192 213,192 213,195 213,195 213,199 212,199 212,203 212,203 212,209 209,209 209,216 207,216 207,219 205,219 205,223 202,223 202,223 202,223 202</inkml:trace>
<inkml:trace contextRef="#ct_0" brushRef="#br_8" duration="1012" timeOffset="16118" >397 214,397 214,395 214,395 214,389 215,389 215,383 215,383 215,373 216,373 216,369 217,369 217,361 218,361 218,353 221,353 221,349 222,349 222,344 223,344 223,336 226,336 226,331 229,331 229,327 230,327 230,323 233,323 233,315 238,315 238,311 240,311 240,301 248,301 248,299 252,299 252,297 255,297 255,293 262,293 262,293 266,293 266,292 269,292 269,292 273,292 273,293 281,293 281,295 285,295 285,295 289,295 289,296 293,296 293,299 297,299 297,300 301,300 301,309 312,309 312,316 319,316 319,320 323,320 323,324 326,324 326,333 332,333 332,337 335,337 335,344 336,344 336,360 342,360 342,367 343,367 343,373 344,373 344,379 345,379 345,404 345,404 345,411 344,411 344,417 344,417 344,424 343,424 343,437 340,437 340,443 340,443 340,449 338,449 338,457 336,457 336,465 333,465 333,475 331,475 331,491 325,491 325,511 319,511 319,516 316,516 316,524 311,524 311,528 310,528 310,531 307,531 307,536 302,536 302,537 299,537 299,541 292,541 292,543 288,543 288,543 286,543 286,544 279,544 279,543 271,543 271,540 270,540 270,539 268,539 268,533 265,533 265,531 263,531 263,528 262,528 262,523 257,523 257,520 254,520 254,515 249,515 249,512 246,512 246,508 243,508 243,501 239,501 239,497 238,497 238,493 237,493 237,484 233,484 233,480 231,480 231,475 230,475 230,465 226,465 226,461 223,461 223,409 214,409 214,403 214,403 214,397 214,397 214,383 214,383 214,379 214,379 214,379 214,379 214</inkml:trace>
<inkml:trace contextRef="#ct_0" brushRef="#br_9" duration="792" timeOffset="18293" >105 375,105 375,111 376,111 376,121 377,121 377,128 378,128 378,139 378,139 378,144 377,144 377,149 376,149 376,171 372,171 372,180 368,180 368,185 366,185 366,193 362,193 362,196 360,196 360,204 355,204 355,207 352,207 352,211 348,211 348,217 341,217 341,220 337,220 337,223 333,223 333,228 325,228 325,229 321,229 321,232 315,232 315,233 311,233 311,232 309,232 309,229 304,229 304,228 303,228 303,221 299,221 299,219 296,219 296,215 295,215 295,204 289,204 289,196 286,196 286,192 285,192 285,187 284,187 284,176 283,176 283,165 283,165 283,159 283,159 283,152 283,152 283,143 283,143 283,137 283,137 283,133 284,133 284,121 287,121 287,116 291,116 291,112 292,112 292,108 295,108 295,101 299,101 299,99 301,99 301,92 306,92 306,88 308,88 308,83 313,83 313,80 316,80 316,79 319,79 319,77 328,77 328,79 331,79 331,81 337,81 337,83 340,83 340,84 343,84 343,87 349,87 349,88 352,88 352,91 357,91 357,92 360,92 360,96 364,96 364,101 367,101 367,107 368,107 368,111 368,111 368,115 367,115 367,120 366,120 366,124 366,124 366,129 365,129 365,139 365,139 365,139 365,139 365</inkml:trace>

```
<inkml:trace contextRef="#ct_0" brushRef="#br_10" duration="598"
timeOffset="20416" >608 218,608 218,604 220,604 220,600 222,600
222,596 224,596 224,592 226,592 226,587 230,587 230,584 233,584
233,583 234,583 234,580 238,580 238,579 240,579 240,577 242,577
242,577 245,577 245,575 250,575 250,575 252,575 252,572 255,572
255,572 260,572 260,572 265,572 265,573 269,573 269,577 272,577
272,583 274,583 274,587 275,587 275,592 276,592 276,596 276,596
276,601 275,601 275,607 272,607 272,609 271,609 271,611 269,611
269,616 265,616 265,617 263,617 263,620 260,620 260,620 256,620
256,620 251,620 251,620 249,620 249,620 245,620 245,620 242,620
242,621 238,621 238,621 236,621 236,621 234,621 234,619 230,619
230,617 229,617 229,601 222,601 222,601 222,601 222</inkml:trace>
</inkml:ink>
```


The captured image and audio recording play while redrawing the ink trace on the image, enabling sharing of rich annotations with care-givers.

9.7 Conclusion

Implementation of the platform using the W3C MMI architecture helped in streamlining the process of structured authoring for application developers. It mainly helped in the separation of concerns between application-design and interaction-design and helped them focus on the application functionality without unduly getting distracted by the differences in the features of underlying device/OS architecture. Further, it helped portability of their applications while shielding authors from nuances of the implementation of modality-components or getting locked-in to a particular vendor's speech-engine or location-service or other modality components. The soundness of the component-architecture of MMI enabled authors in extending modality components and thus the Cue-me platform to suit their needs. As of the date of writing this book, Cue-me applications are deployed on over 3.5 million devices around the world. The types of deployed applications range from banking and financial services, field-force automation, retail-store operations, health-care services, and corporate applications facilitating collaboration and enhancing human interaction, while increasing user-productivity. It helped support distributed deployment of applications across connected devices. Some of the latest features of the Cue-me platform include registration and discovery of modality components, so that authors can now write applications that can dynamically discover and bind to modality components/services.

Acknowledgments Thanks to all the team members of Openstream especially, R. Anthapu, R. Sripada, G. McCobb, K. Patel, N. Shah, T. Sindhaghatta, S. Anthapu, B. Narayana, and S. Khandavilli, who helped in the design and development and the guidance and help provided by D. Dahl, K. Ashimura at W3C MMI.

References

1. Barnett, J., Bodell, M., Dahl, D. A., Kliche, I., Tumuluri, R., Larson, J., Porter, B., et al. (2012). Multimodal architecture and interfaces. World Wide Web Consortium. http://www.w3.org/TR/mmi-arch/. Accessed 20 Nov 2012.
2. Larson, J. A., Raman, T. V., & Raggett, D. (2002). W3C Multimodal Interaction Framework. W3C. http://www.w3.org/TR/mmi-framework/.
3. Watt, S. M., Underhill, T., Chee, Y.-M., Franke, K., Froumentin, M., Madhvanath, S., et al. (2011). Ink Markup Language (InkML). World Wide Web Consortium. http://www.w3.org/TR/InkML. Accessed Nov 2012.
4. Barnett, J. (2016). Introduction to SCXML. In D. Dahl (Ed.), *Multimodal interaction with W3C standards: toward natural user interfaces to everything*. New York, NY: Springer.
5. Barnett, J., Akolkar, R., Auburn, R. J., Bodell, M., Burnett, D. C., Carter, J. et al. (2015) State Chart XML (SCXML): State Machine Notation for Control Abstraction. World Wide Web Consortium. http://www.w3.org/TR/scxml/. Accessed 20 Feb 2016.
6. Johnston, M. (2016). EMMA. In D. Dahl (Ed.), *Multimodal interaction with W3C standards: towards natural user interfaces to everything*. New York, NY: Springer.
7. Johnston, M., Baggia, P., Burnett, D., Carter, J., Dahl, D. A., McCobb, G., et al. (2009). EMMA: Extensible MultiModal Annotation markup language. W3C. http://www.w3.org/TR/emma/. Accessed 9 Nov 2012.
8. Johnston, M., Dahl, D. A., Denny, T., & Kharidi, N. (2015). EMMA: Extensible MultiModal Annotation markup language Version 2.0. World Wide Web Consortium. http://www.w3.org/TR/emma20/. Accessed 16 Dec 2015.

URL's

Easy Health Assistant use case video: http://youtu.be/x2Tte0QyTiA.
Openstream Easy Health Assistant Mindwave Demo: http://youtu.be/aEgdRU-yM2o.
Poor Medication adherence costs $290 Billion Annually: http://mobilehealthnews.com/3901/poor-medication-costs-290-billion-a-year/.
W3C Multimodal Architecture: http://www.w3.org/TR/mmi-arch/.
Cue-me™ Multimodal Authoring Platform: http://www.openstream.com/cueme.
SCXML – State Chart XML: (http://www.w3.org/TR/scxml/).
Neurosky Mindwave: http://neurosky.com/products-markets/eeg-biosensors/hardware/.

Chapter 10
SCXML on Resource Constrained Devices

Stefan Radomski, Jens Heuschkel, Dirk Schnelle-Walka, and Max Mühlhäuser

Abstract Ever since their introduction as a visual formalism by Harel et al. in 1987, state-charts played an important role to formally specify the behavior of reactive systems. However, various shortcomings in their original formalization lead to a plethora of formal semantics for their interpretation in the subsequent years. In 2005, the W3C Voice Browser Working Group started an attempt to specify SCXML as an XML dialect and corresponding semantic for state-charts and their interpretation, promoted to W3C recommendation status in 2015. In the context of multimodal interaction, SCXML derives a special relevance as the markup language proposed to express dialog models as descriptions of interaction in the multimodal dialog system specified by the W3C Multimodal Interaction Working Group. However, corresponding SCXML interpreters are oftentimes embedded in elaborate host environments, are very simplified or require significant resources when interpreted. In this chapter, we present a more compact, equivalent representation for SCXML documents as native data structures with a respective syntactical transformation and their interpretation by an implementation in ANSI C. We discuss the characteristics of the approach in terms of binary size, memory requirements, and processing speed. This will, ultimately, enable us to gain the insights to transform SCXML state-charts for embedded systems with very limited processing capabilities and even integrated circuits.

S. Radomski (✉)
TU Darmstadt, Telekooperation Group, Darmstadt, Germany
e-mail: radomski@tk.tu-darmstadt.de

J. Heuschkel
TU Darmstadt, Telekooperation Group, Darmstadt, Germany
e-mail: heuschkel@tk.tu-darmstadt.de

D. Schnelle-Walka
S1nn GmbH & Co. KG, Stuttgart, Germany
e-mail: dirk.schnelle-walka@s1nn.de

M. Mühlhäuser
TU Darmstadt, Telekooperation Group, Darmstadt, Germany
e-mail: max@tk.tu-darmstadt.de

D.A. Dahl (ed.), *Multimodal Interaction with W3C Standards*,
DOI 10.1007/978-3-319-42816-1_10

10.1 Introduction

The State-Chart eXtensible Markup Language (SCXML) is a W3C recommendation for a specific syntax and semantics of Harel state-charts [4] as a compact visual formalism for state-transitioning systems. It was finalized in September of 2015 [6] and is suggested in the W3C Multimodal Architecture and Interfaces recommendation [1] as a possible description of interaction managers to control modality components in a multimodal user interface. While state-charts, as a visual formalism, were already proposed by Harel in 1987, deficiencies with the initial semantic [5] lead to the development and scientific publication of more than 40 different semantics in the subsequent years [3, 9].

As such, the endeavor of SCXML to standardize the syntax and semantic via the W3C is direly needed to reestablish compatibility of the various tools and platforms available to model and interpret Harel state-charts. However, the syntactical description of SCXML as an XML dialect and several language features implied by tests in the SCXML Implementation Report Plan (IRP) strongly suggest an implementation via interpretation at runtime with a full XML document object model still available. While this overall approach has been spectacularly successful, e.g., with HTML [2] and enables considerable flexibility to dynamically adapt the XML description via scripting during interpretation, it severely limits the applicability of SCXML to platforms with sufficient computing power and memory.

In the following sections, we will describe an approach to preprocess SCXML documents into more suitable data structures and present an implementation of the `microstep(T)` function in ANSI C. This implementation, by a large margin, outperforms the pseudo-code description of the same algorithm in Appendix D of the SCXML recommendation. This is relevant as many SCXML interpreters do indeed align their implementation of this central piece of functionality with the pseudo-code description. Furthermore, by employing the syntax and semantics of ANSI C as a formal programming language, we do address one point of critique with the pseudo-code in Appendix D, that is to provide an actually executable description for `microstep(T)`.

While the evaluation of the ANSI C algorithm will already show general applicability for even the smallest off-the-shelf micro-controllers, the last part of this chapter will describe a first approach for a transformation from SCXML onto VHDL as a hardware description language. Such a description would allow to mold SCXML documents into Field Programmable Gate Arrays (FPGAs) and even Application Specific Integrated Circuits (ASICs), which we expect to gain elevated relevance in the scope of applications for the Internet Of Things (IoT).

10.2 Semantic of SCXML

Before we dive into the actual algorithms, we will need to define some important sets and relations from the SCXML recommendation that are relevant to retrace the algorithms' functionality and convince ourselves of their correctness. This section does assume a passing familiarity with the SCXML recommendation or, at least, with Harel state-charts in general.

The interpretation and execution of an SCXML document at runtime can be conceived as a series of *microsteps* over a set of transitions (T) enabled by an event e. At any point in time, the interpreter is in a given *configuration* as a set of proper states that are said to be *active*. Any change to the configuration of an interpreter is assumed to be instantaneous (perfect synchronicity hypothesis [9]) and always caused by events that enable transitions. A special non-event ε is introduced in the SCXML recommendation to extend this notion for spontaneous transitions. Figure 10.1 summarizes the sequence of activities for an interpreter within a `microstep(T)` iteration. Every iteration starts with establishing the current event as follows:

1. If the previous iteration did not exhaust spontaneous transitions, set the current event to ε as the non-event (**1a**) that only enables event-less (spontaneous) transitions.
2. If there were no more spontaneous transitions enabled by ε in the previous iteration, dequeue an event from the *internal* event queue (**1b**).
3. If there are no events remaining on the internal event queue, attempt to dequeue from the *external* queue (**1c**) or block execution until an event becomes available. After a series of such micro-steps and before dequeuing an event, the interpreter is said to have performed a *macro-step* and reached a new stable configuration. At this point, a compliant interpreter has to make sure that all invocations for external systems specified via <invoke> within states of the active configurations are started and all other such invocations stopped.

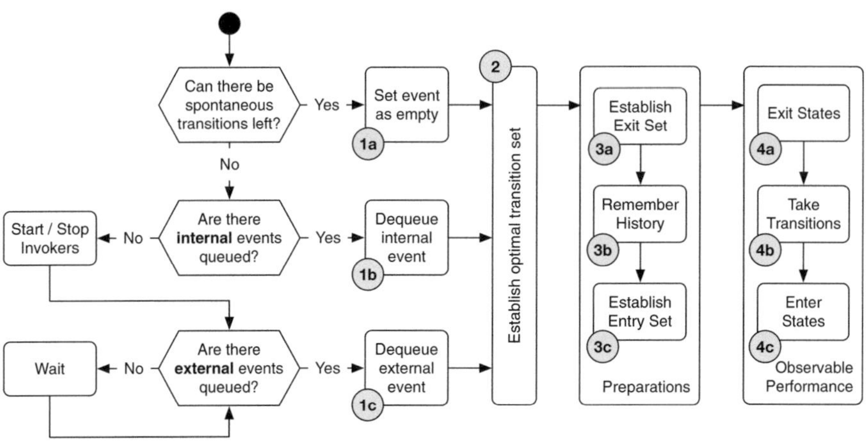

Fig. 10.1 Flow of activities within a micro-step iteration

Now, whenever we are about to proceed to the next activity (**2**), we can assume an event to be set (be it ε or an actual event). For this given event e, we will have to establish the *optimal transition set*. To this effect, the SCXML recommendation defines a containment hierarchy of transition sets as follows:

- **Active Transitions**
  All `<transition>` elements contained as direct children of states in the active configurations are said to be *active*. They form the superset of all other transition sets below.
- **Matched Transitions**
  The subset of active transitions, with at least one event descriptor in their `event` attribute matching the current event's name are said to be *matching*. If the current event is the ε event, all active transitions with no `event` attribute (spontaneous transitions) are matched.
  It is allowed for an event descriptor to have a `.*` suffix for compatibility with CCXML. Furthermore, it is legal for a transition to specify multiple, space-separated event descriptors in their `event` attribute. In this case, an event matches a transition if one of the transition's event descriptors matches the event's name.
- **Enabled Transitions**
  The set of matching transitions is further reduced by requiring an eventual `cond` attribute to evaluate to `true` on the *data-model* (usually an embedded scripting language context). A matched transition with a condition that holds or without a condition is said to be *enabled*.
- **Optimally Enabled Transitions**
  For a transition to be optimally enabled, there can be no earlier enabled transition with the same source state, neither can a transition in a descendant state of our source be enabled. The first criterion provides an ordering for enabled transitions within the same state. The second criterion allows to *specialize* a state-chart's behavior in response to events by overriding behavior in a more deeply nested sub-state of a composite state.
- **Optimal Transition Set**
  Generally, it is not possible for all optimally enabled transitions to be taken in the same micro-step as they might lead to an invalid subsequent configuration. Therefore, the optimal transition set is established as the largest set of non-conflicting, optimally enabled transitions. Here, two transitions are said to be conflicting, if the intersection of their exit sets is nonempty. For any two such transitions, the one with the highest *priority* will be added to the optimal transition set. The priority of a transition is defined very similar to the precedence with the optimally enabled transition set in such that a transition t_1 has a higher priority than t_2 if (1) the source of t_1 is a descendant of the source of t_2, (2) or t_1 precedes t_2 in document order.

The optimal transition set at the end of the above containment hierarchy now contains all the transitions T that are to be performed in response to an event in the current `microstep(T)` iteration. It is crucial for any performant implementation of SCXML to be able to identify this optimal transition set efficiently as it is calculated at least twice for any non-ε event: once as the set of transitions to be taken for the event itself and, subsequently, at least once for the ε event to exhaust any spontaneous optimal transitions in the new configuration.

We will see later that the transitions in the optimal transition set already define the microstep's *exit-set* (**3a**) as the set of active states to be exited within the current micro-step. For any composite state in this set that contains a `<history>` pseudo state as a child we will have to remember its active children or, depending on the histories `type`, even all its active descendants (**3b**) to be reentered when the `<history>` pseudo state is in the target-set of a subsequent iteration.

The optimal transition set also already defines an intermediate *target-set* as the set of states directly referenced in the transitions' `target` attributes. From this target-set, we can establish the *entry-set* of the optimal transition set (**3c**) by calculating its *completion*, which defines the set of states to be actually entered within the current micro-step. The completion of a state in the target-set depends on its type and we will discuss all of them in more detail when we step through the actual `microstep(T)` algorithm below.

After we established the `exit-`, `transition-`, and `entry-set` for a given event in a state-chart's configuration, we can perform the actual micro-step as (**4a**) exiting states, (**4b**) transitioning and (**4c**) entering the new states. This will update the state-chart's active configuration and invoke any*executable content* associated with these activities.

10.2.1 Scope of the Algorithm

The `microstep(T)` function outlined above is at the core of every SCXML interpreter and its description constitutes the bulk of the SCXML recommendation. There are, however, additional responsibilities for a compliant interpreter that we do not address in the algorithm we are about to describe below:

- We do not concern ourselves with invocations of external components via the `<invoke>` element. Such invocations are to be processed prior to dequeueing an external event, right before the interpreter is said to have performed a macro-step. It is perfectly possible to trigger these invocations via our algorithm, but the transformation onto ANSI C we implemented will, currently, only process a single state-chart per file and virtually all tests defined for `<invoke>` in SCXML assume a nested state-chart to be processed.
- We do not support any I/O processor other than the SCXML I/O processor.
- We have not implemented file operations or any retrieval of content referenced via a URL.

We do, however, support the transformation of executable content into semantically equivalent control flow constructs in ANSI C for various callbacks into user-supplied code as well as various datamodel implementations. Both features are required to pass any meaningful subset of the SCXML IRP tests and evaluate the algorithm. The datamodel integration is not part of the actual algorithm but assumed to be available as a set of respective callback functions that will evaluate the various expressions.

10.3 Preparing SCXML Data Structures

If we are to target embedded platforms, it seems wasteful not to preprocess the SCXML documents into a more compact representation. While there are XML parsers available that compile into binary code as small as 30 KB,[1] they only offer a streaming API for XML documents and still require us to establish and maintain a suitable representation at runtime. As such, we might as well preprocess the SCXML documents into a native representation and pre-calculate several sets and relations that will become relevant when we discuss the actual `microstep(T)` algorithm below.

10.3.1 States

When we regard the states of an SCXML document, we can encode all the information required for a semantically equivalent execution of a given state-chart via the compound data structure given in Listing 10.1. During transformation, an array of such structures is defined, containing all the states (along with the pseudo-states and the root state) of an SCXML state-chart. The states in this array are sorted by document-order, which corresponds to entry-order and reverse exit-order for the `microstep(T)` algorithm.

Listing 10.1 Representing a state as a compound data structure

```
struct state {
const char*           name;
const uint8_t         type;
const uint16_t        parent;
const exec_content_t  on_entry;
const exec_content_t  on_exit;
const char            children[STATE_BYTES];
const char            completion[STATE_BYTES];
const char            ancestors[STATE_BYTES];
const elem_data*      data;
};
```

[1]https://dev.yorhel.nl/yxml.

- The **`name`** field contains the eventual identifier of the state or is `NULL` if the state does not specify an identifier. This identifier is only needed for the `In ('state')` predicate and is not used during the actual `microstep(T)` algorithm below.
- The **`type`** field identifies the states type in the original SCXML document and can be one of `{PARALLEL, COMPOUND, ATOMIC, FINAL, INITIAL, HIST_DEEP, HIST_SHALLOW}`. The most significant bit is reserved for the `HAS_HIST` flag, which denotes (1) whether there is a `<history>` child element for a composite state or (2) whether there is another `<history>` element in the descendants of a given history's parent state. This flag will become important later when we complete a history element in the target set onto its entry set.
- The **`parent`** field identifies the index of this state's parent in the array of all states per document.
- The **`on_entry`** and **`on_exit`** fields are pointers to static functions where the respective executable content is generated.
- The next three fields, **`children`**, **`completion`**, and **`ancestors`** are bit arrays with a width sufficient to model every state from the original SCXML document as a single bit. The `children` and `ancestors` bit arrays are initialized such that the bit at index N is set if the state at index N is in the respective relation to the current state. The semantic of the `completion` bit array is more ambiguous and depends on the state's type:

  - For <**`parallel`**> states, it identifies all the direct, proper child states.
  - For **compound** `<state>`s, the completion identifies the first child in document order or the states from the target set identified by the state's `initial` attribute.
  - For <**`final`**> and **atomic** `<state>`s, the completion is empty.
  - For <**`initial`**> pseudo states, it identifies the states in the target set of a contained `<transition>` element.
  - For <**`history`**> pseudo states, its semantic is rather complicated. Essentially, it identifies all the parent's descendant states that are *covered* by the history, i.e. the parent's proper child states for `shallow` histories. For `deep` histories, however, it does not necessarily identify all proper descendant states, but only those that are not already covered by a nested `<history>` pseudo state. We will see later that this construction allows us to model all of the state-chart's history as a single bit-array with a width corresponding to the number of states only.

- Finally, the **`data`** element contains a pointer to an NULL terminated array of compound data structures, representing the optional `<data>` elements for late initialization upon first activation with a late data binding. For an early data-binding, all these `<data>` elements are attached to the state-chart's `<scxml>` root state.

The memory layout of an individual compound data structure for a state is depicted in Fig. 10.2 with its actual size depending on the target architectures bit-width and the total number of states in a state-chart. Figure 10.3 shows the size of a single state structure as a function of the total number of states when assuming a 16-bit target architecture. It is noteworthy that the three bit-arrays (`children`, `completion`, and `ancestors`) are the largest contributors to its size and will completely dominate the required memory for large number of states.

Figure 10.3 also allows to determine the total amount of memory required to represent all of a state-chart's states as an array of these compound data structures by counting the number of all states in an SCXML document and multiplying it with the function value at the given point (e.g., to encode 100 states, we would need round about 100×60 bytes). This is in addition to the memory required for the string literals for the states' identifiers pointed to by the `name` field. If we were required to reduce this memory, we have many options to trade runtime for memory in this data structure, e.g. to

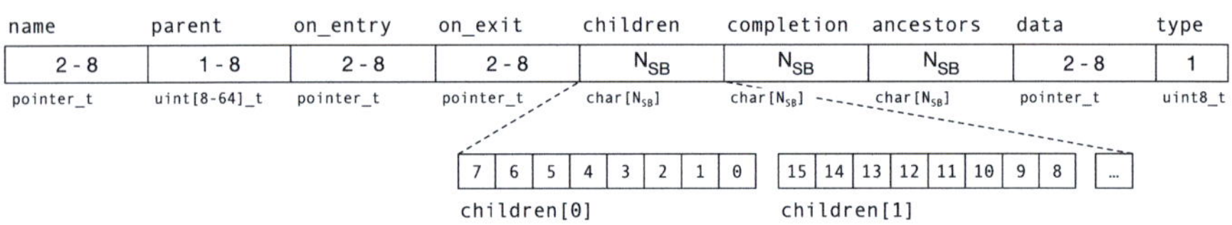

Fig. 10.2 The memory representation of a single state structure depends on the target platform's bit-width, the total number of states and any eventual padding

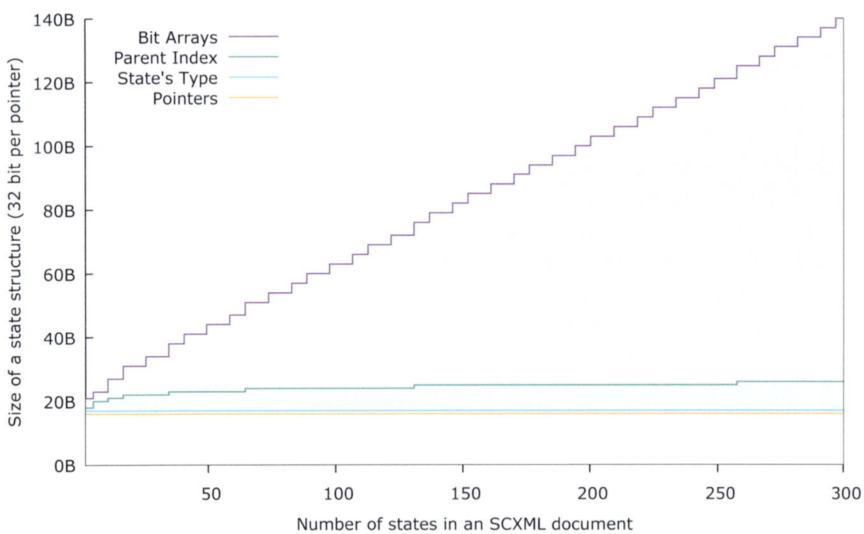

Fig. 10.3 Aggregated size of a single state structure as a function of the total number of states in an SCXML document without alignment padding

- Calculate the `children` via the `parent` relation.
- Calculate the `ancestors` via the reversed `children` relation.
- Calculate the `completion` via the state's `type` along with its `children` relation.
- Calculate the `parent` as the most significant bit in the `ancestors` relation.

However, some of these calculations can be quite expensive (most notably the completion of deep `<history>` states with nested `<history>` elements).

10.3.2 Transitions

Similar to the states above, we can establish an array of structures containing all the relevant information from the `<transition>` elements in the original SCXML state-chart. These compound data structures (Listing 10.2) will also already contain several pre-calculated, important sets and relations that are static with regard to a given SCXML state-chart and relevant for the execution of `microstep(T)`. The array is sorted in post-order traversal of all transitions and we will see later why this is very beneficial.

Listing 10.2 Representing a transition as a compound data structure

```
struct transition {
    const uint16_t          source;
    const char              target[STATE_BYTES];
    const char              exit_set[STATE_BYTES];
    const char              conflicts[TRANS_BYTES];
    const uint8_t           type;
    const char*             event;
    const char*             condition;
    const exec_content_t    on_transition;
};
```

- The **`source`** of a transition identifies its parent state (proper, or otherwise) by the state's index in the array of all states.
- The **`target`** field is a bit-array in which a given bit is set, if the `<transition>` element identified the respective state in its `target` attribute.
- The **`exit_set`** field is a bit-array identifying the transition's *complete* exit set. The definition of the actual exit-set from the SCXML standard is as follows:

  The exit set of a transition in configuration C is the set of states that are exited when the transition is taken when the state machine is in C. If the transition does not contain a `target`, its exit set is empty. Otherwise (i.e., if the transition contains a `target`), if its `type` is `external`, its exit set consists of all active states in C that are proper descendants of the Least Common Compound Ancestor (LCCA) of the source and target states. Otherwise, if the transition has `type internal`, its source state is a compound state, and all its target states are proper descendants of its source state, the exit set consists of all active states in C that are proper descendants of its source state.

Unfortunately, the exit-set depends on the state-chart's configuration C. The implied assumption is that we can calculate the exit-set for the complete configuration in which every state is active and establish the actual exit-set at runtime by intersecting each transition's complete exit-set with the active configuration. We do not have a proof for this assumption, but it makes sense given the definition, the calculation in the pseudo-code from Appendix D in the SCXML recommendation and, indeed, all relevant IRP tests do pass.

- The **`conflicts`** field is a bit-array that identifies other transitions which can, for whatever reason, never occur in an optimal transition set with the given transition. If we look at the definition of the optimal transition set in the containment hierarchy from Sect. 10.2, we can syntactically identify several situations in which two transitions conflict:

  1. For two transitions to be *active* within the same iteration, their source states need to be active at the same time. This can only be the case if their least common ancestor is a `<parallel>` element.
  2. For two transitions to be *matched* at the same time, there has to be an event that matches both transitions. This can never be the case for event-less and eventful transitions or two eventful transitions that have no event descriptor that matches a common event.
  3. We cannot exploit any criteria with regard to the *enabled* transition set as we, usually, cannot make any assumption about the evaluation of an eventual `cond` attribute at transformation time.
  4. For two transitions to be *optimally enabled*, their source states cannot be identical or ancestrally related.
  5. For two transitions to be in the *optimal transition set*, their exit-sets may not overlap.

  This results in quite a number of sufficient criteria for two transitions to conflict and minimizes the amount of transitions to consider when establishing the optimal transition set per micro-step considerably.
- The **`type`** field is interpreted as a bit array that specifies the type of the transition, it might be one or any of `{SPONTANEOUS, TARGETLESS, INTERNAL, HISTORY, INITIAL}`. Not all of these are currently used in the actual algorithm below, though.
- The **`event`** field contains a pointer to the constant string literal with the transition's event descriptor list and is required to establish the *matched* transition set.
- The **`cond`** field contains a pointer to the constant string literal with the transition's condition and is required to establish the *enabled* transition set.
- Finally, the **`on_trans`** field is a pointer to a static function with the transition's executable content.

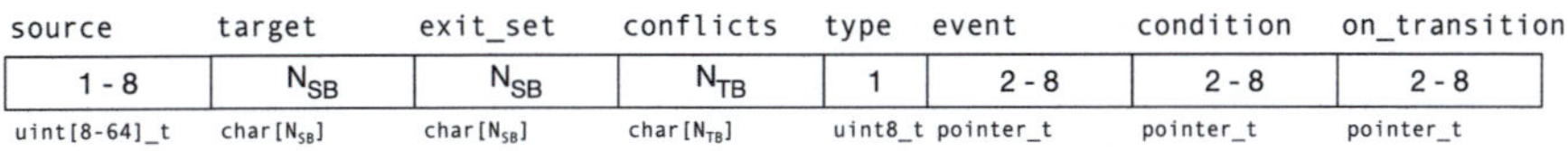

Fig. 10.4 The memory representation of a single transition structure also depends on the target platform's bit-width, the total number of states and transitions, as well as any eventual padding

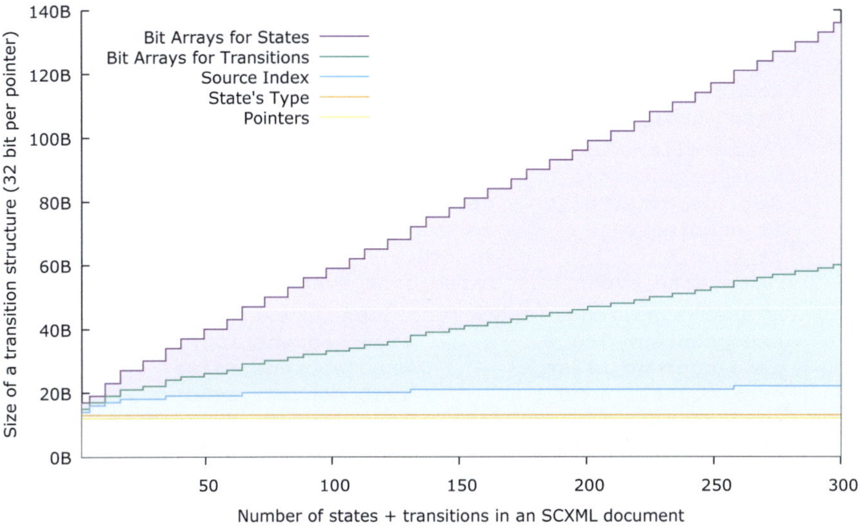

Fig. 10.5 Aggregated size of a single transition structure as a function of the total number of states plus transitions when assuming equal numbers

The memory layout of such a transition structure is given in Fig. 10.4 and, again, its size depends on the bit-width of the target platform and the total number of states and transitions in a given SCXML document. The size of this structure as a function of the total number of states and transitions, when assuming equal numbers and a 16-bit architecture is depicted in Fig. 10.5.

The relation of the structure's size with regard to the complexity of the complete document shows the exact same development as the one for states in Fig. 10.3, though, its increase in size is somewhat dampened if there are more transitions than states as we only need a single bit-array for transitions.

10.3.3 SCXML Context

The states and transitions above represent immutable, constant data for any given SCXML document and can be generated during transformation. But there is also a dynamic part for the interpretation of a state-chart, which we will represent as an SCXML context (see Listing 10.3). This allows us to maintain multiple instances of a state-chart at runtime as distinct contexts that share the states and transitions from above as static data.

Listing 10.3 The context of an SCXML instance at runtime

```
struct ctx {
    uint8_t flags;
    char    config[STATE_BYTES];
    char    history[STATE_BYTES];
    char    initialized_data[STATE_BYTES];

    void* event;
    void* user_data;

    /* miscellaneous user supplied callback functions */
    dequeue_internal_t    dequeue_internal;
    dequeue_external_t    dequeue_external;
    is_enabled_t          is_enabled;
    is_true_t             is_true;
    raise_done_event_t    raise_done_event;

    /* user-supplied callback functions for executable content */
    exec_content_log_t            exec_content_log;
    exec_content_raise_t          exec_content_raise;
    exec_content_send_t           exec_content_send;
    exec_content_foreach_init_t exec_content_foreach_init;
    exec_content_foreach_next_t exec_content_foreach_next;
    exec_content_foreach_done_t exec_content_foreach_done;
    exec_content_assign_t         exec_content_assign;
    exec_content_init_t           exec_content_init;
    exec_content_cancel_t         exec_content_cancel;
    exec_content_script_t         exec_content_script;
};
```

- The **`flags`** field is a generic member of the context to remember various boolean values across invocations of a `microstep(T)`. Currently, it encodes (1) whether the state-chart's context is still in pristine condition and some setup is required (`CTX_PRISTINE`), (2) whether we already exhausted spontaneous transitions (`CTX_SPONTANEOUS`) and need to dequeue an event, and (3) whether the state-chart entered a top-level final state and is done (`CTX_TOP_LEVEL_FINAL`).
- The **`config`** field contains the state-chart's currently active configuration as a bit-array, such that the bit at index i is set if the corresponding state in the array of all states is active.
- The **`history`** field is another bit-array that encodes the valuation of all `<history>` elements (deep or shallow) from the original state-chart. It is not obvious how we can encode all of the history in a single bit-array and we will discuss this point in more detail below.
- The **`initialized_data`** field is a bit-array that encodes which states were already entered. This is only required for SCXML documents with a late data-binding and allows us to perform the initialization of eventual nested `<data>` elements only for the first activation of a state.

- The **event** field is an opaque pointer to a memory region containing the current event. The microstep(T) algorithm will indeed not know any details about the current event as we will just employ the user-supplied is_enabled callback to determine whether an event matched and enabled any transition under consideration.
- The **user_data** field is another opaque pointer, where user-supplied code can register any additional data that might be required per SCXML interpreter instance and has no purpose in the scope of the microstep(T) algorithm below.
- All other fields are callbacks into user-supplied code. Most notably:
  - The **dequeue_internal** and **dequeue_external** functions will return an opaque pointer for the current event.
  - The **is_enabled** callback is called with a transition structure and the opaque event pointer to determine whether the given transition is matched and enabled by the given event.
  - The **is_true** callback determines whether a given expression evaluates to true on the data-model.
  - The **raise_done_event** is called with a state structure, and the information from an optional <donedata> element to raise the respective done.state.ID event on the internal queue.

The other callbacks are invoked as part of the executable content in the various, generated on_entry, on_exit, and on_trans functions referenced from the respective function callbacks in the state and transition structures above.

The total size of the context structure as a function of the number of states in the SCXML document is given in Fig. 10.6. For documents with only a few states, the

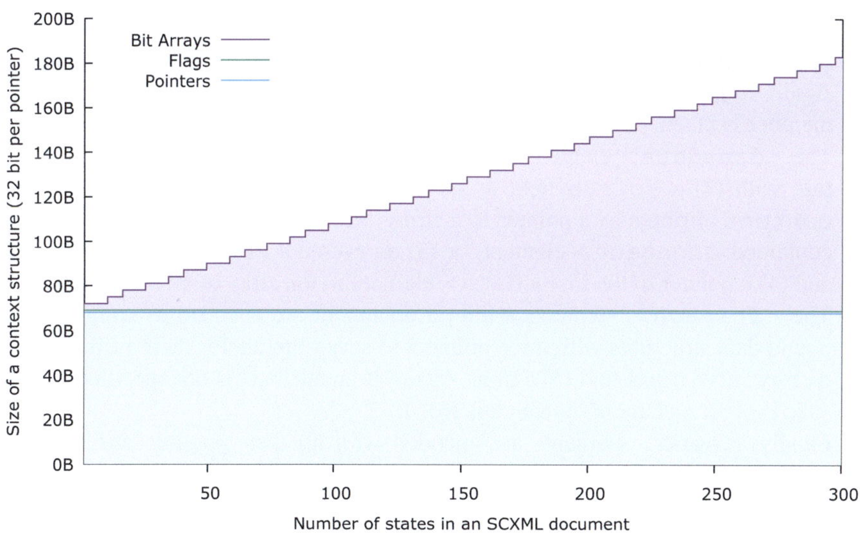

Fig. 10.6 Aggregated size of a single context structure as a function of the total number of states

size is dominated by the memory required to store the addresses of the callback functions for executable content and other datamodel queries. When the number of states increases, however, the size of the bit-arrays required to model the history, the active configuration and already initialized states start to become the major contribution. Ultimately, another such bit-array for current invocations per `<invoke>` would likely have to be added as well.

10.3.4 Other Elements

In order to enable the processing of executable content, we also need to encode various other SCXML elements into compound data structures. These merely encode the information contained within the respective elements and, as opposed to the states and transitions above, there are no special considerations with regard to their representation other than to make the information available to user-supplied code.

- The <**data**> elements, as children of the `<datamodel>` elements, are represented as compound data structures with four pointers to string literals for (1) their `id` attribute, (2) the `src` attribute, (3) the `location` attribute, and eventual content. If any of those attributes is unspecified, its value will be initialized as `NULL`.

  All compound data structures for `<data>` elements are contained within an array, with `NULL` entries in between as delimiters. To reference a consecutive set of data structures, the address of the first member is taken and a macro is provided to iterate all subsequent members until the next `NULL` delimiter.
- A similar approach is taken for all <**param**> elements. These are encoded in compound data structures, each with pointers to three string literals as (1) their `name` attribute, (2) the `location` attribute, and (3) their `expr` attribute. Again, to reference a consecutive set of such elements, the address of the first member is taken with macros to test for additional structures.
- The <**donedata**> elements are also just encoded as a compound data structure with (1) a `source` field as the index of the containing state, (2) their `content` attribute as a pointer to a string literal with the textual value of any contained `<content>` element, or (3) an eventual `contentexpr` attribute and (4) a pointer to the first `<data>` element in the array of all data structures.
- The <**foreach**>, elements within a document are encoded as simple compound data structures with three pointers to string literals for their attributes (1) `array`, (2) `index` and (3) `item`. Again, if an attribute is not specified with a `<foreach>` element, its value is `NULL`.
- Finally, <**send**> elements are encoded with all their possible attributes as compound data structures as well. Eventual `<content>` children are given as string literals in their textual representation and `<param>` elements are given as a reference to their first entry in the array of all `param` data structures.

10.3.5 Executable Content

As part of the transformation from SCXML onto C, we will also transform the executable content contained as children of the `<onexit>`, `<onentry>` and `<transition>` elements. This is not strictly required for an implementation of the `microstep(T)` algorithm, but it is easily done and extends the domain of the transformation.

To transform the executable content is to invoke the various callbacks in the SCXML context (cf. Listing 10.3) in the correct order and under the correct conditions. Furthermore, we will have to regard the error semantics of the various blocks. If an error occurs within a block of executable content, a compliant interpreter is required to raise a respective event and continue processing with the next block of executable content. To this effect, we encode every individual block of executable content as a static function that is exited if an error occurs and call each block sequentially from within a general `[ID]_on_entry`, `[ID]_on_exit`, `[ID]_on_trans` function. Here, the `ID` of an element is either the value of its eventual `id` attribute or a unique identifier derived from its position in the SCXML document.

With the callbacks given in the SCXML context and the representation of the various elements above, it is straightforward to see how we can generate C code to model the behavior of the executable content. The remaining SCXML elements of executable content for which we did not define a compound data structure above are merely passed via their various attributes into the user-supplied callbacks.

10.4 A Compact Algorithm for Interpretation

Now that we have all the data structures and control flow for executable content from an SCXML document defined, we can present the actual algorithm for `microstep(T)`. The algorithm is closely aligned with the sequence of activities from Fig. 10.1 and we will, indeed, describe its workings by presenting each activity in turn.

10.4.1 Preparations

Currently, the arrays of compound data structures for the transitions, states, and other SCXML elements are modeled as static global variables accessible throughout the compilation unit and their identifiers hard-coded into the algorithm. As such, there is very little preparation required but to allocate memory for a `ctx` and register the various callback functions:

Listing 10.4 Instantiating a state-machine context

```
int main(int argc, char** argv) {
   int err;
   ctx ctx;
   memset(&ctx, 0, sizeof(ctx));

   /* register callbacks */
   ctx.is_enabled = &is_enabled;
   ctx.is_true = &is_true;
   ...

   /* run interpreter until done */
   while(true) {
     err = microstep(&ctx);
     if (err == ERR_DONE)
       break;
     if (err != ERR_OK)
   return EXIT_FAILURE;
   }
   return EXIT_SUCCESS
}
```

The callbacks are not shown, but we did indeed implement them in order to pass the SCXML IRP tests.

10.4.2 Dequeuing Events

The first thing to do within the `microstep` function is to transition into the initial configuration if the state-machine is still pristine or to set the current event (Listing 10.5).

Listing 10.5 Initialization the state-chart and dequeing events

```
size_t i, j, k;
int err = ERR_OK;
char conflicts[TRANS_BIT_ARRAY] = TRANS_BIT_ARRAY_INIT;
char target_set[STATE_BIT_ARRAY] = STATE_BIT_ARRAY_INIT;
char exit_set[STATE_BIT_ARRAY] = STATE_BIT_ARRAY_INIT;
char trans_set[TRANS_BIT_ARRAY] = TRANS_BIT_ARRAY_INIT;
char entry_set[STATE_BIT_ARRAY] = STATE_BIT_ARRAY_INIT;
char tmp_states[STATE_BIT_ARRAY] = STATE_BIT_ARRAY_INIT;

if (ctx->flags & CTX_TOP_LEVEL_FINAL)
    return ERR_DONE;
if (ctx->flags == CTX_PRISTINE) {
    global_script(ctx, &states[0], NULL);
    bit_or(target_set, states[0].completion);
    ctx->flags |= CTX_SPONTANEOUS | CTX_INITIALIZED;
    goto ESTABLISH_ENTRY_SET;
}
```

```
20 if (ctx->flags & CTX_SPONTANEOUS) {
       ctx->event = NULL;
       goto SELECT_TRANSITIONS;
   }
   if ((ctx->event = ctx->dequeue_internal(ctx)) != NULL) {
25     goto SELECT_TRANSITIONS;
   }
   if ((ctx->event = ctx->dequeue_external(ctx)) != NULL) {
       goto SELECT_TRANSITIONS;
   }
```

Each iteration of a micro-step starts by allocating memory on the stack for all variables required in the function's scope as is required in ANSI C. Afterwards, we check to see whether the state-chart already entered a top-level final state (line 10–11), which signifies the end of all processing. If this is not given, we check whether the state-chart is still in pristine condition, in which case we execute any eventual global script elements and set the target set to the root state's completion as defined for compound states in Sect. 10.3.1 before we continue processing with the completion of the target set as the entry set (line 13–18). Otherwise we establish the current event (Fig. 10.1 **(1a–c)**) as `NULL` if spontaneous transitions were not yet exhausted per `ctx->flags` or attempt to dequeue an event and continue to establish the optimal transitions set (line 20–29).

10.4.3 Selecting Transitions and Establishing the Exit-Set

The next activity is to establish the optimal transition set (Fig. 10.1 **(2)**). This is a crucial section of the algorithm as it will be executed at least twice for any non-null event. The corresponding pseudo-code from Appendix D in the SCXML recommendation is rather obscure and very elaborate. For the ANSI C implementation in Listing 10.6, a single iteration of all transitions is sufficient with the majority of iterations skipped very early.

Listing 10.6 Establishing the optimal transition set

```
SELECT_TRANSITIONS:
for (i = 0; i < NUMBER_TRANSITIONS; i++) {
    if (transitions[i].type & (TRANS_HIST | TRANS_INITIAL))
        continue;

if (BIT_HAS(transitions[i].source, ctx->config)) {
    if (!BIT_HAS(i, conflicts)) {
        if (ctx->is_enabled(ctx, &transitions[i], ctx->event) > 0) {
            bit_or(conflicts, transitions[i].conflicts);
            bit_or(target_set, transitions[i].target);
            bit_or(exit_set, transitions[i].exit_set);
            BIT_SET_AT(i, trans_set);
            ctx->flags |= CTX_TRANSITION_FOUND;
```

```
            }
15      }
    }
}
```

To understand this piece of the algorithm, it is important to realize that the transitions in `transitions` are sorted from a post-order traversal of all `<transition>` elements in the original SCXML document. This corresponds to the *priority* of a transition from the definition for the optimal transition set in Sect. 10.2: Transitions with the same source state are given in document order and transitions within descendant source states precede those in ancestor source states. This means that the first enabled transitions (t_1) is necessarily in the optimal transition set and its *conflicts* will exclude all other transitions that cannot form an optimal transition set with t_1 included. By iterating all other transitions, we can successively establish the optimal transition set by adding enabled transitions that are not conflicting and skipping those that are.

We start the iteration (line 2) and skip any `<transition>` elements that originate in an `<initial>` or `<history>` element (line 3–4) as we will handle them differently when completing the target-set as the entry-set below. For the remaining transitions, we check whether they are *active*, *non-conflicting*, and *enabled* (line 6–8). The order of these conditions is arbitrary, though, the check for the enabled status of a transition is potentially expensive and thus the last condition.

If all these conditions hold for the current transition, we add its conflicts to the set of conflicting transitions, its targets to the intermediate target-set and its exit-set to the complete exit-set (line 9–11). Finally we remember the transition as being part of the optimal transition set in `trans_set` (line 12) and the fact that we found a transition at all as flag in the context (line 13).

Finally, we need to intersect the optimal transition set's exit set with the active configuration to arrive at the set of states actually exited (Fig. 10.1 **(3a)**) and determine whether we need to perform another round of spontaneous transitions or dequeue an event in the next round (Listing 10.7).

Listing 10.7 Establishing the actual exit-set and determining whether spontaneous transitions are exhausted

```
bit_and(exit_set, ctx->config);
if (ctx->flags & CTX_TRANSITION_FOUND) {
    ctx->flags |= CTX_SPONTANEOUS;
    ctx->flags &= ~CTX_TRANSITION_FOUND;
} else {
    ctx->flags &= ~CTX_SPONTANEOUS;
}
```

Now we have already established (1) the optimal transition set, (2) the exit set, and (3) an intermediate target set that we will have to complete below.

10.4.4 Remembering the History

Before we can establish the missing entry-set as the completion of the target-set, we will have to process any <history> elements (Fig. 10.1 **(3b)**) whose parent states are in the exit-set as they might eventually be entered within the same micro-step again (Listing 10.8).

Listing 10.8 Remembering the history

```
1  REMEMBER_HISTORY:
   for (i = 0; i < NUMBER_STATES; i++) {
    if (STATE_MASK(states[i].type) == STATE_HIST_SHALLOW ||
        STATE_MASK(states[i].type) == STATE_HIST_DEEP) {
5       if (BIT_HAS(states[i].parent, exit_set)) {
            bit_copy(tmp_states, states[i].completion);
            bit_and(tmp_states, ctx->config);
            bit_and_not(ctx->history, states[i].completion);
            bit_or(ctx->history, tmp_states);
10      }
    }
   }
```

If control flow reaches the inner-most block, *i* contains the index of a history state whose parent is about to be exited within the current micro-step and we have to remember its history. We defined a history state's completion as the set of states that are *covered* by the history and are not already covered by a nested history (note that a state can still be covered by more than one history elements with the same parent state). To remember a histories active states, we set all bits from the histories completion within the temporary state bit-array (line 6–7). Then, we intersect the states covered by the history with the active configuration and reset the context's history with the new history for the states covered (line 8–10).

Here, the fact that we excluded those states already covered by nested histories from the histories completion will ensure that no states covered by more deeply nested history elements are reset. If they are to be reset, we will pass the respective history element in a later step of the iteration.

10.4.5 Establishing the Entry Set

The next activity to perform is to complete the target-set as the actual entry-set (Fig. 10.1 **(3c)**). To this effect we, again, first define the complete entry set and later intersect it with the non-active states to arrive at the actual entry-set. The first thing to realize is that if a state (proper or otherwise) is in the target-set, all its ancestors will necessarily be active in the next configuration. As such, we can just add all ancestors of states in the target-set (Listing 10.9) and, subsequently complete them.

Listing 10.9 Extending the target set with all ancestors

```
ESTABLISH_ENTRY_SET:
bit_copy(entry_set, target_set);
for (i = 0; i < NUMBER_STATES; i++) {
    if (BIT_HAS(i, entry_set)) {
        bit_or(entry_set, states[i].ancestors);
    }
}
```

To complete the target-set and its ancestors more efficiently, we have to make sure that the states in the completion of a given state *s* always succeed *s* in document order. This seems obvious but is actually not necessarily the case if we targeted an `<initial>` or `<history>` pseudo-state as their completion might be siblings. As such, we have to postulate that for all children of a given parent, all `<initial>` elements precede `<history>` elements precede proper `<state>`s. This is merely a syntactic transformation of the SCXML document that we will have to perform prior to establishing the array with all state structures above. If this is given, we can iterate the set of all states in document order and dispatch on their type to add their completion to the entry set (Listing 10.10).

Listing 10.10 Adding the completion of all states into the entry set

```
for (i = 0; i < NUMBER_STATES; i++) {
    if (BIT_HAS(i, entry_set)) {
        // mask the MSB with the HAS_HIST flag
        switch (STATE_MASK(states[i].type)) {
            ...
        }
    }
}
```

The actual completion of a state from the preliminary uncompleted entry-set depends on its type as follows:

- `case STATE_PARALLEL:`

```
bit_or(entry_set, states[i].completion);
break;
```

  If a `<parallel>` element is in the entry set, all of its child states will have to be in the complete entry set.
- `STATE_INITIAL:`

```
for (j = 0; j < NUMBER_TRANSITIONS; j++) {
    if (transitions[j].source == i) {
        BIT_SET_AT(j, trans_set);
        CLEARBIT(i, entry_set);
        bit_or(entry_set, transitions[j].target);
        for (k = i + 1; k < NUMBER_STATES; k++) {
```

```
              if (BIT_HAS(k, transitions[j].target)) {
                 bit_or(entry_set, states[k].ancestors);
              }
10          }
       }
  }
break;
```

If a transition or the completion of another state targeted an <initial> state, we search for its default <transition> and add the transition's target state and its ancestors to the complete entry-set. We do know that the initial transition's target state succeeds the initial state in document order (as we sorted the array of states accordingly), and can start the search for the target at the state succeeding the initial state (line 6).

- STATE_COMPOUND:

```
1 if (!bit_has_and(entry_set, states[i].children) &&
      (!bit_has_and(ctx->config, states[i].children) ||
       bit_has_and(exit_set, states[i].children)))
  {
5     bit_or(entry_set, states[i].completion);
      if (!bit_has_and(states[i].completion, states[i].children)) {
          for (j = i + 1; j < NUMBER_STATES; j++) {
            if (BIT_HAS(j, states[i].completion)) {
               bit_or(entry_set, states[j].ancestors);
10             break;
            }
          }
      }
  }
15 break;
```

When we encounter a compound state while completing the target set and its ancestors, we first have to check whether it is already complete (line 1–3) in which case we do not do anything. Otherwise, we add its completion and check (line 6) whether its completion is referencing a state more deeply nested (e.g., via an initial attribute into a non-child descendant), in which case we have to add the completion ancestors as well (line 8–11).

- case STATE_HIST_SHALLOW:
  case STATE_HIST_DEEP:

  Completing history states is the most complicated case as we have to account for various situations and take deep nested histories into account. We can differentiate two general cases first:

  – The history is empty:

```
for (j = 0; j < NUMBER_TRANSITIONS; j++) {
   if (transitions[j].source == i) {
     bit_or(entry_set, transitions[j].target);
```

```
     if(STATE_MASK(states[i].type) == STATE_HIST_DEEP &&
5       !bit_has_and(transitions[j].target, states[i].children))
     {
       for (k = i + 1; k < NUMBER_STATES; k++) {
         if (BIT_HAS(k, transitions[j].target)) {
           bit_or(entry_set, states[k].ancestors);
10         break;
         }
       }
     }
     BIT_SET_AT(j, trans_set);
15   break;
   }
 }
```

If we never before exited the history's parent state, we merely need to add its default transition to the transition-set (to process its eventual executable content later) and the default transition's target to the entry-set. If the history is `deep`, its default *default history configuration* may be a descendant of a sibling, in which case we have to add its ancestors as well (line 4–13). For `shallow` histories, the standard mandates that the target is a sibling of the history.

– We already remembered states for the history:

```
1 bit_copy(tmp_states, states[i].completion);
  bit_and(tmp_states, ctx->history);
  bit_or(entry_set, tmp_states);
  if (states[i].type == (STATE_HAS_HIST | STATE_HIST_DEEP)) {
5   for (j = i + 1; j < NUMBER_STATES; j++) {
      if (BIT_HAS(j, states[i].completion) &&
          BIT_HAS(j, entry_set) &&
          (states[j].type & STATE_HAS_HIST)) {
        for (k = j + 1; k < NUMBER_STATES; k++) {
10        if (BIT_HAS(k, states[j].children) &&
              (STATE_MASK(states[k].type) == STATE_HIST_DEEP ||
               STATE_MASK(states[k].type) == STATE_HIST_SHALLOW)) {
            BIT_SET_AT(k, entry_set);
          }
15      }
      }
    }
  }
```

In this case, we need to add the states we remembered earlier which are covered by the history to the entry-set (line 1–4). If the current history element has nested history elements (line 5) and their parents were added to the entry-set via our coverage (line 7–9), we need to add them as well, to be processed likewise in a later iteration (line 11–15). Here, we can again exploit the fact that they will necessarily succeed the current history element in document order and start iteration at the state succeeding the current history pseudo state.

Now we have all the sets in place to perform the actual transitions and call executable content in the following sections.

10.4.6 *Exiting States*

Listing 10.11 Exiting states in reverse document order

```
1  size_t i = NUMBER_STATES;
2  while(i-- > 0) {
3      if (BIT_HAS(i, exit_set) && BIT_HAS(i, ctx->config)) {
4        if (states[i].on_exit != NULL) {
5          err = states[i].on_exit(ctx, &states[i], ctx->event);
6          if (err != ERR_OK)
7            return err;
8        }
9        CLEARBIT(i, ctx->config);
10     }
11 }
```

To exit states during a microstep (Fig. 10.1 **(4a)**) is merely to iterate all states from the complete exit-set (line 1–2) that are active (line 3) in reverse document order, invoke their `on_exit` handlers (line 4–8), and remove them from the active configuration (line 9).

10.4.7 *Taking Transitions*

Listing 10.12 Taking transitions in document order

```
for (i = 0; i < NUMBER_TRANSITIONS; i++) {
    if (BIT_HAS(i, trans_set) &&
        (transitions[i].type & (TRANS_HIST | TRANS_INITIAL)) == 0) {
      if (transitions[i].on_transition != NULL) {
        err = transitions[i].on_transition(
                ctx,
                &states[transitions[i].source],
                ctx->event);
        if (err != ERR_OK)
          return err;
      }
    }
}
```

After we exited all states from the intersection of the complete exit-set with the currently active configuration, we need to perform any eventual executable content associated with transitions in the optimal transition set in document order

(Fig. 10.1 **(4b)**). We do iterate the array with the transition structures in a post-order sequence, though, the optimal transition subset of all transitions is implicitly ordered in document-order. This becomes clear if we consider that for a transition t_1 in the optimal transition set, no other transition t_2 in the optimal transition set can precede t_1 in post-order and succeed t_2 in document-order as it would never be optimally enabled with its source state being ancestrally related to the source of t_1.

In this step, we will not yet perform executable content associated with transitions whose parent is an `<initial>` or `<history>` element (line 3) as these are to be processed after the `<onentry>` elements of their parent states.

10.4.8 Entering States

As the last activity within a micro-step, we need to enter all states from the intersection of the complete entry-set with the negated active configuration (Fig. 10.1 **(4c)**). There are, however, quite some additional activities associated with the entering of states that are outlined in Listing 10.13 and detailed below.

Listing 10.13 Entering states in document order.

```
for (i = 0; i < NUMBER_STATES; i++) {
  if (BIT_HAS(i, entry_set) && !BIT_HAS(i, ctx->config)) {
    if (STATE_MASK(states[i].type) == STATE_HIST_DEEP ||
      STATE_MASK(states[i].type) == STATE_HIST_SHALLOW ||
        STATE_MASK(states[i].type) == STATE_INITIAL)
      continue;
    BIT_SET_AT(i, ctx->config);
    // 1. Initialize data
    // 2. Perform executable content for on_entry
    // 3. Process history and initial transitions
    // 4. Raise done events
  }
}
```

1. **Initialize Data**
   After we added the new state to the active configuration, we need to initialize its associated `<data>` elements if the document has a `late` data binding. We do keep a bit-array of states that were already initialized in the interpreter's context and did transform all `<data>` elements accordingly.

Listing 10.14 Initializing nested data elements for late data binding

```
if (!BIT_HAS(i, ctx->initialized_data)) {
    if (states[i].data != NULL && ctx->exec_content_init != NULL) {
        ctx->exec_content_init(ctx, states[i].data);
    }
    BIT_SET_AT(i, ctx->initialized_data);
}
```

(2) **Perform Executable Content**
To perform the executable content is merely to invoke the states `on_entry` callback function for the generated code as introduced in Sect. 10.3.5.

Listing 10.15 Calling executable content for the entry of states

```
if (states[i].on_entry != NULL) {
  err = states[i].on_entry(ctx, &states[i], ctx->event);
  if (err != ERR_OK)
    return err;
}
```

(3) **Process History and Initial Transitions**
When we completed the target-set as the entry-set above, we did remember all initial and history transitions that would have to be performed, but ignored them when we performed the transitions' executable content after exiting the states from the exit-set above. Their respective bits are still set in the transition-set and the standard mandates to invoke their executable content after the parent states on-entry handlers.

Listing 10.16 Calling executable content for history and initial transitions

```
for (j = 0; j < NUMBER_TRANSITIONS; j++) {
   if (BIT_HAS(j, trans_set) &&
       (transitions[j].type & (TRANS_HIST | TRANS_INITIAL)) &&
       states[transitions[j].source].parent == i) {
     if (transitions[j].on_transition != NULL) {
       err = transitions[j].on_transition(ctx,
                                           &states[i],
                                           ctx->event));
       if (err != ERR_OK)
         return err;
       }
   }
}
```

(4) **Raise Done Events**
Special considerations have to be given when entering <`final`> states as part of a microstep.

```
if (STATE_MASK(states[i].type) == STATE_FINAL) {
   ...
}
```

If the parent of the final state is the <`scxml`> element itself, the interpreter is done and we set the `CTX_TOP_LEVEL_FINAL` flag in the interpreter's context (Listing 10.17).

Listing 10.17 Top-level final state reached

```
if (states[i].ancestors[0] == 0x01) {
  ctx->flags |= CTX_TOP_LEVEL_FINAL;
}
```

Otherwise, if the final state is the child of a compound state, we need to raise a `done.state.[ID]` event on the interpreter's internal queue and attach any eventual `<donedata>` with the event (Listing 10.18).

Listing 10.18 Final state of a compound state entered

```
else {
  const elem_donedata* donedata = &elem_donedatas[0];
  while(ELEM_DONEDATA_IS_SET(donedata)) {
    if unlikely(donedata->source == i)
      break;
    donedata++;
  }
  ctx->raise_done_event(ctx,
                        &states[states[i].parent],
              (ELEM_DONEDATA_IS_SET(donedata) ? donedata : NULL));
}
```

In this last case, we also need to check whether the current final state is the last one to finalize all children of a parallel ancestor, in which case we need to raise a `done.state.[PARALLEL_ID]` event in addition (Listing 10.19).

Listing 10.19 Raising done events for finalized parallel states

```
1  for (j = 0; j < NUMBER_STATES; j++) {
     if (STATE_MASK(states[j].type) == STATE_PARALLEL &&
         BIT_HAS(j, states[i].ancestors)) {
       bit_and_not(tmp_states, tmp_states);
5      for (k = 0; k < NUMBER_STATES; k++) {
         if (BIT_HAS(j, states[k].ancestors) &&
             BIT_HAS(k, ctx->config)) {
           if (STATE_MASK(states[k].type) == STATE_FINAL) {
             bit_and_not(tmp_states, states[k].ancestors);
10         } else {
             BIT_SET_AT(k, tmp_states);
           }
         }
       }
15     if (!bit_any_set(tmp_states)) {
         ctx->raise_done_event(ctx, &states[j], NULL);
       }
     }
   }
```

We start by iterating all states and search for parallel states which are ancestrally related to the current final state (line 1–3). If we found such a state, we clear out the temporary bit array of states to remember any active descendant of the parallel (line

5–7). If we found an active state in the descendants of the parallel state under consideration and it is a final state itself, we clear all its ancestors in the temporary bit-array, if it is anything else, we set its ancestor's bits. After we processed all active descendant states of the parallel in this manner and the temporary bit-array is empty (line 15–17), all of the parallel's child states have also entered a final state and we need to raise the `done.state.[PARALLEL_ID]` event for the parallel state.

This concludes the description of the `microstep(T)` algorithm in ANSI C and we will evaluate its performance and memory consumption in the following sections.

10.5 Evaluating the ANSI C Implementation

In this section, we will evaluate the ANSI C algorithm presented above with regard to its runtime, binary size, and memory consumption. As a baseline, we took the `microstep(T)` implementation from our `uSCXML` implementation, which is relevant as it, rather literally, employs the pseudo-code from Appendix D of the SCXML recommendation. Though, even with this baseline implementation, we already employ some caching, e.g. for state look-ups by identifier, the exit- and target-set of transitions and proper ancestors of two states. As such, it establishes a lower bound for any implementation that approaches the `microstep(T)` algorithm as specified in the recommendation. We are aware that it was never the intention of said pseudo-code to be performant or small, but many SCXML interpreters do, indeed, implement the `microstep(T)` algorithm very similarly.

10.5.1 Methodology

For all our measurements, we transformed all SCXML IRP tests for the ECMAScript datamodel and generated the compound data structures as introduced above. We wrote the callback functions as required for the SCXML context connecting to the respective functionality in the `uSCXML`[2] interpreter and explicitly excluded:

- 37 tests due to missing support for the <invoke> element.
- 17 tests due to missing support for anything but the SCXML I/O processor.
- 4 tests that attempt to retrieve data from a URL.
- 1 test with an XML node in a variable.
- Some more manual tests.

[2]https://github.com/tklab-tud/uscxml (accessed January 26th, 2015).

This set of tests forms the basis for all subsequent measurements below. All measurements are taken on a MacBook 13" (early 2015) with Intel Core i7 CPU @ 3.1 GHz. While this is not exactly a *resource constrained device*, the actual values measured give every reason to assume that the implementation is perfectly suited for resource constrained devices.

10.5.2 Compliance

The set of SCXML IRP tests that is passed by our implementation for the ECMAScript data-model is given in Table 10.1. Do note that we pass all tests for core constructs but the one for invocation order (`test422`) as we did not implement `<invoke>` yet. Even though the tests are merely an enumeration of correct behavior for a compliant interpreter and no proof of compliance, it is a good indicator of a *largely correct* implementation of `microstep(T)`.

We also wrote initially failed and ultimately passed three additional tests for deep completions via the initial attribute and nested history pseudo-states (deep and shallow) to account for some border cases we realized when designing the algorithm above.

Table 10.1 Number of tests in the SCXML Implementation and Report Plan with corresponding section from specification

Class	#Pass	#Total	Class	#Pass	#Total
Core constructs		*40 (1)*	*Data model and manipulation*		*50 (4)*
General	2	2	Data	5	7
State	1	1	Assign	4	4
Final	2	2	Donedata	1	1
OnEntry	2	2	Content	3	3
OnExit	2	2	Param	3	3
History	4	4	Script	3	4 (1)
Events	4	4	Expressions	7	8 (3)
Transition selection	22	23 (1)	System variables	19	20
Executable content		*13*	*External communications*		*51 (3)*
Raise	1	1	Send	16	19 (1)
If	3	3	Cancel	2	3
Foreach	7	7	Invoke	0	29 (2)
Evaluation	2	2	*Data models*		*51*
Event I/O processors		*28 (1)*	NULL	1	1
SCXML	10	16	ECMAScript	15	20
Basic HTTP	0	12 (1)	XPath	0	30
			Total	140	*233 (9)*

Brackets indicate manual tests

10.5.3 Performance

For the performance measurements, we instrumented the code-base with timers using `mach_absolute_time` as the highest precision monotonic clock available. It is difficult to get any reliable information about its precision, accuracy or resolution. However, an example in the official technical QA1398[3] from Apple does convert its return value into nanoseconds, suggesting a sufficient granularity for the measurements. Furthermore, all measurements were averaged over 1.000 iterations and the methodology was the same for the baseline. Still, the approach of measuring the performance of a given piece of code by averaging its runtime is far from objective as seemingly unrelated changes in the runtime can have a considerable effect on the measurements [7]. As such, the numbers below are to be interpreted with some reservations.

Using this approach, we were able to measure the performance of 132 individual tests, with the remaining 8 tests relying on the timeout of an event, which prevented us from measuring. We did measure the time to completion for a single interpretation per test excluding and subtracted the time spent in the data-model's functions. The difference was divided by the number of iterations for the `microstep(T)` algorithm described above.

Figure 10.7 depicts a distribution for the average duration of such an iteration per SCXML IRP test with 5 us bins. We can see that for the majority of tests, their microsteps averaged to about 5–15 us, which translates to `650.000 - 2.000.000`

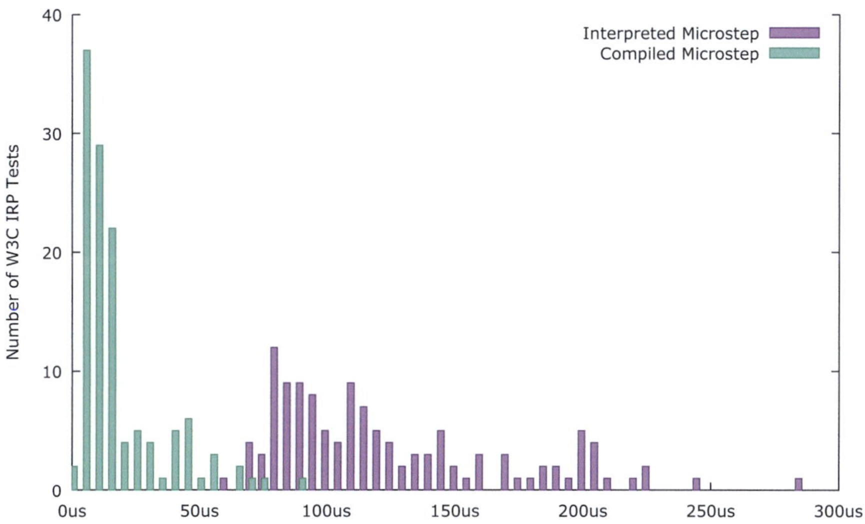

Fig. 10.7 Distribution of the execution speed of a single microstep for the interpreted and compiled case (averaged per SCXML IRP test)

[3]https://developer.apple.com/library/mac/qa/qa1398/_index.html (accessed January 26th, 2016).

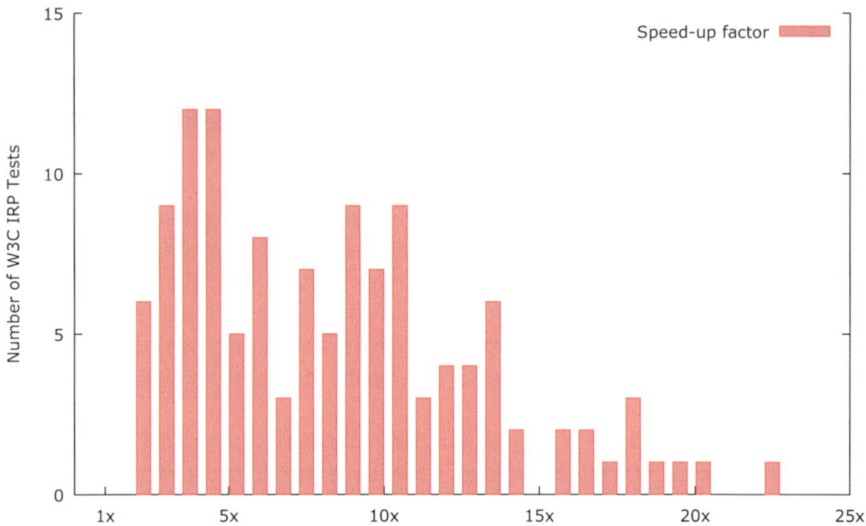

Fig. 10.8 Distribution of the speed-up for a single microstep (averaged per SCXML IRP test)

iterations per second or, at most, `300.000-1.000.000` events per second (when assuming no spontaneous transitions).

We also did a direct comparison for the average duration of a microstep iteration per test and Fig. 10.8 depicts a distribution of the speed-up factor when using the algorithm described above compared to the baseline. We can see that the proposed algorithm always outperforms a more literal implementation of the pseudo-code from Appendix D in the SCXML recommendation and, on occasion, is more than 20 times as fast.

10.5.4 Binary Size and Memory

An important consideration when targeting a resource constrained platform is the memory available. For example, the ATmega8 from Atmel only features 8 KB of flash memory with 512 Byte SRAM for dynamic data, its more powerful counterparts up to 256 KB flash memory with 8 KB SRAM. As such, a compact representation for the logic representing the control flow from the SCXML document has a direct consequence for its applicability in this domain. A major problem in this regard is the employed data-model: A single instance of the JavaScriptCore ECMAScript implementation will, regardless of actual usage, allocate 8 MB of memory upon instantiation on top of its already considerable binary size; orders of magnitude more than what would be available on an ATmega8. One scripting language explicitly touted for scripting on resource constrained devices is Lua with a binary size of round about 80 KB and very conservative memory usage and, indeed, the `uSCXML` platform does support a Lua datamodel.

While the SCXML recommendation does provide a normative specification for an ECMAScript data model and a supplementary W3C note for an XPath data model, there is no mandatory requirement for a compliant interpreter to implement either. This offers considerable flexibility, but comes with the cost of reduced interoperability.

For our measurements, we explicitly excluded the size of the data-model. If one were to seriously target a resource constrained platform, ultimately, a data-model that can syntactically transformed onto ANSI C seems most suited as it can directly be subjected to the compiler for the respective platform without any requirement for runtime interpretation. As such, we only measured the size of the compiled control flow logic with all required static data and executable content functions introduced above and excluded anything linked from the user-supplied callback functions. The distribution of binary sizes for the 140 IRP tests from Table 10.1 is given in Fig. 10.9.

Two distributions for different compiler switches are displayed. When optimizing for speed (`-Ofast`), the resulting binaries will be anywhere from 3 to 6 KB. When optimizing for size, the resulting binaries are round about the same size, at times even somewhat larger. With the possible options for reducing the required memory by dropping some of the bit-arrays in the static data structures introduced in Sects. 10.3.2 and 10.3.1, this size is perfectly suited to run on a device with 8–16 KB of memory. However, as it is, the size for compiled binaries grows quadratic with the size of the input SCXML document as each additional state or transition will increase each relation modeled in the bit-arrays (cf. Figs. 10.3 and 10.5).

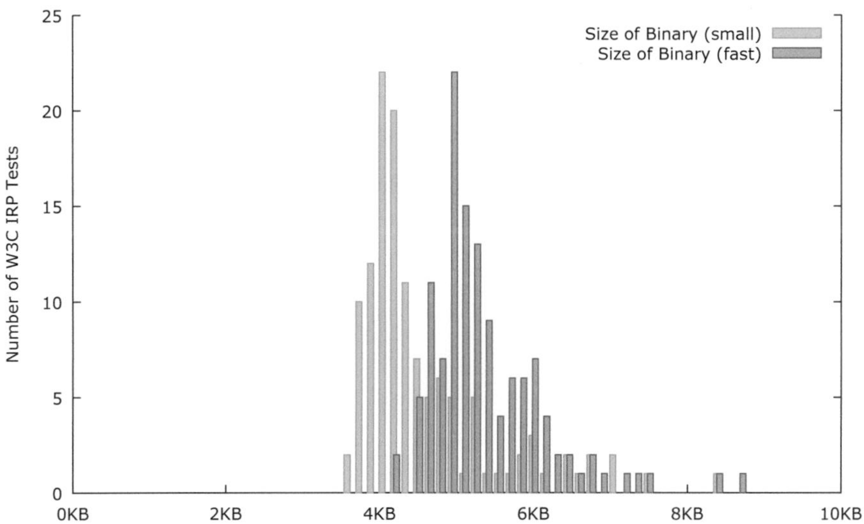

Fig. 10.9 Distribution of the 16-bit binary size for the compiled SCXML IRP tests (state-transitioning and executable content calls only, horizontal lines denote base size for *empty* statechart)

With regard to the dynamic memory requirements at runtime, it is noteworthy that the algorithm above does, at no point, allocate memory on a heap structure (`malloc`) and all allocations are performed on the stack. The sum of memory required per iteration depends on the number of states and transitions in the original SCXML document. If we assume an original SCXML document with 50 states and transitions each, we can calculate its dynamic memory requirements as follows: Each instance of an interpretation will require round about 90 bytes for its SCXML context structure (Fig. 10.6) and every iteration of a microstep will instantiate

- **3 unsigned integer variables** as indices during iteration for a total of 6 bytes on a 16-bit architecture.
- **2 bit-arrays for transitions set**, namely for the optimal transition set in `trans_set` and for conflicting transitions in `conflicts`, amounting to 2 ×`ceil`(NUMBER_TRANS/8) bytes.
- **4 bit-arrays for transitions set**, namely for the set of states targeted by transitions in the optimal transition set as `target_set`, the entry-set in `entry_set` and the exit-set in `exit_set`. One more bit array is allocated as a general, temporary bit-array `tmp_states` and used to remember and reenter history states and when raising the `done.state.[PID]` event for parallel states. This amounts to a total of 4 ×`ceil`(NUMBER_STATES/8) bytes.
- **A single additional byte** for the return value in `err`.

If we, again, assume a state-chart with 50 transitions and states, any bit-array will consist of 7 bytes for a total of 49 bytes allocated on the stack per invocation (not accounting for alignment padding). The development for the memory requirements of static and dynamic memory is depicted in Fig. 10.10, excluding memory for code and additional elements other than transitions and states.

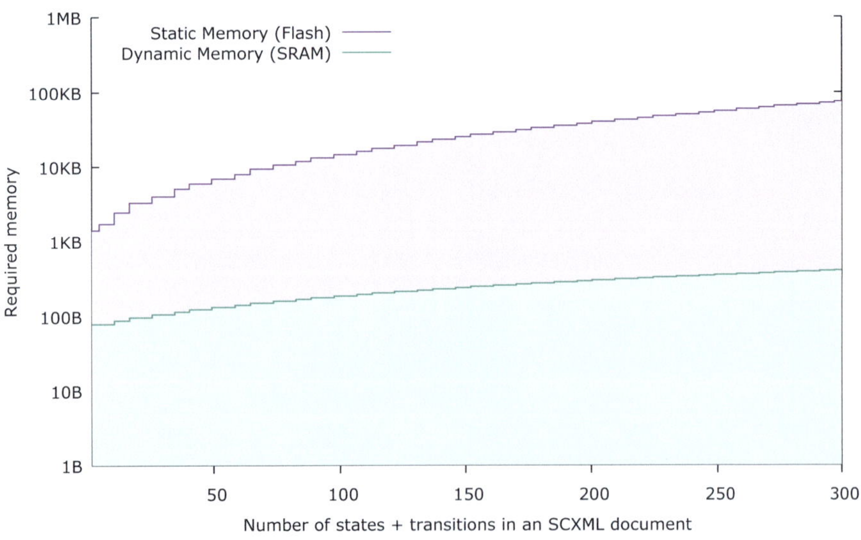

Fig. 10.10 Static memory required for the data structures (transitions and states) and dynamic memory (context and microstep stack)

10.6 Transformation for VHDL

While the C implementation of `microstep(T)` described above will already allow to address a wide range of off-the-shelf micro-controllers, a single iteration will still require tens of thousands of cycles. In this section we present a hardware realization for a subset of SCXML state-charts and discuss its possible performance. To this effect, we will not generate C code, but descriptions for hardware building blocks, expressed in the widely used hardware description language VHDL. With VHDL, it is possible to program FPGA logic blocks for dynamic hardware state machines and even to design custom ASICs.

The general description is, again, aligned with the set of steps depicted in Fig. 10.1 and based on the pre-calculated sets and relations already introduced as part of the C implementation above. While we already excluded the `<invoke>` element, custom I/O processors and some other features for the description of the C algorithm, the domain of the VHDL transformation will be even more restricted:

- We will not concern ourselves with any data-model, but only describe the transitioning of active configurations and the corresponding entry-, exit-, and transition-sets.
- We do not yet address the semantic of the `<history>` element nor are `<initial>` elements supported. The `initial` attribute is supported though.
- Events are enumerated and expressed as individual lines. Any data attached to an event would be inaccessible anyway as we do not provide a data-model.
- No executable content other than `<raise>` and `<send>` are supported and these can only address simple events to either the internal or external event queue.

The general architecture of the hardware realization is depicted in Fig. 10.11 and consists of a *microstepper* with an attached *event controller*. Each microstep is performed in a single cycle and several outputs are available

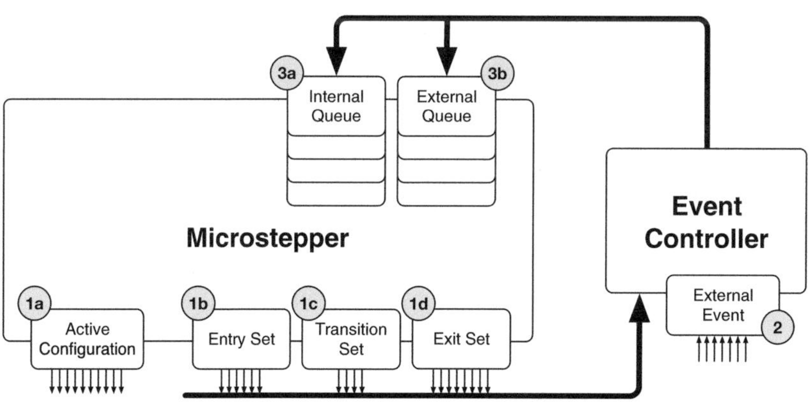

Fig. 10.11 A high level overview of the generated hardware architecture

- The **Active Configuration (1a)** provides the valuation of active states in the current configuration. It's realized as a parallel bus, where every line signifies the activation status of one state, indicating an active state with HIGH and an inactive state with LOW. The bus width equates to the number of proper states in the SCXML document.
- The **Entry Set (1b)** pins provide the information, which states were in the entry-set when the microstepper transitioned to the active configuration by setting the corresponding pin to signal HIGH. To save some of the rare I/O pins, we just generate pins for states, that have defined an <onentry> child element.
- Analogously, the **Exit States (1d)** pins provide the information, which states were in the micro-step's exit-set by setting the corresponding pin to signal HIGH. Again, we just generate pins for states, that have an <onexit> child element.
- The **Transition Set (1c)** pins provide the information, which <transition> elements with executable content were in the optimal transition set.
- For the **Internal Queue (3a)** and the **External Queue (3b)**, the microstepper offers a writing interface, that provides enough pins to differentiate the individual events specified in the SCXML document.

  Since we cannot, in the general case, give an upper-bound for the maximum length of either event queue at transformation time [8], it is important to take carenot to overflow them. If an event is about to be enqueued on a full queue, the microstepper will, for now, just assume an error state readable through an interface pin.

The event-controller will, depending on the occurrence of <raise> and <send> elements in executable content, deliver these events in accordance with the valuation of the entry-, exit-, and transition-set bus. It is also available to, asynchronously, deliver additional external events not enqueued by the state-chart itself **(2)**.

Figure 10.12 illustrates the inside architecture of the microstepper component. It mainly consists of the event queues **(3a–b)** and an elaborate Moore state machine **(4a–c)**, to perform the actual micro-steps. The most relevant parts of the state machine are the transition logic **(4a)** and the state memory **(4b)**. These are described more detailed in the following chapter. As the current state configuration and relevant sets are available as interface busses, the output logic **(4c)** is just responsible for setting the completed signal, which indicates that the state-chart is in a top-level final state.

With the general architecture of the hardware established, we can now describe its actual implementation to realize the steps from Fig. 10.1 in more detail.

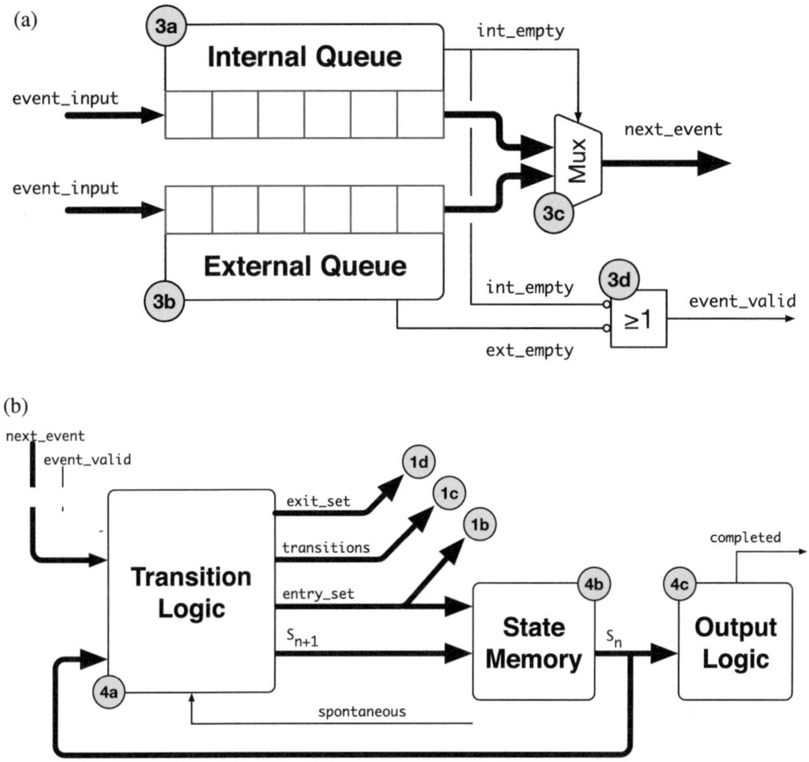

Fig. 10.12 Architecture of the microstepper. (**a**) Internal and external event queue with bus selection. (**b**) Finite-State-Machine Implementation

10.6.1 Dequeuing Event

The logic depicted in Fig. 10.12 **(3a–d)** shows how non-null event dequeuing is implemented: If there is an event enqueued at the internal queue, the `int_empty` signal is `LOW` and the multiplexer **(3c)** connects the `next_event` bus to the internal queue or external queue otherwise. If both the `int_empty` and `ext_empty` signals are `HIGH`, the `event_valid` signal goes to `LOW` to indicate that no events are available. Both signals along with the `spontaneous` signal from the state memory can, subsequently, be used to perform transition selection.

10.6.2 Selecting Transitions

Just as with the C implementation, the next step is to establish the optimal transition set for the current event (Fig. 10.13). We have already described, in the scope of the

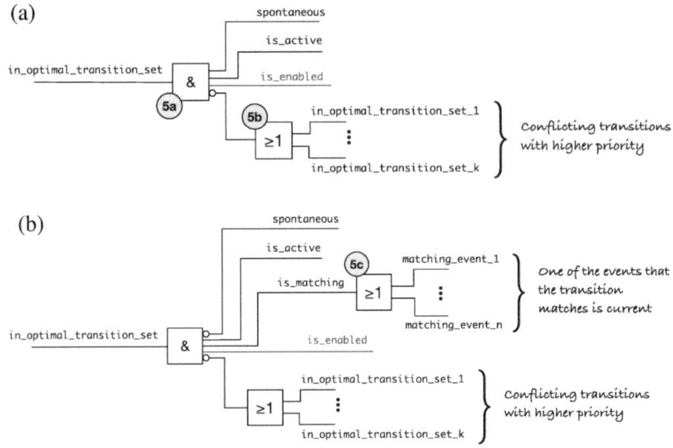

Fig. 10.13 Establishing the optimal transition set. (**a**) Transition selection for spontaneous transitions. (**b**) Transition selection for non-spontaneous transitions

C implementation above, how we can employ a post-order traversal of all transitions to have higher priority transitions precede those with a lower priority. Furthermore, we did introduce a *conflicts* relation of transitions to prevent the selection of invalid transition sets. Both are also relevant to select the optimal transition sets with dedicated hardware.

If the last micro-step did not exhaust spontaneous transitions, the `spontaneous` is still set to `HIGH` in the state memory and the logic in Fig. 10.13a is applied. For any given spontaneous transition, we will set its `in_optimal_transition_set` line to `HIGH` **(5a)** if its parent state is active as per configuration in the state memory and no other spontaneous transition with a higher priority conflicts **(5b)**. Here, we can just connect all `in_optimal_transition_set` lines for conflicting transitions with a higher priority as they are known at transformation time. We also included an eventual `is_enabled` signal, which would need to be set by some external component that would realize the data-model.

The case for non-spontaneous transitions, selected for a non-null event, is very similar, but an enumeration of matching events would need to set the `is_matching` signal **(5c)** to `HIGH` as well. This will give us the valuation of signals for the external interface bus at **(1c)** above.

10.6.3 Establishing the Exit-Set

When we identified the set of `in_optimal_transition_set` signals that are set to `HIGH` for transitions in the optimal transition set, we can instantaneously establish the exit-set of the current micro-step. In the scope of the C implementation, we argued that we can identify a transition's complete exit-set as the exit-set

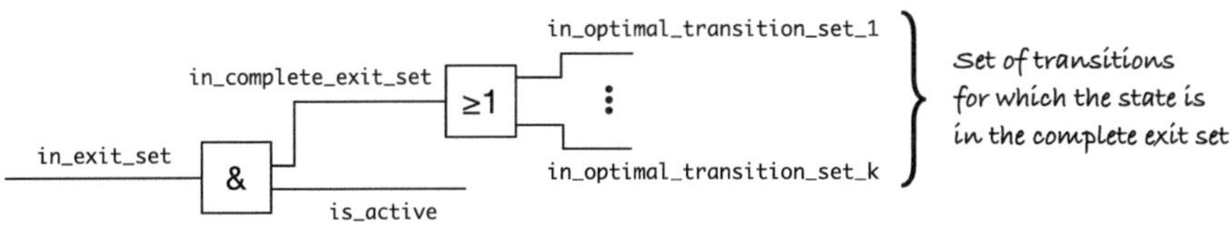

Fig. 10.14 Establishing the exit-set by intersecting the complete exit-set with the active configuration

when assuming the complete configuration. Now, if we intersect the complete exit-set with the active configuration, we arrive at the micro-step's actual exit-set (Fig. 10.14) and can set the respective signals to high on the external interface bus **(1d)**.

10.6.4 Establishing the Entry-Set

As we do not support `<history>` elements yet, the next step is to establish the entry-set of the current micro-step. This is by far the most complicated step, but by regarding the implementation in C, we can gain some insights that help us to understand the respective logic.

In the C implementation, there were three general situations for any given proper state to become part of the complete entry set:

1. The state is **targeted directly** by a transition in the optimal transition set.
2. The state is added as an ancestor of a targeted state (**ancestor completion**).
3. The state is added as the completion of a parent state (**descendant completion**).

If any state is targeted directly, it will set its `in_complete_entry_set_up` signal to `HIGH`, which causes ancestor completion for all its ancestor states (Fig. 10.15a, b). This signal is received by composite parent states and recursively passed to their parents causing all targeted states and their ancestors to have the respective signal set to `HIGH`. In order to arrive at a valid completion, any composite states added via ancestor completion will have to be completed as well (descendant completion). Composite parents of type `PARALLEL` will, unconditionally, add all their child states to the complete entry set (Fig. 10.15c). Composite states of type `COMPOUND` are more complicated: They will only need to be completed if they are not already complete, that is, if none of their children are already active and not exited during the current micro-step (Fig. 10.15d) and the given child state is the default completion per document order or `initial` attribute.

This will, recursively, establish the complete entry set which has to be intersected with the set of states that are active and not in the exit-set (Fig. 10.16) to arrive at the actual entry set for the external interface bus at **(1b)**.

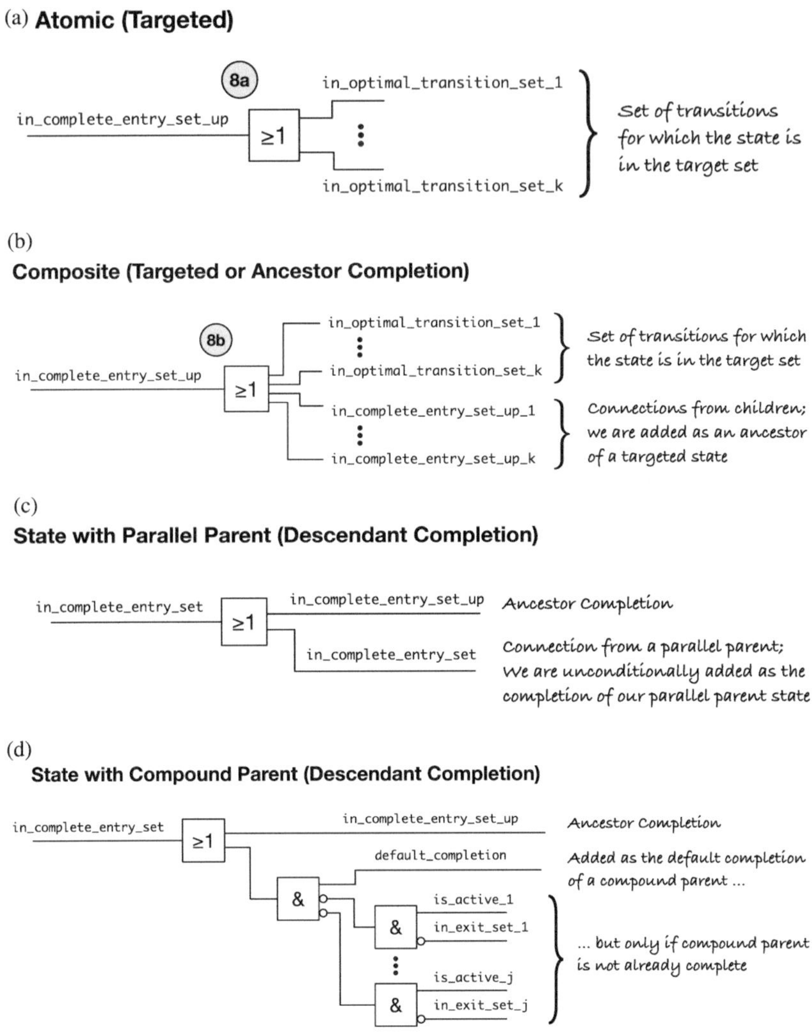

Fig. 10.15 Completing the target set as the complete entry set. (**a**) An atomic state added by being targeted directly. (**b**) A composite state added by being targeted directly or via ancestor completion. (**c**) A state added by its parallel parent state. (**d**) A state added as the default completion of a compound parent state

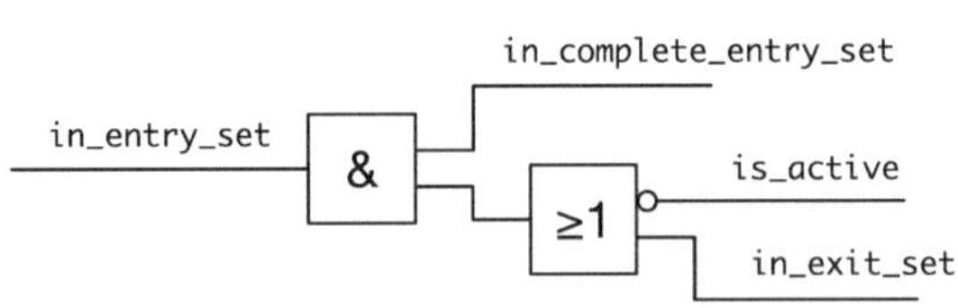

Fig. 10.16 Establishing the entry-set by intersecting the complete entry-set with the subset of the active configuration that is not exited

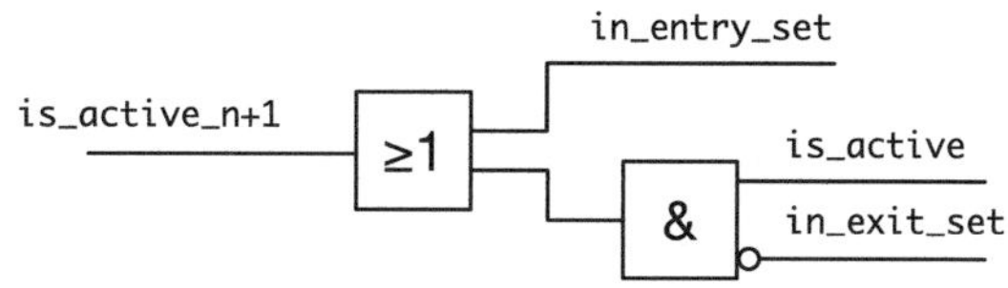

Fig. 10.17 The new active configuration is the set of states already active and not exited together with the set of states entered

10.6.5 Observable Performance

In the C implementation above, the preparations above triggered the actual exiting, transitioning, and entering of states. For the hardware implementation, these activities are to be performed by components connected to the external interface bus and dispatching over the various sets. As such, we just need to set the follow-up configuration (Fig. 10.17) as the set of states already active and not exited together with the set of states entered in the state-memory and process the next microstep.

10.6.6 Evaluation

The transformation from SCXML to VHDL is still very raw with only a select few language features implemented. As such, it is futile to evaluate its compliance with regard to the SCXML IRP tests. We did, however, write two simple SCXML statecharts that we successfully simulated to pass.

The first test in Listing 10.20 employs a parallel state with an atomic and a compound child state, with the compounds child, in turn, an atomic event raising an event upon entry on the internal queue that matches a transition in the other atomic state.

Listing 10.20 VHDL test for a parallel state with nested compound state

```
<parallel id="p1">
    <state id="p1.1">
      <transition event="foo" target="pass"/>
    </state>
   <state id="p1.2">
   <state id="p1.2.1">
  <onentry>
    <raise event="foo" />
    </onentry>
    </state>
  </state>
 <final id="pass" />
 <final id="fail" />
</parallel>
```

The second test in Listing 10.21 relies on transition preemption of the first transition by the more deeply nested second transition to pass.

Listing 10.21 VHDL test for transition preemption

```
<state id="s1">
  <transition event="foo" target="fail"/>
  <state id="s1.1">
    <onentry>
      <raise event="foo" />
    </onentry>
    <transition event="foo" target="pass"/>
  </state>
  <final id="pass" />
  <final id="fail" />
</state>
```

Ultimately, it is definitely desirable to get a more rigorous evaluation of the VHDL description's compliance and we are confident that, by aligning the VHDL description with the C implementation, we will be able to pass a similar subset.

10.6.6.1 Performance

Since all dynamic functions such as transition-, entry-, and exit-set generation or the calculation of the next configuration are implemented via combinatorial logic, the hardware performs one microstep per clock cycle. The maximum frequency for such a hardware component depends on several properties of the original SCXML state-chart and the hardware employed:

- **Critical Path:** The critical path is the longest combinatorial path in the design. A clock cycle has to be long enough, for a signal to pass through this path and stabilize. In our implementation it highly depends on the interleaving depth of the state machine and, as such, the complexity of the original SCXML document.
- **Hardware Specifics:** In particular the signal propagation time, which depends, among other things, on fabrication node and core voltage, is an important factor for the pass-through time of the longest path.

We expect, in any case, that the state controller's speed will not be the limiting factor for the overall system, external components like sensors and actuators are orders of magnitude slower and would stall the microstepper most of the time.

10.6.6.2 Hardware Costs

If we are to mold the VHDL description into an ASIC, we need to estimate the chip area required for the various transistors in our solution. Since this number depends on many factors we present a worst case estimation, wherein we treat every state in the SCXML document as an atomic state, which is the default completion of a `COMPOUND` state. We will further assume all of these states to have two transitions event driven transitions, and to be the target of two transitions each. This hypothetical set of states represents the worst “transistors per transition” relation.

From this assumptions we get

- `04` transistors for the buffer to save the state,
- `16` transistors for the atomic state function,
- `06` transistors for the activation set of the atomic state,
- `12` transistors for the exit set and the exclusive exit line,
- `06` transistors for the additional pins at the parent gates,
- `18` transistors for the transition function,
- `16` transistors for the interface buffers for the sets on the external interface bus,

for a total of 78 transistors per state. For a comparison, an Intel i7 Haswell-E has around 2.6 billion transistors.

If we are to implement the VHDL on an FPGA we need to estimate the required logic cells and flip-flop memory cells. Since an FPGA can build logic cells via its architecture, most vendors give “logic cell equivalent” numbers for their products. For our assumed worst case scenario above, we get

- `01` flip-flop to save the state,
- `03` gates for the atomic state function,
- `01` gates for the activation set of the atomic state,
- `02` gates for the exit set and the exclusive exit line,
- `03` gates for the additional pins at the parent gates,
- `05` gates for the transition function,
- `05` flip-flops for the interface buffers for the sets on the external interface bus,

for a total of 20 logic cells per state. As a consequence, a Xilinx Spartan 6 SLX9 would hold about 450 states. We choose this FPGA as comparison, because it is the smallest FPGA which could hold the AX8 as a VHDL description of the AVR architecture, which would be able to run the C implementation.

10.7 Conclusion

In this chapter we presented two implementation of the `microstep(T)` algorithm, central to every SCXML interpreter, one in ANSI C, another in VHDL. In the scope of the ANSI C implementation, we introduced several sets and relations that can be derived syntactically from a given SCXML document along with a few important observations:

- Most of the criteria for an optimal transition sets can be derived syntactically and encoded in a static $\texttt{conflicts}(t_1, t_2) \subseteq T \times T$ relation to identify pairs of transitions that can, for whatever reason, never be part of the same optimal transition set.

- The post-order traversal sorting for transitions is equivalent to the *priority* of a transition. Together with the `conflicts(t1,t2)` relation, this allows to identify the optimal transition set in a single iteration of transitions with most steps skipped very early.
- The complete exit set of a transition, as a superset of the actual exit set can be calculated at transformation time. The actual exit set is the intersection of this complete exit set with the active configuration. This notion extends to sets of transitions, i.e. the optimal transition set.
- The same is true for the complete entry set and the actual entry set of a transition set.
- Sorting the states such that the states in a given state's completion will always succeed the given state in document order allows to identify the entry set in a single iteration after we identified the target sets' ancestors.
- All of a state-chart's history can be encoded in a single bit per state.

By exploiting these techniques, we were able to improve the performance for a `microstep(T)` implementation considerably. Along with a transformation of an SCXML document onto a set of native data-structures, we managed to provide semantically equivalent object code with a size suitable to be deployed for even the smallest of micro-controllers.

The insights gained from the ANSI C implementation were subsequently applied for a transformation from SCXML onto VHDL to implement SCXML as dedicated hardware elements, be it by programming FPGAs or even to mold custom ASICs on a die. Even though the VHDL transformation was not evaluated with the same scientific rigor as the ANSI C implementation, we are confident that it provides an excellent starting point to support a larger set of SCXML language features.

Acknowledgements This work was supported by Institute for Information & communications Technology Promotion (IITP) grant funded by the Korea government (MSIP) (No. B0101-15-1292, Development of Smart Space to promote the Immersive Screen Media Service) and by the German Research Foundation (DFG) as part of the projects B02 within the Collaborative Research Center (CRC) 1053 – MAKI.

References

1. Barnett, J., Bodell, M., Dahl, D., Kliche, I., Larson, J., Porter, B., et al. (2012). *Multimodal architecture and interfaces*. W3C recommendation, W3C. http://www.w3.org/TR/2012/REC-mmi-arch-20121025/.
2. Berjon, R., Faulkner, S., Leithead, T., Pfeiffer, S., O'Connor, E., & Navara, E. D. (2014). *HTML5*. Candidate recommendation, W3C. http://www.w3.org/TR/2014/CR-html5-20140731/.
3. Crane, M. L., & Dingel, J. (2005). *On the Semantics of UML State Machines: Categorization and Comparison*. Technical Report 2005-501, School of Computing, Queen's.
4. Harel, D. (1987). Statecharts: A visual formalism for complex systems. *Science of Computer Programming, 8*(3), 231–274.

5. Harel, D., Pnueli, A., Schmidt, J. P., & Sherman, R. (1987). On the formal semantics of statecharts (extended abstract). In *Proceedings of the Symposium on Logic in Computer Science*, Ithaca, NY, USA (pp. 54–64).
6. Hosn, R., Carter, J., Burnett, D., Lager, T., Barnett, J., Raman, T., et al. (2015). *State chart XML (SCXML): State machine notation for control abstraction*. W3C recommendation, W3C. http://www.w3.org/TR/2015/REC-scxml-20150901/.
7. Mytkowicz, T., Diwan, A., Hauswirth, M., & Sweeney, P. F. (2009). Producing wrong data without doing anything obviously wrong! *ACM Sigplan Notices, 44*(3), 265–276.
8. Radomski, S. (2015). *Formal Verification of Multimodal Dialogs in Pervasive Environments*. Ph.D. thesis, Technische Universität Darmstadt. http://tuprints.ulb.tu-darmstadt.de/5184/
9. von der Beeck, M. (1994). A comparison of statecharts variants. In H. Langmaack, W. P. de Roever, & J. Vytopil (Eds.), *Formal techniques in real-time and fault-tolerant systems. Lecture notes in computer science* (Vol. 863, pp. 128–148). Berlin, Heidelberg: Springer.

Chapter 11
Standard Portals for Intelligent Services

Deborah A. Dahl

Abstract Some multimodal interpretation services natively support W3C multimodal standards, but most still use their own proprietary formats and protocols. This makes it much more difficult for developers to use different systems because they have to learn and program to a new API for each vendor. This paper describes how standards-based servers can wrap proprietary systems in the W3C MMI Architecture and EMMA 2.0 to allow developers to interact with modality interpretation services in a standard way, even if the service that they are using does not natively support the standards.

11.1 Introduction

Multimodal technology that supports forms of input (modalities) such as natural language processing, speech recognition, handwriting recognition, and object recognition from images is becoming increasingly powerful and is being employed in a wide variety of useful applications. However, it is currently typical for each vendor to have its own proprietary application programming interface (API). Because of this, developing multimodal applications requires mastering a different API for each vendor. Furthermore, these API's differ for different vendors' versions of each modality. The result is that developing multimodal applications becomes unnecessarily complex and difficult. Developers require extensive expertise and experience in order to master all of these API's. Acquiring this expertise is especially difficult for developers at small companies. This situation slows down the rate at which multimodal applications can be implemented and makes them more expensive than they would be if API's were uniform.

Standards such as the W3C Multimodal Architecture and Interfaces specification (MMI Architecture) [1–3] define a generic modality API, but the adoption of this standard across many vendors and modalities will take time. In the interim, an

D.A. Dahl (✉)
Conversational Technologies, Plymouth Meeting, PA, USA
e-mail: dahl@conversational-technologies.com

D.A. Dahl (ed.), *Multimodal Interaction with W3C Standards*,
DOI 10.1007/978-3-319-42816-1_11

alternative approach for developers who would like to take advantage of the standards would be to use portals that reformat proprietary results in standard formats.

By providing a standard, generic, modality, and vendor independent API in the form of the MMI Architecture, standard portals greatly simplify the learning process for developers. This approach is also much more extensible to new modalities than proprietary approaches. Furthermore, it makes it easier to change modality vendors if another vendor offers a superior product.

11.2 Overview of a Portal

As stated above, multimodal technology that supports such capabilities as natural language processing, speech recognition, handwriting recognition, and object recognition from images is becoming increasingly powerful and is being used in many applications. There are many products available in this space. Just looking at natural language processing offerings alone, some examples are wit.ai (Facebook) [4], api.ai [5], Microsoft LUIS (Language Understanding Intelligent System) [6], and Amazon Alexa Skills Kit [7], to name just a few of the systems available in 2016. Similarly, there are a number of API's for emotion recognition, including affectiva [8], EmoVu [9], Microsoft Emotion Recognition [10], Kairos [11], and nViso [12].

However, currently all of these systems have their own proprietary API's. Because of this, developing multimodal applications requires mastering a different API for each modality, and each vendor or even multiple API's for one modality, if the application supports multiple modality services.

This problem can be addressed through a standard multimodal web service portal. A standard portal can provide access to many types of modalities through a standard API; specifically, the W3C Multimodal Architecture and Interfaces (MMI Architecture) specification [1–3], as shown in Fig. 11.1. A standard portal serves as a layer of middleware between client applications and modalities. Developers of client applications only need to code to the standard MMI Architecture and the multimodal web service portal will provide the interface to the vendor-specific API, shielding developers from the details of the proprietary API and simplifying development. The standard portal is in fact an MMI Architecture Modality Component, communicating with clients using MMI Architecture Life Cycle events. A

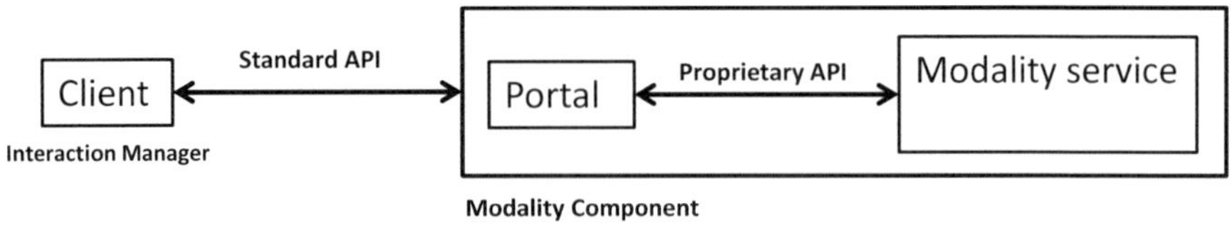

Fig. 11.1 Portal wrapping a standard API around a proprietary API

uniform API also makes it significantly easier to integrate, or fuse, inputs from multiple components. For example, it would be very useful to integrate speech and geolocation inputs in order to respond to user questions such as "where is the nearest Chinese restaurant" or "How far am I from home?" It is easy to see that as mobile devices add capabilities the problem of integrating multiple API's becomes very complex very quickly. While the problem of integrating inputs from multiple device capabilities is to some extent addressed by standard device API's such as the Media Capture and Streams API [13] these API's are still modality-specific, so that cross-modality integration of inputs is still up to the developer.

11.3 The Standard API

11.3.1 MMI Architecture

The standard API discussed in this paper consists of two components:

1. The MMI Architecture Life Cycle events for communication between an Interaction Manager (IM) and the Modality Components (MC's) that support the application.
2. Extensible Multimodal Annotation markup (EMMA 2.0) [14–16] for representing user input and system output.

The MMI Architecture includes both components and events. The components are (1) the Interaction Manager (IM), which coordinates the interaction, and (2) Modality Components (MC's). MC's both interpret multimodal inputs (from users as well as sensors) and create multimodal outputs. Modality Components communicate only with the Interaction Manager, they do not communicate directly with each other.

In addition to the components, the MMI Architecture also includes a set of high level Life Cycle events for communication between the IM and the MC's. Life Cycle events focused on controlling components include **StartRequest**, **PauseRequest**, **ResumeRequest**, and **CancelRequest**. These are messages sent from the IM to MC's. MC's, upon receiving one of these messages, respond with **Response** events, such as **StartResponse** and **PauseResponse**, for acknowledging receipt of the **Request** events and reporting errors. In addition, MC's can send a **DoneNotification** event when the requested processing is completed. Either the IM or an MC can also send an **ExtensionNotification** event at any time. **ExtensionNotification** events can contain arbitrary, application-specific data. No specific syntax is required for Life Cycle events, but XML is used in the examples in the specification, and will be used in this chapter.

Every Life Cycle event can optionally contain a **Data** field with additional information about the event. In the cases where the event pertains to user input or system output, the **Data** field contains Extensible Multimodal Annotation (EMMA) [14–16] data, which represents the user input and/or system output.

11.3.2 EMMA

EMMA is an XML language that is especially appropriate for representing semantically complex information. The semantics of the information itself is contained in the `<emma:interpretation>` element for user input or the `<emma:output>` element for system output. In addition to the actual semantics of the information, EMMA is also able to represent a rich set of metadata related to the context of the input or output. EMMA metadata includes, for example, processor confidence, timestamps, alternatives (nbest), medium and mode, the process that produced the EMMA result, tokens of input and pointers to the original signal (such as an audio file or image), among many other types of metadata.

In effect, the standard API referred to in this chapter consists of MMI Life Cycle events containing EMMA to represent interpreted inputs from users or sensors and system outputs.

11.4 Details of Multimodal Interaction with the Portal

An example architecture of a standard multimodal portal is shown in Fig. 11.2.

Interaction is initiated in the client-side components (1) by the user. Interaction modalities may include, for example, speech, typing, or mouse input, but may potentially include many other forms of input. The client-side components include application logic (2) implemented, for example, in HTML and JavaScript in

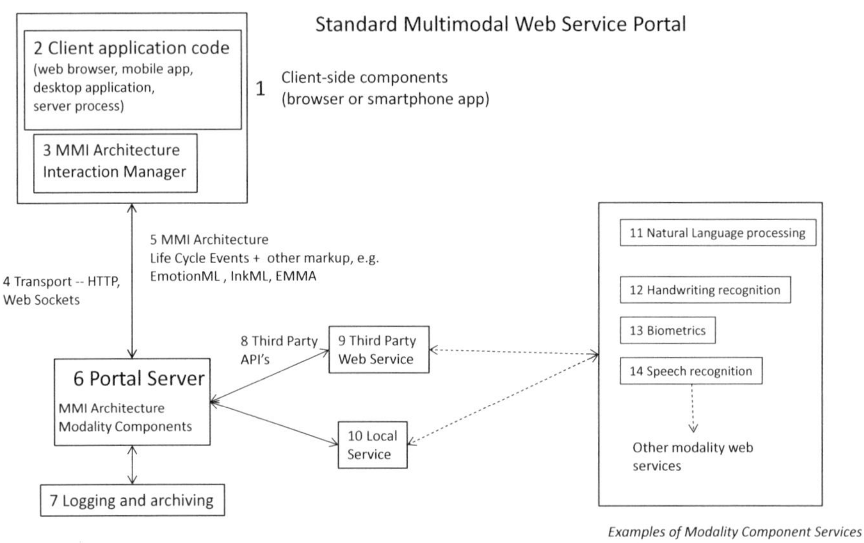

Fig. 11.2 Architecture of an MMI portal

browser-based implementations. The MMI Architecture Interaction Manager (3) sends over a transport mechanism such as HTTP (4) an MMI Architecture compliant Life Cycle event (5).

The Life Cycle event instructs the Portal Server (6) to process the user's input as required by the nature of the input (natural language understanding for language input, for example). (7) Logging and archiving of Life Cycle events may optionally occur at any point in processing.

Most critically, once the Portal Server has determined which third party services (if any) are required to process the event, it creates an API call (8) to that service (9). Although these API calls themselves may be proprietary, knowledge of any proprietary details is restricted to the Portal Server and is therefore isolated from the application developer, who only has to be concerned with sending and receiving standard MMI Architecture Life Cycle events. Within this architecture, it is also possible for services to be provided locally, within the portal (10).

Examples of possible (remote or local) modality services include but are not limited to natural language processing (11), handwriting recognition (12), biometric processing (13), and speech recognition (14). Once the appropriate service is contacted, its result is transmitted back to the Portal Server (6), reformulated into standard Life Cycle events (5), and sent back to the client-side components (1), specifically to the client Interaction Manager (3). Finally, application-specific code in the client (2) executes the appropriate action as determined by the processing result.

11.5 Implementing a Portal

Developing a standard portal requires developing several components. Going back to Fig. 11.2, the first component (2) is an application using client-side code (running in a web browser or as native code) which captures user input in modalities that are appropriate to the application. For example, a hand-held translation system requires speech to be captured for speech to speech translation, or keystrokes to be captured for translation from typing.

In addition, the client-side code will include functionality that controls the components with standard MMI Architecture Life Cycle events (that is, it will include an Interaction Manager (3)). The Interaction Manager can be implemented as a reusable library (for example, a Javascript library for browser clients) that can be used in many applications. SCXML [17, 18] is a suggested choice for Interaction Managers in the MMI Architecture. SCXML as a choice for Interaction Managers is especially efficient because the SCXML interpreter itself need only be implemented once for each platform, with the Interaction Managers for specific applications being implemented in SCMXL markup.

The Portal Server, which processes the Life Cycle events receives events using a standard transport such as HTTP [19] or Web Sockets [20, 21] (see Fig. 11.3 for an example of an actual HTTP POST message). The Portal Server is the key to the portal, because it serves to isolate proprietary API's (8) from the developer and enables the developer to access modality component services (11–14) entirely

```
POST https://proloquia-nlservice.rhcloud.com/rest/processmessage HTTP/1.1
Host: proloquia-nlservice.rhcloud.com
Connection: keep-alive
Content-Length: 725
Origin: https://proloquia-nlservice.rhcloud.com
User-Agent: Mozilla/5.0 (Windows NT 6.1; WOW64) AppleWebKit/537.36 (KHTML, like Gecko)
Chrome/47.0.2526.106 Safari/537.36
Content-Type: text/xml
Accept: */*
Referer: https://proloquia-nlservice.rhcloud.com/understanding.html
Accept-Encoding: gzip, deflate
Accept-Language: en-US,en;q=0.8

<mmi:StartRequest xmlns:mmi="http://www.w3.org/2008/04/mmi-arch" mmi:Context="nlClient0395"
mmi:RequestID="requestID0249" mmi:Source="ctNLClient"
mmi:Target="ctNLServer"><mmi:Data><function>understanding</function><emma:emma
xmlns:emma="http://www.w3.org/2003/04/emma" version="1.1"><emma:interpretation id="initial1"
emma:function="understanding" emma:tokens="put a couple cans of tomato soup on the
shopping list" emma:medium="tactile" emma:mode="keys" emma:verbal="true" emma:device-
type="keyboard" emma:end="1452629301869" emma:lang="en-US" emma:expressed-
through="text"><emma:literal>put a couple cans of tomato soup on the shopping
list</emma:literal></emma:interpretation></emma:emma></mmi:Data></mmi:StartRequest>
```


Fig. 11.3 HTTP POST request with MMI StartRequest event for "put a couple cans of tomato soup on the shopping list"

through standard mechanisms. Implementing the Portal Server requires developing code that can (1) interpret MMI Life Cycle events, (2) determine what services are being requested, (3) translate the user's request to the native API used by the service, (4) call the required services, and (5) reformat the results back into standard MMI Life Cycle events. In addition, a Portal Server can optionally perform other useful functions such as logging and archiving the event traffic, providing information as to what services are available, acting as a security gateway, format conversions, and managing user credentials.

The Interaction Manager (3) and the Portal Server (6) are essential parts of the portal. The Interaction Manager creates the Life Cycle events from the user's input and interprets the Life Cycle events sent back from the Portal Server. A transport mechanism (4) is required for the portal, but it is not necessary for the developer to implement the transport mechanism because a number of standard transport mechanisms are already available and are appropriate for use in this architecture, including HTTP or WebSockets. The Portal Server can be used on its own, without the ability to access third party components (8, 9, 11–14), just using local services (10); however, the portal is far more useful if translation from standard Life Cycle events to third party API's (8) is implemented for accessing existing third party services. In addition, logging and archiving services (7), while not necessary, are extremely useful in production systems for monitoring usage and debugging problems. Another aspect of a portal that would be very useful, although not required, is a way for clients to query the Portal Server in order to discover available services. Discovery and Registration functionality of this kind could be implemented using the W3C Discovery and Registration approach discussed in [22–24].

11.6 An Example: Home Control

The Internet of Things (IoT) has enormous potential for adding convenience, comfort, safety, and efficiency to everyday life as well as for supporting larger scale enterprise applications. However, there will soon be too many items in the IoT to realistically expect conventional graphical interfaces to support all the ways in which users might want to interact with them. For this reason, natural language using a standard API will become very important for these types of interactions. This section discusses an IoT example in the area of home control.

Home control is a common use case for the IoT. Home control includes control of lighting, appliances, heating and air conditioning, entertainment and security, among many other possibilities. Even limiting consideration to items that users will want to interact with in the home still leaves the possibility of interaction with hundreds of devices. If each device, or even each vendor of a connected home system, has its own API, this will quickly become unmanageable for developers who wish to integrate many devices into an application. Here we will describe an MMI Architecture approach for controlling lighting with a standard portal.

Figure 11.4 shows a web page with a user interface for natural language control of lighting. The user can click "Start Recognition" to start recognition and speak, or the user can type the request into a text box. In this case the user has typed "It's dark in here." The web page Javascript wraps the input in EMMA and the **StartRequest** LifeCycle event to produce the event shown in Fig. 11.5. Application-specific information is contained in "`<mmi:Data>`". In this example there is an application-specific field "`function`" which determines which function the data pertains to, in this case "`lightControl`". The user input itself, expressed in EMMA, is also contained in the "`<mmi:Data>`" field.

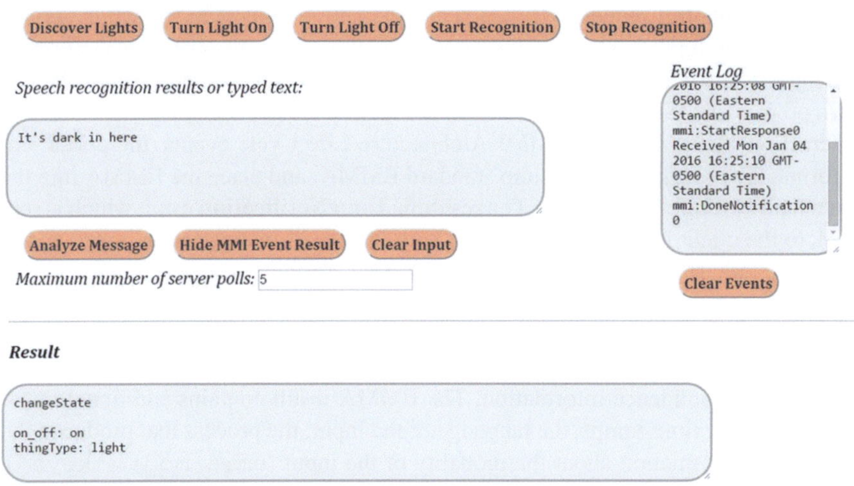

Fig. 11.4 Web page for home control

```
<mmi:StartRequest xmlns:mmi="http://www.w3.org/2008/04/mmi-arch"
          mmi:Context="nlClient0515"
          mmi:RequestID="requestID1841"
          mmi:Source="ctNLClient"
          mmi:Target="ctNLServer">
  <mmi:Data>
    <function>lightControl</function>
    <emma:emma xmlns:emma="http://www.w3.org/2003/04/emma" version="2.0">
      <emma:interpretation
        id="initial2"
        emma:function="lightControl"
        emma:tokens="It's dark in here"
        emma:medium="acoustic"
        emma:mode="voice"
        emma:verbal="true"
        emma:device-type="microphone"
        emma:end="1451946835723"
        emma:lang="en-US"
        emma:expressed-through="text">
        <emma:literal>It's dark in here</emma:literal>
      </emma:interpretation>
    </emma:emma>
  </mmi:Data>
</mmi:StartRequest>
```

Fig. 11.5 StartRequest Life Cycle event for "it's dark in here"

The **StartRequest** event is sent to the portal via HTTP POST and the portal is polled using AJAX [25] for information returned in response to the **StartRequest**. The first event returned from the portal is a **StartResponse** which simply acknowledges that the **StartRequest** was received. The portal then creates an API request to a wit.ai [4] natural language processing endpoint which has been trained to understand home control requests. It then sends the native request to the wit.ai service endpoint. Wit.ai interprets "it's dark in here" to mean that the user wants to turn on the light. The wit.ai endpoint returns natural language understanding results in a its own proprietary JSON format, as shown in Fig. 11.6. However, since the web client Interaction Manager expects MMI Architecture Life Cycle events, the portal will reformat the proprietary result into standard EMMA, and place the EMMA into the Data field of a Life Cycle event. The resulting **DoneNotification** event which is sent back to the client is shown in Fig. 11.7, with the actual interpretation boxed and in bold (see [14, 16] for details of the EMMA XML format).

Comparing the native API result in Fig. 11.6 with the MMI Architecture/EMMA result in Fig. 11.7, we can note that the semantic information contained in the result is the same—"it's dark in here" is interpreted as "turn the light on." Both formats also include confidence information. The EMMA result contains additional metadata, including timestamps, the language of the input, the process that produced the result, and information about the modality of the input (`emma:mode="keys"`). While some of this information is optional in EMMA, including the richer metadata can become very important for debugging and tuning large-scale, enterprise

```
{
 "msg_id": "a81ffcef-4606-4a93-8f63-0ccf9d8a5b05",
 "_text": "it's dark in here",
 "outcomes": [
  {
   "_text": "it's dark in here",
   "confidence": 0.782,
   "intent": "changeState",
   "entities": {
    "thingType": [
     {
      "type": "value",
      "value": "light"
     }
    ],
    "on_off": [
     {
      "value": "on"
     }
    ]
   }
  }
 ]
}
```


Fig. 11.6 Native wit.ai JSON output

applications. It is also possible to retain the complete EMMA data on the server (where it can be used in debugging and tuning) while sending only the minimum amount of data to a client (for use in interactive dialogs), using mechanisms that have been newly introduced in EMMA 2.0 [16]. While in this case the native format of wit.ai is JSON and the MMI/EMMA format is in XML, there are many software tools available for converting between these formats.

11.7 Existing Portals

A very experimental MMI Architecture client and portal has been implemented by the author. Please contact the author for access to the portal. This portal includes demos of emotion recognition from language, natural language understanding, and part of speech tagging, among others. The portal accepts MMI Architecture Life Cycle events over HTTP with user inputs represented in EMMA. The examples in this chapter were produced by this portal.

For emotion recognition, an EmotionML wrapper for the Microsoft Project Oxford Emotion Recognizer is also available [26]. While not a full MMI Architecture portal, it does wrap a proprietary API with a standard, EmotionML [27, 28], which is very much in the spirit of providing standard API's to otherwise proprietary services.

```
<mmi:mmi
  xmlns:mmi="mmi">
  <mmi:DoneNotification mmi:Context="nlClient0515" mmi:RequestID="requestID0217" mmi:Source="ctNLServer"
mmi:Status="success" mmi:Target="ctNLClient">
    <mmi:Data>
      <emma:emma
  xmlns:emma="http://www.w3.org/2003/04/emma"
  xmlns:xsi="http://www.w3.org/2001/XMLSchema-instance" version="2.0"
xsi:schemaLocation="http://www.w3.org/2003/04/emma  http://www.w3.org/TR/2009/REC-emma-
20090210/emma.xsd">
        <emma:interpretation emma:confidence="0.744" emma:process="wit.ai" emma:tokens="It's dark in here"
id="interp12">
          <emma:derived-from composite="false" resource="#initial1"/>
          <nlResult>
            <_text>It's dark in here</_text>
            <msg_id>6a24c2e5-a904-4f9d-95fd-a0dc892d9460</msg_id>
            <outcomes>
              <e>
                <_text>It's dark in here</_text>
                <confidence>0.744</confidence>
                <entities>
                  <on_off>
                    <e>
                      <value>on</value>
                    </e>
                  </on_off>
                  <thingType>
                    <e>
                      <type>value</type>
                      <value>light</value>
                    </e>
                  </thingType>
                </entities>
                <intent>changeState</intent>
              </e>
            </outcomes>
          </nlResult>
        </emma:interpretation>
        <emma:derivation>
          <emma:interpretation emma:device-type="keyboard" emma:end="1451942708899" emma:expressed-
through="text" emma:function="lightControl" emma:lang="en-US" emma:medium="tactile" emma:mode="keys"
emma:tokens="It's dark in here" emma:verbal="true" id="initial1">
            <emma:literal>It's dark in here</emma:literal>
          </emma:interpretation>
        </emma:derivation>
      </emma:emma>
    </mmi:Data>
  </mmi:DoneNotification>
</mmi:mmi>
```


interpretation

Fig. 11.7 DoneNotification event for the interpretation of "it's dark in here" as "turn the light on"

11.8 Integrating Portals with Other MMI-Standards Compliant Components

As the standards become more widely integrated into modality services, there will be increasing native support for EMMA and the MMI Architecture. This development will be completely compatible with the portal model. Components supporting the standards natively will be fully interoperable with a standard portal. For example, an application for emotion recognition might fuse results from language

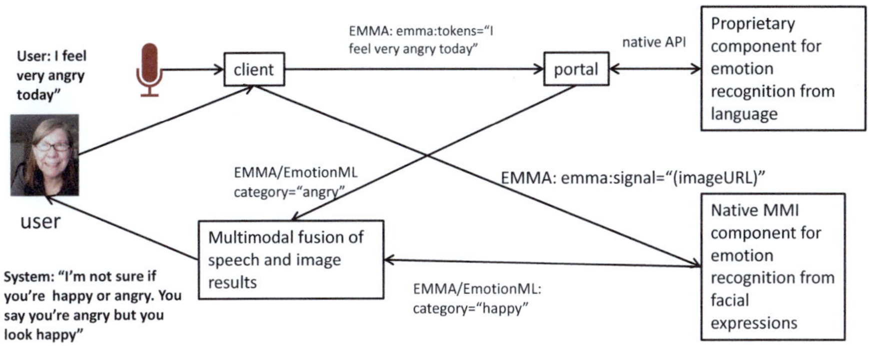

Fig. 11.8 Mixing portals with MMI native components

and facial expressions to improve the accuracy of the emotion recognition result. The language recognition could come from a service provided by a portal, while the facial expression analysis could come from a service that supports the MMI Architecture natively. Integration of information from different modalities (fusion) would be provided by a fusion component, as shown in Fig. 11.8. Of course, systems can include more than one standard portal, where each portal provides different modality services.

11.9 Developing Standard Modality Components and Portals

Given an existing modality processor (for example, handwriting recognition, speech recognition, object recognition, or emotion recognition) developing a standard component is straightforward. Following the requirements and documentation guidelines in [29], the developer provides access to the native capabilities of the component through MMI Life Cycle events. Thus, the user of the component will use a standard API call such as the one shown in Fig. 11.5, rather than the corresponding native call, the HTTP GET message https://api.wit.ai/message?v=20141022&q=it%27s%20dark%20in%20here.

Clearly, the native call is less verbose, but much of the detailed information in the standard API call is optional. In addition, the additional standard information, if used, can provide a great deal of detail that is valuable for logging, archiving, and tuning applications. This kind of information is especially important in large scale commercial applications.

Multiple modality components can be aggregated into a portal by providing a single REST endpoint and including information in the `mmi:Data` field to indicate which modality component is being requested.

11.10 Conclusions

In summary, standards-based multimodal portals can provide standard interfaces to otherwise proprietary services, providing a way for developers to use standards with proprietary systems.

In doing so, they provide the following advantages over proprietary approaches:

1. They reduce the need for developers to learn proprietary API's.
2. They can foster the adoption of standards by supporting a phased implementation approach.
3. They increase vendor-independence.
4. They can simplify logging and analysis of inputs for debugging and tuning because processing results from different vendors' services will be in the same format.
5. They simplify adding new modalities to an existing application because inputs from different modalities will be in the same format.
6. They simplify integration of inputs from components using the MMI Architecture API's natively with information produced by proprietary systems.

References

1. Barnett, J., Bodell, M., Dahl, D. A., Kliche, I., Larson, J., Porter, B., et al. (2012). Multimodal architecture and interfaces. World Wide Web Consortium. http://www.w3.org/TR/mmi-arch/. Accessed 20 Nov 2012.
2. Dahl, D. A. (2013). The W3C multimodal architecture and interfaces standard. *Journal on Multimodal User Interfaces*, 1–12 (2013). doi:10.1007/s12193-013-0120-5.
3. Barnett, J. (2016). Introduction to the multimodal architecture. In D. Dahl (Ed.), *Multimodal interaction with W3C standards: Towards natural user interfaces to everything*. New York, NY: Springer.
4. wit.ai (2015). wit.ai. https://wit.ai/. Accessed 17 Mar 2015.
5. api.ai (2015). api.ai. http://api.ai/. Accessed 17 Mar 2015.
6. Microsoft (2015). Language Understanding Intelligent Service (LUIS). Microsoft. http://www.projectoxford.ai/luis. Accessed 5 June 2015.
7. Amazon (2016). Alexa Skills Kit. Amazon. https://developer.amazon.com/public/solutions/alexa/alexa-skills-kit. Accessed 6 Jan 2016.
8. affectiva (2016). Affdex emotion sensing and analytics. affectiva. http://www.affectiva.com/solutions/apis-sdks/. Accessed 11 Jan 2016.
9. EmoVu (2016). EmoVu Cloud API. Eyeris. http://emovu.com/e/developers/api/. Accessed 12 Jan 2016.
10. Microsoft (2016). Project oxford emotion recognition. Microsoft. https://www.projectoxford.ai/demo/emotion. Accessed 12 Jan 2016.
11. Kairos (2016). Emotion analysis API. Kairos. https://www.kairos.com/emotion-analysis-api. Accessed 11 Jan 2016.
12. nViso (2016). nViso emotion recognition. nViso. http://www.nviso.ch/index.html. Accessed 11 Jan 2016.
13. Burnett, D., Bergkvist, A., Jennings, C., & Narayanan, A. (2015). *Media capture and streams* (14th ed.). Boston, MA: World Wide Web Consortium.

14. Johnston, M. (2016). Extensible multimodal annotation for intelligent interactive systems. In D. Dahl (Ed.), *Multimodal interaction with W3C standards: Towards natural user interfaces to everything*. New York, NY: Springer.
15. Johnston, M., Baggia, P., Burnett, D., Carter, J., Dahl, D. A., McCobb, G., et al. (2009). EMMA: Extensible MultiModal Annotation markup language. W3C. http://www.w3.org/TR/emma/. Accessed 9 Nov 2012.
16. Johnston, M., Dahl, D. A., Denny, T., & Kharidi, N. (2015). EMMA: Extensible MultiModal Annotation markup language Version 2.0. World Wide Web Consortium. http://www.w3.org/TR/emma20/. Accessed 16 Dec 2015.
17. Barnett, J. (2016). Introduction to SCXML. In D. Dahl (Ed.), *Multimodal interaction with W3C standards: Toward natural user interfaces to everything*. New York, NY: Springer.
18. Barnett, J., Akolkar, R., Auburn, R. J., Bodell, M., Burnett, D. C., Carter, J., et al. (2015). State Chart XML (SCXML): State machine notation for control abstraction. World Wide Web Consortium. http://www.w3.org/TR/scxml/. Accessed 20 Feb 2016.
19. Fielding, R. T., Gettys, J., Mogul, J., Frystyk, H., Masinter, L., Leach, P., et al. (1999). RFC 2616 hypertext transfer protocol—HTTP/1.1. Internet Engineering Task Force (IETF). https://tools.ietf.org/html/rfc2616. Accessed 12 Jan 2016.
20. Fette, I., & Melnikov, A. (2011). RFC 6455 The WebSocket Protocol. Internet Engineering Task Force (IETF). https://tools.ietf.org/html/rfc6455. Accessed 12 Jan 2016.
21. Hickson, I. (2012). The WebSocket API. The World Wide Web Consortium. http://www.w3.org/TR/websockets/. Accessed 20 Nov 2012.
22. Rodríguez, B. H., Barnett, J., Dahl, D., Tumuluri, R., Kharidi, N., & Ashimura, K. (2015). Discovery and registration of multimodal modality components: State handling. World Wide Web Consortium. https://www.w3.org/TR/mmi-mc-discovery/.
23. Rodriguez, B. H., & Moissinac, J.-C. (2016). Discovery and registration—finding and integrating components into dynamic systems. In D. A. Dahl (Ed.), *Multimodal interaction with W3C standards: Toward natural user interfaces to everything*. New York, NY: Springer.
24. Rodriguez, B. H., Wiechno, P., Dahl, D. A., Ashimura, K., & Tumuluri, R. (2012). Registration & discovery of multimodal modality components in multimodal systems: Use cases and requirements. World Wide Web Consortium. http://www.w3.org/TR/mmi-discovery/. Accessed 26 Nov 2012.
25. Garrett, J. J. (2005). Ajax: A new approach to web applications. Adaptive Path. https://web.archive.org/web/20080702075113/http://www.adaptivepath.com/ideas/essays/archives/000385.php. Accessed 14 Jan 2016.
26. Hilton, A. (2015). EmotionAPI 0.2.0. Coolfire solutions. https://github.com/Felsig/Emotion-API. Accessed 11 Jan 2016.
27. Schröder, M., Baggia, P., Burkhardt, F., Pelachaud, C., Peter, C., & Zovato, E. (2014). Emotion Markup Language (EmotionML) 1.0 World Wide Web Consortium. http://www.w3.org/TR/emotionml/.
28. Burkhardt, F., Pelachaud, C., & Schuller, B. (2016). Emotion markup language. In D. Dahl (Ed.), *Multimodal interaction with W3C standards: Toward natural user interfaces to everything*. New York, NY: Springer.
29. Kliche, I., Dahl, D. A., Larson, J. A., Rodriguez, B. H., & Selvaraj, M. (2011). Best practices for creating MMI modality components. World Wide Web Consortium. http://www.w3.org/TR/2011/NOTE-mmi-mcbp-20110301/. Accessed 20 Nov 2012.

Chapter 12
Applications of the Multimodal Interaction Architecture in Ambient Assisted Living

António Teixeira, Nuno Almeida, Carlos Pereira, Miguel Oliveira e Silva, Diogo Vieira, and Samuel Silva

Abstract Developing applications for ambient assisted living (AAL) scenarios requires dealing with diverse user groups, heterogeneous environments, and a large plethora of devices. These requirements pose several challenges on how to design and develop user interaction with the proposed applications and services. In this context, the versatility provided by multimodal interaction (MMI) is paramount and the adopted architecture should be instrumental in harnessing its full potential. This chapter offers an insight on how AAL challenges can be tackled by multimodal-based solutions. It presents the authors' views and research outcomes in multimodal application development for AAL grounded on an architecture for MMI aligned with the W3C recommendations.

12.1 Introduction

In our aging society, ambient assisted living (AAL) is defined as an "aim to extend the time people can live in a decent way in their own home by increasing their autonomy and self-confidence, the discharge of monotonous everyday activities, to monitor and care for the elderly or ill person, to enhance security and to save resources" [1]. AAL is a natural extension of the ambient intelligence paradigm [2, 3], whose main objective is the integration of a broad range of technologies, namely smart materials, microelectromechanical components, sensor technologies, embedded devices, ubiquitous communications, or intelligent interfaces to proactively support people in their daily lives.

In a similar way, as many Internet services are now part of everyday life of technological non-impaired people, one can envision AAL services as also playing an important role in the future living environment of everyone, with a special relevance for those with physical limitations or health problems.

A. Teixeira (✉) • N. Almeida • C. Pereira • M. Oliveira e Silva • D. Vieira • S. Silva
DETI—Department of Electronics, Telecommunications and Informatics, IEETA—Institute of Electronics and Informatics of Aveiro, University of Aveiro, Campus Universitário de Santiago, Aveiro, Portugal
e-mail: ajst@ua.pt

D.A. Dahl (ed.), *Multimodal Interaction with W3C Standards*,
DOI 10.1007/978-3-319-42816-1_12

Despite its origins, AAL does not need to be restricted to household scenarios. Assistance is needed outside and in many other activities. Examples of Pervasive AAL—or ubiquitous AAL—include assistance for: (a) using new social networks [4]; (b) patients with their medication [5, 6]; (c) firefighters in emergency situations; (d) drivers in their cars; (e) tourists in new places [7]; and (f) older or disabled within environments such as their neighborhood, shopping mall, and other public places [8]. Many of these new application scenarios are mobile and require distributed heterogeneous solutions.

The creation of AAL applications and services, particularly Pervasive AAL, presents a number of challenges to researchers and developers alike. An AAL scenario with a simple application typically includes several sensors (e.g., [9]), communication networks, computational devices, and other technologies [10]. With the fast evolution of concepts such as the Internet of Things (IoT) [11, 12], these scenarios will rapidly grow and become even more complex to manage [13]. Testing and fine-tuning a service or an application in one environment does not guarantee the same performance in another. Environments and contexts of use change within and among users and this must be taken into consideration when developing new applications or services.

Within AAL's objectives [14], multimodality can provide an important push in shortening the gap between the user and the ambient. By offering different interaction methods, developers are not only creating redundancy, but they are also increasing usability rates. Given the volatility of the environment itself, doing so is fundamental to ensuring that, regardless of the context, users are able to access applications and services when needed.

At the same time, several situations point to the need to explore more natural ways of interaction, usually associated with human to human communication, such as speech or gestures. Speech, despite the far-from-perfect performance of the state-of-the-art technologies, has a strong potential in areas such as robot assistants [15] or assistive technologies [16], particularly by allowing hands-free interaction, at a distance, and benefiting from its intuitive use, potentially requiring minimal learning. Use of hands and body gestures can also be very useful, but, in general, hands will not be free for other tasks and more learning is needed.

Our long-term involvement in designing and developing for generalized assistance scenarios offers exciting opportunities and challenges regarding interaction: the existence of many sensors capable of providing information on the environment, user activity, and even user mood, the need to interact at a distance, and to provide information to the user through small devices, for example. Good AAL applications require constant adaptation to the user, not only to his/her position, time of day, or task and environment characteristics, but also to the user's mood, preferences, or even disabilities.

In addition to this, the heterogeneity of environments in AAL is a serious issue both from the technical [17] and user experience perspective. From early on, developers may try to pay special attention to aspects such as usability or user experience, and rapidly become frustrated as their application eventually ceases to operate properly due to changes in the environment. At the extreme, the application may not be compatible at all due to irreconcilable differences in the data formats or adopted infrastructures. In this regard, standardization is crucial not only in fostering an easier design and development of novel applications [18], but also in enabling interactively rich AAL services that harness the full capabilities of the environment (devices, sensors, etc.) in favor of higher user acceptance [19] and improved life experiences.

In line with these views, the remainder of this chapter is organized as follows. In Sect. 12.2, we start by presenting the context in which our work evolved, resulting in our adoption of a multimodal interaction (MMI) architecture based on the W3C recommendations, followed by a brief presentation of the architecture we adopted for AAL MMI. Section 12.3 is then devoted to presenting two practical examples of AAL works adopting the proposed architecture, highlighting their main features and how the architecture enabled fulfilling the requirements. Finally, the chapter ends with some discussion and conclusions regarding the accomplished work and some views on future lines of development.

12.2 Multimodal Interaction Architecture for AAL Based on the W3C Standards

Our adoption of the W3C MMI architecture for AAL was the result of a long-term effort in designing and developing systems and applications for AAL. Therefore, in what follows, we first provide a brief account of how our approach evolved, considering several of the challenges faced, followed by a characterization of our current W3C-based architecture for MMI, including our proposal of generic interaction modalities.

12.2.1 Integrating Support for Multimodality in AAL Architectures

To address the interaction requirements for the AAL scenarios established by project LUL—Living Usability Lab[1] and AAL4ALL[2] [20] a novel architecture to support MMI was proposed, favoring autonomy and decentralization, based on a distributed paradigm grounded on modules and services. This offered enough flexibility to encompass scalability, and adaptive capabilities (e.g., to dynamic environments and user characteristics), particularly relevant for the envisaged environments. Likewise, a distributed architecture offered an easier integration and testing of novel components.

[1] Project LUL—Living Usability Lab, http://www.livinglab.pt.

[2] Project AAL4ALL—Ambient Assisted Living for All, http://www.aal4all.org.

To instantiate the architecture, we considered different alternatives. In the first iteration, we adopted agents and the Java Agent Development Framework (JADE), allowing the development of agent-based systems, supporting multiple communication methods among agents, and an agent registration and indexing service. Considering this first implementation, a method was proposed to adapt the output to the user and context—codenamed AdaptO—showing the overall capabilities of the proposed architecture [21].

Since service-oriented architectures (SOA) offer adequate solutions for the main requirements of AAL services and applications, such as heterogeneity, they have been widely adopted [22]. Nevertheless, they do not directly address interaction and it was important to support MMI on the proposed AAL architecture. As a result, a novel version of the architecture was proposed, adopting web services and HTTP-based communication.

When deploying this new version of the architecture it was clear that several aspects required special attention, such as the use of a standard communication language and the existence of a base set of modalities. The use of a non-standard communication language made it almost impossible to connect new components from the different stakeholders, turning development into a hard task, and delaying the deployment of novel multimodal systems [23]. The lack of support for a basic modality set would impact negatively on users' decision to adopt the architecture at all.

The effort to tackle these challenges naturally resulted in the consideration of several of the W3C recommendations as an integral part of our architecture, namely: Extensible MultiModal Annotation markup language (EMMA) [24] for the representation of data transferred among entities of a multimodal system; life cycle events, to deal with the communication between modalities and the Interaction Manager (IM); and the modality definition, considering the use of external services. Beyond enabling us to address these aspects, the adoption of the W3C standards is also well aligned with the principles of standardization and open systems that should foster further advances in AAL [18]. This novel iteration of the architecture is described in what follows.

12.2.2 Multimodal Architecture for AAL: Overview

Our AAL architecture is strongly based on the separation of modalities from the application and on the use of an IM, as in the W3C architecture. The communication between the modalities and the IM, and between the IM and the application, uses the key ideas of the W3C regarding life cycle events, and encoding of information using EMMA. Figure 12.1 presents a simple example of the life cycle events generated after an event in the speech recognition, part of the speech modality.

Due to our goal of enabling simpler and faster MMI, assigning a high priority to the integration of speech and gesture-based interaction modalities, we consider as

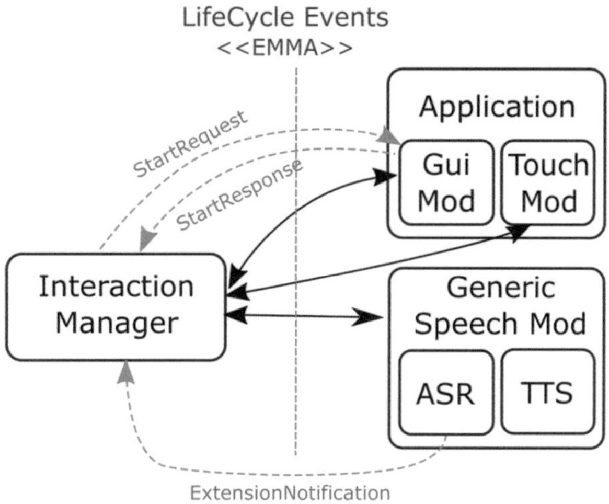

Fig. 12.1 Simple example of exchanged life cycle events

an integral part of our AAL MMI architecture a set of complex modalities supporting these interaction features. These modalities should support application developers in creating MMI interfaces without the need to master modality complexity and, therefore, we opted, in our architecture, for a set of comprehensive modalities that could be easily configurable for many situations. We consider these "Generic Modalities" (i.e., complex modalities developed with the purpose of being easily integrated in any MMI application) as key elements of our architecture.

The alignment with the W3C MMI architecture presents itself as a good solution to some AAL issues:

- Heterogeneity becomes a smaller issue. The inclusion of a new modality does not imply revisions to existing applications or in the AAL environment.
- Autonomy can be increased. Given its focus on interaction, the W3C MMI architecture allows including into input/output modalities some degree of autonomy and intelligence. Modalities are capable of receiving updates and adapting themselves to the present conditions.
- Usability through choice. Applications can communicate with multiple modalities such as speech, gestures, keyboard, and touch, without changes to the core programming. New modalities can be added at any time. Users can now use several methods to interact with applications, potentially increasing their overall usability.

Also, to achieve truly adaptable solutions, we believe that the environment must constantly evolve. This obviously includes input and output characteristics of our interfaces. With the adoption of the MMI architecture, especially through the usage of MMI life cycle events and the EMMA standard, we have been able to make advancements in tackling some of the difficulties.

The implementation of the multimodal framework implementing the envisaged architecture follows the W3C standard for multimodal architecture [25].

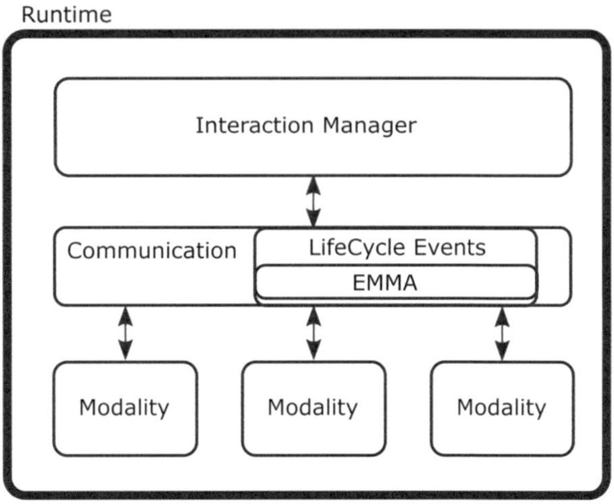

Fig. 12.2 Overview of the adopted multimodal architecture comprising its main components

The framework instantiating our MMI architecture for AAL includes an IM, a set of modalities, and methods for communication between modalities and IM. Figure 12.2 presents an overview of the multimodal architecture.

The IM operates as a central service accessible to all modalities. At its core, the IM operates using a State Chart extensible Markup Language (SCXML) definition loaded during initialization. Using this definition, the IM parses and answers modality requests.

All communication is specified using MMI life cycle events and the EMMA language. As such, the IM receives MMI life cycle Events coming from modalities via HTTP. However, because communication is bidirectional and can be started by either the IM or the modalities, the latter can also operate as HTTP servers.

When an MMI life cycle event is received, the IM or the modality parses the message and analyzes its content. In the case of the IM, depending on the result, it might trigger an event depending on what is defined in the SCXML file. This usually involves responding to the incoming message, but it can also involve modifications to the data model, or contacting other modalities. Modality components act according to their own logics and implementation.

Figure 12.3 illustrates the communication between the IM and several modalities.

12.2.3 Generic Modalities

Considering the multimodal architecture recommendations, modalities should be decoupled and communicate with the IM using standard MMI life cycle events. This enables two important features: (1) easily registering novel modalities with existing applications; and (2) allowing developers with no expertise in particular

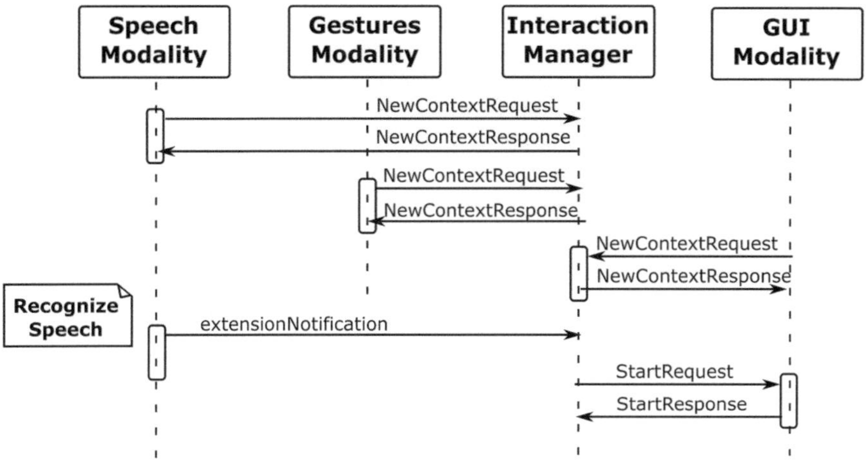

Fig. 12.3 Illustrative example of the communication between modules for the adopted MMI architecture

technologies to easily support them in their applications, keeping their focus solely on the application development. With this in mind, we have worked on the proposal of several modalities that could serve any AAL application adopting the architecture.

A basic set of such generic modalities for AAL needs to include support for touch, body gestures (particularly hand gestures), speech input and output, and classical graphical output. Due to the potential of speech for input and output in AAL scenarios—freeing hands and eyes and allowing interaction at a distances—a particularly important generic modality was designed and developed for speech.

The generic speech modality follows the W3C multimodal architecture recommendation, implementing the standard MMI life cycle events and EMMA to communicate with the IM. Since it is a decoupled modality, it is easily improved and added to other systems using the multimodal framework.

The generic modality can be configured to support speech input and/or output, resorting to speech recognition and synthesis technologies, and is capable of recognizing or synthesizing speech for multiple languages, making the modality useful for multilingual systems. A notable characteristic is that, for example, for speech input, the modality uses a webservice, capable of translating a given grammar to other languages and performing semantic parsing to the output of a speech recognizer [26]. This use of a service for complex processing, providing a centralized way for handling different languages, is very important in AAL scenarios, since devices running the modality can have limitations regarding memory and computational power. In addition, it provides a way for manufacturers to deploy updates to the grammars used by a large set of applications with little effort. Eventually, the service could also run the speech recognizer (ASR) and receive a stream with recorded speech, but this would entail higher bandwidth demands.

To create the first instantiation of this generic modality, the Microsoft Speech Platform[3] was used as the speech engine, as it supports a large set of languages and is simple to expand by adding additional language packs [27].

12.3 AAL Applications

At this time, we have used the described multimodal architecture in several AAL projects. The following sections present information regarding two application contexts and how the architecture served their requirements: a telerehabilitation service and a personal life assistant (PLA).

12.3.1 A New Telerehabilitation Service for the Older Adults

The percentage of people over 60 is increasing more rapidly than any other age group. By 2025, it is anticipated that there will be 1.2 billion humans over the age of 60 and this will continue to rise to about two billion in 2050. With ageing comes a decrease in functioning associated with a stronger incidence of multiple chronic diseases, which motivates an increasing need for healthcare services [28]. This constitutes a challenging scenario for the traditional healthcare system [29], both regarding its ability to face the larger number of potential patients and the need to provide patient specific approaches and follow up. In this context, telerehabilitation has some advantages [30] for older adults by working as facilitators for operational optimization of care services [31], increasing, for example, the availability of therapists, allowing rehabilitation at home, reducing therapist cost, and fighting isolation.

The new telerehabilitation Service [22, 32] supporting MMI, developed in the scope of projects LUL—Living Usability Lab[4] and AAL4ALL,[5] enables patients to perform rehabilitation sessions, at home or at community centers, under remote supervision by a health professional (e.g., a physiotherapist).

12.3.1.1 Requirements

In order to develop the telerehabilitation service, the following main requirements were considered:

---

[3] Microsoft Speech Platform, https://msdn.microsoft.com/en-us/library/office/hh361572(v=office.14).aspx, accessed March, 2016.

[4] See footnote [1].

[5] See footnote [2].

- Functional requirements:
  - Remotely manage and monitor the rehabilitation session including the definition of an exercise plan;
  - Remotely provide and receive feedback from the user in order to adjust the rehabilitation session;
  - Enable the simultaneous management of a large number of patients by a single supervisor.
- Non-functional requirements:
  - Guarantee the reliability of the system, including fault-tolerance capabilities and the ability to recover from communication errors.
  - Support the distributed execution of the service, in order to support the heterogeneous nature of different environments (such as houses or community centers).
  - Scalable and extensible, thus allowing future iterations and growth.
  - Provide high usability rates.

From this list, it is important to highlight the last requirement, high usability, as it is crucial to the success of the service. Since the user is provided with instructions by the interface and there is constant interaction between himself and the health professional, the user must feel comfortable when interacting with the system, and its operations and functionalities should feel simple and natural to him/her. To achieve this goal, a new set of requirements exclusively based on HCI were considered:

- Modality redundancy: The system must be prepared to offer complementary modalities to improve the chance of message delivery.
- Modality adaptation: Environmental conditions (e.g., light, noise, and distance) as well as user preferences (e.g., font size, resolution) should be considered as contextual information and directly influence the behavior of the modalities.
- Multimodal support for the health professional—Multi-touch input should be available on the health professional side in order to allow him to easily access information or quickly select a course of action regarding the rehabilitation sessions.

12.3.1.2 Service Architecture

Taking into account the functional requirements, the health professional can remotely monitor the patient using video and biosensors (e.g., surface electromyography to monitor muscle activity during exercises); plan, apply, and control an exercise program; provide feedback regarding their performance. The requirement for the service to allow a single health professional to supervise a large number of patients resulted in the option for not having a video feed from supervisor to patient, being a text-based chat the adopted solution. Since using the chat by the

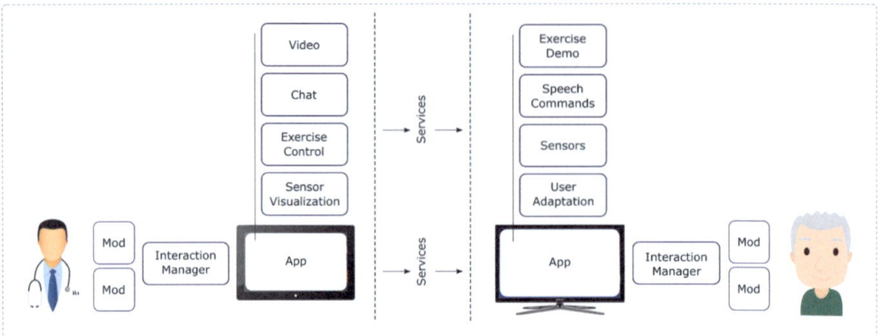

Fig. 12.4 Global view of the telerehabilitation service

patient—reading and writing messages—is often impossible, when doing the exercises, speech input and output is used to complement keyboard, mouse, and display.

To improve its usability, the service includes output adaptation (e.g., font size, speech volume) to the user based on distance, ambient noise, and lighting of the room. In what follows we provide an overall view of the service and its features. For more details regarding the service the reader is referred to [22, 32].

A broad overview of the service is shown in Fig. 12.4. The service provides the health professional with information from the house and control over the rehabilitation session. The user at home receives indications for performing the exercises.

The service depends on two applications, one for the user at home, and the other for the health professional planning, monitoring, and evaluating the session. Both were developed supporting MMI tailored for each specific application goal and to the expected different capabilities of its users.

The health professional application (in Fig. 12.5) is composed by four components. The top left window contains the exercise selection panel. This panel provides the health professional with the ability to create and manage the list of exercises that the user should perform. On the top right is the biosensor information panel, containing the sensor data regarding the use of rehabilitation service. On the bottom left, a messaging window was included with the purpose of allowing direct communication with the patient via text messages. Finally, the bottom right window displays a live video feed of the user performing the session.

The second application is for the patient at home. Its interface, presented in Fig. 12.6, also contains four major components. The top left panel displays visual illustrations on how to perform the current exercise using a step-by-step approach. On the top right, an indication of the remaining time for the current exercise is shown to the user. On the bottom left, a messaging panel is included to allow communication with the health professional. The bottom right panel displays the live video feed of the user to allow him/her to analyze and correct their posture.

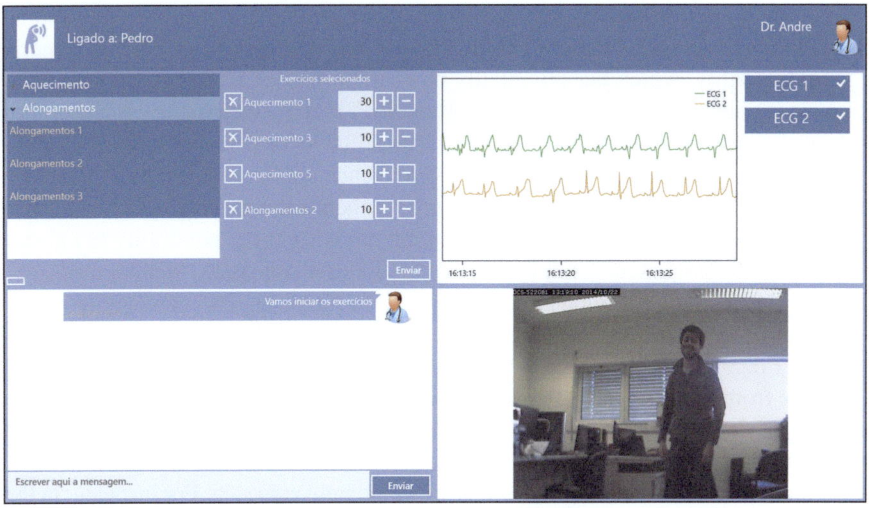

Fig. 12.5 Screen of the health professional application (from [33])

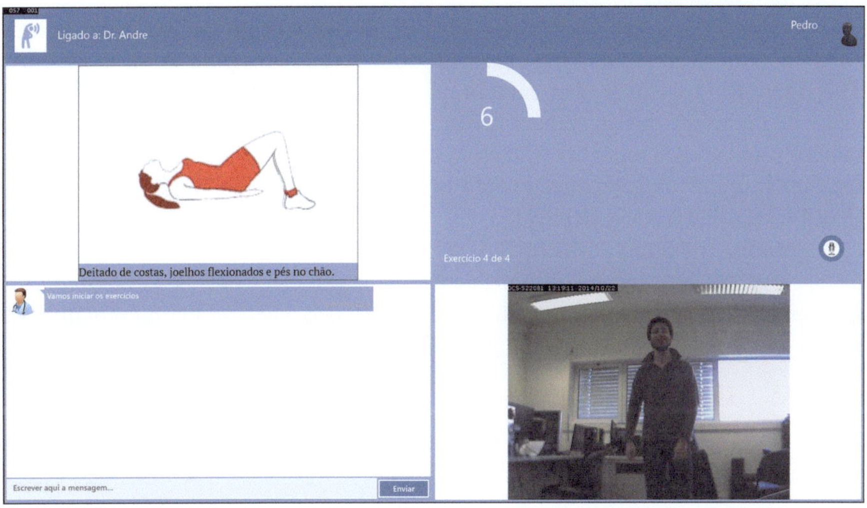

Fig. 12.6 Screen of the patient application (from [33])

12.3.1.3 How the Multimodal Architecture Was Used

Both applications of the telerehabilitation system use the framework in its operation. The two applications are executed and they communicate remotely using services to exchange information related to the components. As the video from the patient to the supervisor is only for human use, it is not handled as a modality.

For each application, an IM runs in the same machine, providing multimodal capabilities to the applications. Both devices are configured to run the generic speech modality, which has potentially more importance for the patient's application since they are making exercises in front of a TV connected to the computer. For the health professional, touch may be more suitable to perform tasks. Notwithstanding, speech is also available. The login to each application can be initiated by pressing the button "Login" or by speaking "authenticate" or "sign in."

An important use of the speech recognition modality is the possibility to interact with the chat component (both applications), the user can speak "dictate message" to activate the dictation mode and then dictate a message. The message is recognized and shown in the chat component, the user verifies that the message was correctly recognized, and speaks "send message." Every time a user receives a chat message from the other user, an event is generated and sent to the IM, which sends it to the speech synthesis to be read. With this type of feedback the user does not need to read the message on the screen.

Many other tasks can be accomplished using touch or speech. For instance, in the patient application the user can choose to pause or move to the next exercise. On the health professional side, it is possible to select the sensor in which to focus, or move the camera direction.

The events generated in the modalities are encoded using the life cycle events, which contain the markup language with the information of the event in EMMA. The messages generated also contain attributes relative to the modality, but the name of the event that generates the same output is the same either in the touch modality and speech modality. If a user swipes the finger to the right or speaks "turn the camera right" the event inside the EMMA will always be [ACTION].[VIDEO].[SWIPE].[DIRECTION].[RIGHT].

A gestures modality, which uses the Kinect sensor, is used to calculate the distance to the user and, based on thresholds, it generates events so the interface can be adapted and the areas with more importance, such as the presentation of exercises, can be presented with a zoom factor.

12.3.1.4 Discussion

This application was our first application considering the MMI architecture. At this stage, we started to develop the multimodal framework, starting with the IM and simple modalities to serve only the purpose of this application. While developing the framework and modalities, the requirements needed to create generic modalities were acquired, enabling us to design and develop the generic framework allowing a simpler integration of MMI to new applications.

12.3.2 AALFred: The Personal Assistant of Project PaeLIFE

In the European AAL Joint Program project PaeLIFE,[6] the W3C MMI architecture was adopted by the consortium as the basis for the MMI. This project developed a personal assistant named AALFred, that aggregates different social and messaging services such as email, Facebook®, and Twitter®, and provides news, weather forecasts, and nearby places of interest. Using a multimodal framework, several modalities are included, offering users a wider range of ways to interact, readily available to any module added to the system by any of the partners. With a strong emphasis on the speech modality, most of the content of the assistant can be accessed through speech interaction, even dynamic content, and the decoupled nature of modalities allowed deploying expedite methods to ensure the support of several languages (Portuguese, English, French, Polish, and Hungarian).

12.3.2.1 Requirements

In order to better perceive the fundamental requirements for the development of a user friendly and multimodal application, a brainstorm session was conducted with the main stakeholders. By focusing on the end user itself and the characteristics of the AAL environments, the following requirements were defined:

- MMI support: With a special focus on natural language via speech;
- Multilingual application: Given the focus of the project as a European project, it was decided to that each partner's native language should be supported;
- Decoupled solution: To allow the distribution of modalities across different devices and to allow a smoother integration of modules developed by partners;
- Modular solution: Not only create modules to support different social services and other information services, but also allow the future inclusion of third party modules, if needed, through the creation of a hub for this purpose.

12.3.2.2 General Presentation of AALFred

The AALFred system is a personal assistant application containing a collection of information and social services developed through the collaboration of multiple partners from the PaeLIFE consortium. The application integrates the developed multimodal framework at its core with the main objective of allowing users to interact with the application using any available modality.

Figure 12.7 illustrates the initial view of the application, the main menu, listing all accessible options within the application. The menu is based on a set of modules added through a developed hub, and offers options such as managing the agenda or

[6] Project PaeLIFE—Personal Assistant to Enhance the Social Life of Seniors, http://www.paelife.eu.

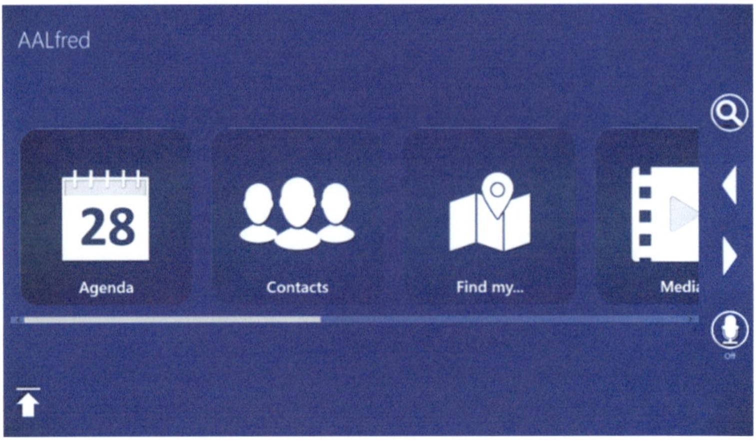

Fig. 12.7 The initial screen of AALFred shows the list of available modules

the user's contacts, check the weather and the most recent news, communicate with others via messaging or access media content such as videos or photos.

Figure 12.8 shows an example of using the agenda. Given the multimodal nature of the application, users, in accordance with their preference, are able to use either touch or speech to navigate. During interaction, the AALFred assistant provides the user with suggestions of possible options using speech synthesis. For instance, it may inform the user on how to create a new appointment using speech instructions ("Do you want to create a new appointment? Open a day and say create a new appointment").

Through the indication of the user localization (or language preference), the system automatically adopts the corresponding language for the speech modality. In the same manner, the interfaces automatically translate existing text to the correct language.

12.3.2.3 System Architecture

The AALFred architecture features two main parts, one on the cloud and the other on a local computational device such as a tablet, smart TV, or a personal computer. The architecture is illustrated in Fig. 12.9. The left side illustrates the user and the available devices. The right side illustrates the cloud services including a database for user data management and several modules for accessing external services such as Facebook or Youtube as well as services to allow improved speech recognition. Connecting the two sides is an API using web services through the PLA.

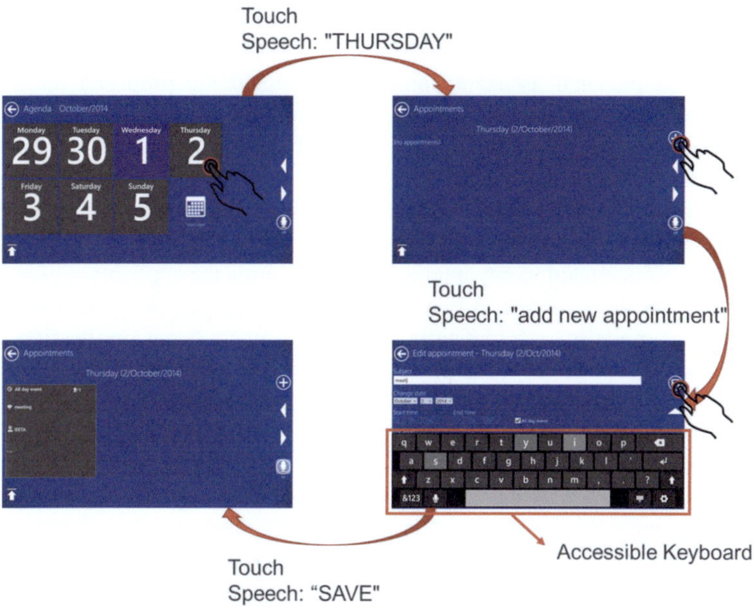

Fig. 12.8 Interactions flow to create a new appointment in the agenda, the user starts by choosing the preferred way to interact with the application (touch or speech) to select the day and then select the option to add a new appointment, then fill the subject with the virtual keyboard, and again select to save the appointment by touch or speech

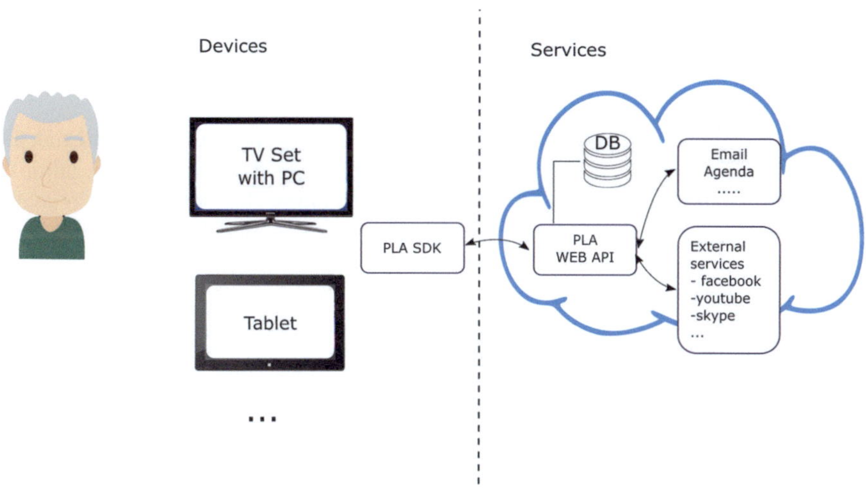

Fig. 12.9 AALFred personal life assistant (PLA) system architecture

12.3.2.4 How the Architecture Was Used

The AALFred personal assistant application adopted the proposed multimodal framework entirely and, therefore, all events related to the interaction are managed, without exception, by the IM. An overview of the connected modalities and their integration is shown in Fig. 12.10.

The graphical user interface mainly encompasses the entire application itself and as such, it is tightly connected to the application's business core. The IM controls the flow of events related to any interaction and informs the application of any new event. Each time an event occurs in a modality, using speech, touch, or gestures, the correspondent modality automatically sends a life cycle event notification to the IM. When received, the event is processed in the IM and sent to the application which may cause changes in the graphical user interface. It is noteworthy that from its part, the application may also notify modalities (such as the speech synthesis), through the IM, to output information to the user.

Regarding the usage of the application, it is possible to perform certain operations using complementary or redundant actions. For instance, in order to operate AALFred's main menu, it is possible to: (1) use touch to drag the content from one side to the other; (2) use Kinect to scroll horizontally using swipe gestures; or (3) use simple commands such as "left" or "right" through speech. These options are shown in Fig. 12.11.

AALFred was the first application in which we made full use of the generic speech modality and its automatic translation feature using grammars [13].

In addition to the automatic translation feature, speech support for AALFred allows the users to use different sentences to produce the same result. For instance, saying "my agenda" or "open agenda" produces the same output thus providing diversity to the user and promoting a more natural interaction instead of a more robotic interaction using fixed commands.

After project conclusion, a modality based in eye-tracking was developed and, as it produced semantic information compatible with the existing modalities, AALFred could be controlled by gaze without complex changes [34]. Also some initial work was performed regarding fusion of this new modality with the speech modality [35].

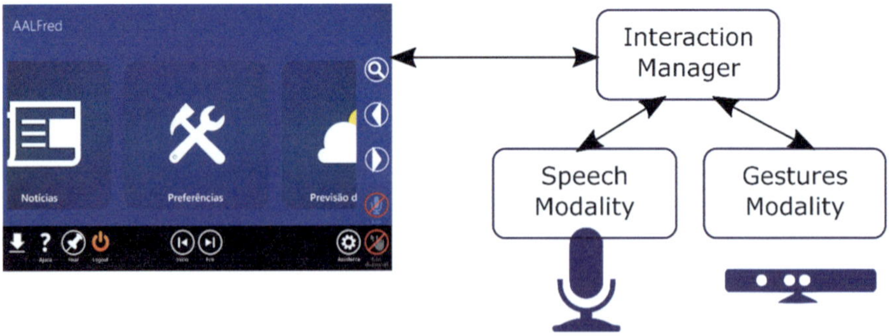

Fig. 12.10 Integration of AALFred with the Multimodal Framework. All interaction related events are managed by the Interaction Manager

Fig. 12.11 Interaction with the application AALFred is possible through different modalities that can be used interchangeably

12.3.2.5 Discussion

The decoupled nature of the architecture allowed the development of the modules without having to concern with the complexity of modalities. At the end, by following the proposed methodology, it is possible to use different modalities, even modalities not considered during the first stages of the application development.

Also, the framework simplified the work needed to support different languages, on the speech modality, due to the automatic translation of grammars.

12.4 Conclusion

This chapter presents the overall characteristics of our W3C-based MMI architecture, developed to serve multimodal application design and development in AAL scenarios. These scenarios require that users are able to access systems and applications in different situations, using different devices, and the adopted MMI architecture enables an easy and scalable solution to adapt to specificities of the users, tasks, and environments.

The adopted multimodal framework enabled the creation of several multimodal applications, for which two examples are provided. Reusing modalities was an important factor, which decreased the development time for each of them, allowing developers to focus on the features of the application and not worry much with the interaction aspects. This result is a strong outcome in favor of our inclusion of a basic set of generic modalities in the architecture, reducing one of the potential barriers for adopting the W3C architecture.

The adoption of the W3C standards in the AAL context is also an important step towards increasingly useful and open AAL solutions by addressing standardization, which is pointed out as one of the key issues to tackle [18] to further develop the field.

The positive impact of adopting a standards-based, loosely coupled architecture is that it also enables a faster response to technological evolution. For example, with the current low cost of eye-tracking technology, gaze interaction will surely be a common feature in the near future that can easily be added, as a new modality, to existing applications. One such example is our recent work on expanding AALFred with a gaze modality [34, 35].

In line with the previous point, one of our visions for the use of the MMI architecture in AAL encompasses its consideration in the increasingly prominent IoT scenarios, with the different sensors and devices seen as modalities. These can be interaction modalities, used directly, or passive modalities, used indirectly (e.g., a mood sensor providing data to adjust speech recognition or decide when to present the user with a warning) or as means for providing context-awareness (e.g., a sensor providing a "low ambient light" input to the environment). In this scope, advances in the standardization of aspects such as discovery and registration of modalities and synchronization among modalities (e.g., an avatar modality synchronized with a speech output modality) are particularly relevant and will further increase the applicability of the MMI architecture in AAL.

Finally, providing a basic set of generic modalities, shareable among different devices, with many of their components running on the cloud, can also strongly benefit the MMI user experience. In fact, considering their decoupled nature, these modalities, e.g., the proposed generic speech modality, can be used independently of the device operating system, or computational resources. Therefore, the modalities, and the data processing associated with them, can be common among devices, ensuring a uniform user experience. For example, the speech synthesizer for the media center can be the same as the one used on the smartphone, or watch, and an improvement on the synthesizer will automatically be propagated to all devices.

12.4.1 Future Work

Following on the positive outcomes of adopting the MMI architecture and proposing generic modalities, there are several aspects that deserve further attention and should serve as novel research goals. To allow full exploration of multimodality, one of the key aspects we deem relevant to address in more detail is fusion and fission. Although they have been addressed for the different applications presented in this chapter, in our view we still lack a more systematic and versatile approach to fusion and fission, namely regarding the best placement of these modules. Another important aspect needing more work is the set of generic modalities that must be expanded.

Acknowledgements The research was partially funded by IEETA Research Unit funding (Incentivo/EEI/UI0127/2014), Marie Curie IAPP project IRIS (ref. 610986, FP7-PEOPLE-2013-IAPP), project Smart Phones for Seniors (S4S), a QREN project (QREN 21541), co-funded by COMPETE and FEDER, project PaeLIFE (AAL-08-1-2001-0001), and project AAL4ALL (AAL/0015/2009).

The authors thank all W3C MMI recommendations contributors for their insightful and inspiring approaches to MMI.

References

1. Steg, H., Strese, H., Loroff, C., Hull, J., & Schmidt, S. (2006). *Ambient assisted living–European overview report, March 2006*. EU Specific Support Action.
2. Kung, A., & Jean-Bart, B. (2010). Making AAL platforms a reality. *Ambient Intelligence, 6439*, Lecture Notes in Computer Science, 187–196.
3. Mikulecký, P., Lišková, T., Čech, P., & Bureš, V. (2009). *Ambient intelligence perspectives: Selected papers from the First International Ambient Intelligence Forum 2008*. Amsterdam: IOS Press.
4. Hämäläinen, A., Teixeira, A., Almeida, N., Meinedo, H., Fegyó, T., & Dias, M. S. (2015). Multilingual speech recognition for the elderly: The AALFred personal life assistant. Proceedings of the 6th International Conference on Software Development and Technologies for Enhancing Accessibility and Fighting Infoexclusion (DSAI 2015) June 10–12, 2015 Fraunhofer FIT, Sankt Augustin, Germany
5. Teixeira, A., Ferreira, F., Almeida, N., Silva, S., Rosa, A. F., Pereira, J. C., et al. (2016). Design and development of Medication Assistant: older adults centred design to go beyond simple medication reminders. *Universal Access in the Information Society*.
6. Jara, A. J., Zamora, M. A., & Skarmeta, A. F. G. (2011). An internet of things–based personal device for diabetes therapy management in ambient assisted living (AAL). *Personal and Ubiquitous Computing, 15*(4), 431–440.
7. Signoretti, A., Martins, A. I., Almeida, N., Vieira, D., Rosa, A. F., Costa, C. M. M., & Teixeira, A. (2015). Trip 4 all: A gamified app to provide a new way to elderly people to travel. *Procedia Computer Science, 67*, 301–311.
8. Li, R., Lu, B., & McDonald-Maier, K. D. (2015). Cognitive assisted living ambient system: A survey. *Digital Communications and Networks, 1*(4), 229–252.
9. Dasios, A., Gavalas, D., Pantziou, G., & Konstantopoulos, C. (2015). Hands-on experiences in deploying cost-effective ambient-assisted living systems. *Sensors, 15*(6), 14487–14512.
10. Blackman, S., Matlo, C., Bobrovitskiy, C., Waldoch, A., Fang, M. L., Jackson, P., Mihailidis, A., Nygård, L., Astell, A., & Sixsmith, A. (2016). Ambient assisted living technologies for aging well: A scoping review. *Journal of Intelligent Systems, 25*(1), 55–69.
11. Gubbi, J., Buyya, R., Marusic, S., & Palaniswami, M. (2013). Internet of Things (IoT): A vision, architectural elements, and future directions. *Future Generation Computer Systems, 29* (7), 1645–1660.
12. Feki, M. A., Kawsar, F., Boussard, M., & Trappeniers, L. (2013). The internet of things: The next technological revolution. *Computer (Long Beach California), 46*(2), 24–25.
13. Gomes, B., Muniz, L., da Silva e Silva, F. J., Ríos, L. E. T., & Endler, M. (2015). A comprehensive cloud-based IoT software infrastructure for ambient assisted living. In *2015 International Conference on Cloud Technologies and Applications (CloudTech)*, pp. 1–8.
14. Moschetti, A., Fiorini, L., Aquilano, M., Cavallo, F., & Dario, P. (2014). Preliminary findings of the AALIANCE2 ambient assisted living roadmap. In S. Longhi, P. Siciliano, M. Germani, & A. Monteriù (Eds.), *Ambient assisted living: Italian forum 2013* (pp. 335–342). Cham: Springer.
15. Teixeira, A. (2014). A critical analysis of speech-based interaction in healthcare robots: Making a case for the increased use of speech in medical and assistive robots. In A. Neustein (Ed.), *Speech and automata in health care* (pp. 1–29). Boston: De Gruyter.
16. Teixeira, A., Braga, D., Coelho, L., & Fonseca, A. (2009). Speech as the basic interface for assistive technology. In *International Conference on Software Development for Enhancing Accessibility and Fighting Info-Exclusion*. Lisbon: UTAD.
17. Wu, D., Cai, Y., & Guizani, M. (2015). Asynchronous flow scheduling for green ambient assisted living communications. *IEEE Communications Magazine, 53*(1), 64–70.
18. Memon, M., Wagner, S. R., Pedersen, C. F., Beevi, F. H. A., & Hansen, F. O. (2014). Ambient assisted living healthcare frameworks, platforms, standards, and quality attributes. *Sensors, 14* (3), 4312–4341.

19. Gövercin, M., Meyer, S., Schellenbach, M., Steinhagen-Thiessen, E., Weiss, B., & Haesner, M. (2016). SmartSenior@home: Acceptance of an integrated ambient assisted living system. Results of a clinical field trial in 35 households. *Informatics for Health and Social Care, 41*(4), 430–447.
20. Teixeira, A., Almeida, N., Pereira, C., Silva, M. O., & Pereira, J. C. (2013). Serviços de Suporte à Interação Multimodal. In A. Teixeira, A. Queirós, & N. Rocha (Eds.), *Laboratório Vivo de Usabilidade* (pp. 151–165). Portugal: ARC Publishing.
21. Teixeira, A., Pereira, C., e Silva, M. O., Pacheco, O., Neves, A., & Casimiro, J. (2011). AdaptO—Adaptive multimodal output. In *Proceedings of PECCS*. Vila Moura, Algarve: INSTICC.
22. Teixeira, A., Pereira, C., e Silva, M. O., Alvarelhão, J., Silva, A., Cerqueira, M., et al. (2013). New telerehabilitation services for the elderly. In M. M. Cruz-Cunha (Ed.), *Handbook of research on ICTs for healthcare and social services: Developments and applications* (pp. 109–132). IGI Global, Hershey, Pennsylvania.
23. Johnston, M. (2009). Building multimodal applications with EMMA. In *Proceedings of the 2009 International Conference on Multimodal Interfaces—ICMI-MLMI'09, Cambridge, MA, USA* (p. 47).
24. Baggia, P., Burnett, D. C., Carter, J., Dahl, D. A., McCobb, G., & Raggett, D. EMMA: Extensible MultiModal Annotation markup language. W3C Recommendation. http://www.w3.org/TR/emma/. Accessed 10 Feb 2009.
25. Dahl, D. A. (2013). The W3C multimodal architecture and interfaces standard. *Journal on Multimodal User Interfaces, 7*(3), 171–182. doi:10.1007/s12193-013-0120-5.
26. Teixeira, A., Francisco, P., Almeida, N., Pereira, C., & Silva, S. (2014). Services to support use and development of speech input for multilingual multimodal applications for mobile scenarios. In *The Ninth International Conference on Internet and Web Applications and Services (ICIW 2014), Track WSSA—Web Services-based Systems and Applications*. IARIA, Paris.
27. Almeida, N., Teixeira, A., Rosa, A. F., Braga, D., Freitas, J., Dias, M. S., et al. (2015). Giving voices to multimodal applications. In M. Kurosu (Ed.), *Human–computer interaction: Interaction technologies is a LNCS volume of the Proceedings of the 17th International Conference HCI International Los Angeles, CA, USA, August 2–7, 2015* (pp. 273–283). Cham: Springer.
28. World Health Organization (2002). Active ageing: A policy framework. A contribution of the World Health Organization to the Second United Nations World Assembly on Ageing, Madrid, Spain, April 2002, WHO/NMH/NPH/02.8. http://www.who.int/ageing/publications/active_ageing/en/
29. Kairy, D., & Lehoux, P. (2009). A systematic review of clinical outcomes, clinical process, healthcare utilization and costs associated with telerehabilitation. *Disability and Rehabilitation, 31*(6), 427–447.
30. Burdea, G. (2002). Keynote address: Virtual rehabilitation-benefits and challenges. In *1st International Workshop on Virtual Reality Rehabilitation (Mental Health, Neurological, Physical, Vocational) VRMHR, Lausanne, Switzerland* (pp. 1–11). Vila Moura, Algarve, INSTICC Portugal.
31. Siegel, C., Hochgatterer, A., & Dorner, T. E. (2014). Contributions of ambient assisted living for health and quality of life in the elderly and care services—a qualitative analysis from the experts' perspective of care service professionals. *BMC Geriatrics, 14*(1), 112.
32. Teixeira, A., Pereira, C., Oliveira e Silva, M., Almeida, N., Pinto, J., Teixeira, C., et al. (2012). Health@home scenario: Creating a new support system for home telerehabilitation. In *Proceedings of the 2nd International Living Usability Lab Workshop on AAL Latest Solutions, Trends and Applications, AAL 2012, in Conjunction with BIOSTEC 2012* (pp. 37–47). INSTICC: Vila Moura.
33. Pereira, C., Almeida, N., Martins, A. I., Silva, S., Rosa, A. F., Silva, M. O., et al. (2015). Evaluation of complex distributed multimodal applications evaluating a telerehabilitation system when it really matters. In J. Zhou & G. Salvendy (Eds.), *Human aspects of IT for the aged population: Design for everyday life (part II)* (Proceedings of the 1st International

Conference on Human Aspects of IT for the Aged Population, part of HCI International 2015). Los Angeles, CA: Springer. published as LNCS volume 9194.

34. Vieira, D. (2015). *Enhanced multimodal interaction framework and applications.* Master thesis, Universidade de Aveiro, Aveiro.
35. Vieira, D., Freitas, J. D., Acartürk, C., Teixeira, A., Sousa, L., Silva, S., et al. (2015). Read that article: Exploring synergies between gaze and speech interaction. In *Proceedings of the 17th International ACM SIGACCESS Conference on Computers and Accessibility* (pp. 341–342). New York, NY.

Part III
Applications

Chapter 13
Assembling the Jigsaw: How Multiple Open Standards Are Synergistically Combined in the HALEF Multimodal Dialog System

Vikram Ramanarayanan, David Suendermann-Oeft, Patrick Lange, Robert Mundkowsky, Alexei V. Ivanov, Zhou Yu, Yao Qian, and Keelan Evanini

Abstract As dialog systems become increasingly multimodal and distributed in nature with advances in technology and computing power, they become that much more complicated to design and implement. However, open industry and W3C standards provide a silver lining here, allowing the distributed design of different components that are nonetheless compliant with each other. In this chapter we examine how an open-source, modular, multimodal dialog system—HALEF—can be seamlessly assembled, much like a jigsaw puzzle, by putting together multiple distributed components that are compliant with the W3C recommendations or other open industry standards. We highlight the specific standards that HALEF currently uses along with a perspective on other useful standards that could be included in the future. HALEF has an open codebase to encourage progressive community contribution and a common standard testbed for multimodal dialog system development and benchmarking.

13.1 Introduction

Dialog systems nowadays are becoming increasingly multimodal. In other words, dialog applications, which started off mostly based on voice and text [15], have increasingly started to encompass other input–output (I/O) modalities such as

V. Ramanarayanan (✉) • D. Suendermann-Oeft • P. Lange • A.V. Ivanov • Y. Qian
Educational Testing Service (ETS) R&D, San Francisco, CA, USA
e-mail: vramanarayanan@ets.org

R. Mundkowsky • K. Evanini
Educational Testing Service (ETS) R&D, Princeton, NJ, USA

Z. Yu
Carnegie Mellon University, Pittsburgh, PA, USA

D.A. Dahl (ed.), *Multimodal Interaction with W3C Standards*,
DOI 10.1007/978-3-319-42816-1_13

video [3], gesture [3, 17], electronic ink [10, 11], avatars or virtual agents [6, 25, 26], and even embodied agents such as robots [7, 29], among others. While the integration of such technologies provides a more immersive and natural experience for the users and enables an analysis of their non-verbal behaviors, it also makes the design of such multimodal dialog systems more complicated. This is because, among other things, one needs to ensure a seamless user experience without any reduction in quality of service—this includes issues such as latency, accuracy, and sensitivity—while transporting data between each of these multimodal (and possibly disparate) I/O endpoints and the dialog system. In addition, dialog systems consist of multiple subsystems; for example, automatic speech recognizers (ASRs), spoken language understanding (SLU) modules, dialog managers (DMs), and speech synthesizers, among others, interacting synergistically and often in real-time. Each of these subsystems is complex and brings with it design challenges and open research questions in its own right. As a result, development of such multi-component systems that are capable of handling a large number of calls is typically done by large industrial companies and a handful of academic research labs since they require individual maintenance of multiple individual subsystems [5]. In such scenarios, it is essential to have industry-standard protocols and specification languages that ensure interoperability and compatibility of different services, irrespective of who designed them or how they were implemented. Designing systems that adhere to such standards also allow generalization and accessibility of contributions from a large number of developers across the globe.

The popularity of commercial telephony-based spoken dialog systems—also known as interactive voice response (IVR) systems—especially in automating customer service transactions in the late 1990s, drove industry developers to start working on standards for such systems [18]. As a core component of an IVR, the voice browser, essentially responsible for interpreting the dialog flow while simultaneously orchestrating all the necessary resources such as speech recognition, synthesis, and telephony, was one of the early components subject to standardization resulting in the VoiceXML standard dating back to 1999[1] (see Sect. 13.4.1.1 for more details on VoiceXML). Since the vast majority of authors responsible for creating standards such as VoiceXML come from the industry, most implementations of spoken dialog systems adhering to these standards are commercial, proprietary, and closed-source applications. Examples of voice browser implementations include

- Voxeo Prophecy[2]
- TellMe Studio[3]
- Plum DEV[4]

[1] http://www.w3.org/TR/2000/NOTE-voicexml-20000505.

[2] https://voxeo.com/prophecy/.

[3] https://studio.tellme.com/.

[4] http://www.plumvoice.com/products/plum-d-e-v/.

- Cisco Unified Customer Voice Portal[5]
- Avaya Voice Portal[6]

In addition to over 20 commercial closed-source voice browsers,[7] we are aware of a single open-source implementation that has been actively developed over the past few years:

- *JVoiceXML*.[8]

We adopted this voice browser for the creation of the multimodal spoken dialog system HALEF (Help Assistant–Language-Enabled and Free), which serves as an example of a standards-based architecture in this chapter.

Note that in addition to industrial implementations of spoken and multimodal dialog systems, there exists an active academic community engaging in research on such systems. Prominent examples include

- CMU's Olympus [4]
- Alex,[9] by the Charles University in Prague [12]
- InproTK,[10] an incremental spoken dialog system
- OpenDial[11]
- the Virtual Human Toolkit [9]
- Metalogue,[12] a multimodal dialog system
- IrisTK,[13] a multimodal dialog system

Many of these examples, along with other (multimodal) dialog systems developed by the academic community, are built around very specific research objectives. For example, Metalogue provides a multimodal agent with metacognitive capabilities; InproTK was developed mainly for investigating the impact of incremental speech processing on the naturalness of human–machine conversations; OpenDial allows one to compare the traditional MDP/POMDP[14] dialog management paradigm with structured probabilistic modelling [14]. Due to their particular foci, they often use special architectures, interfaces, and languages paying little attention to existing speech and multimodal standards (e.g., see the discussions in [2]). For example, none of the above research systems implements VoiceXML, MRCP, or EMMA (see Sect. 13.4 for more details on these standards).

---

[5]http://www.cisco.com/c/en/us/products/customer-collaboration/unified-customer-voice-portal.

[6]https://support.avaya.com/products/P0979/voice-portal.

[7]Find a comprehensive list at https://www.w3.org/Voice/voice-implementations.html.

[8]https://github.com/JVoiceXML/JVoiceXML.

[9]https://github.com/UFAL-DSG/alex.

[10]https://bitbucket.org/inpro/inprotk.

[11]http://www.opendial-toolkit.net.

[12]http://www.metalogue.eu.

[13]http://www.iristk.net.

[14]Partially Observable Markov Decision Processes.

In this chapter, we describe a system that was designed to bridge the gap between the industrial demand for standardization and the openness, community engagement, and extensibility required by the scientific community. This system, HALEF, is an open-source cloud-based multimodal dialog system that can be used with different plug-and-play back-end application modules [21, 24, 30]. In the following sections, we will first describe the overall architecture of HALEF (Sect. 13.2) including its operational flow explaining how multimodal interactions are carried out (in Sect. 13.3). We will then review major components of multimodal dialog systems that have previously been subject to intensive standardization activity by the international community and discuss to what extent these standards are currently reflected (or are planned in the future) in the HALEF framework. These include

- standards for **dialog specification** describing system prompts, use of speech recognition and interpretation, telephony functions, routing logic, etc. (primarily VoiceXML), see Sect. 13.4.1.1 (also see [1]);
- standards controlling properties of the **speech recognizer**, primarily grammars, statistical language models, and semantic interpretation (e.g., JSGF, ARPA, WFST), see Sect. 13.4.1.2;
- standards controlling properties of the **speech synthesizer** (primarily SSML);
- standards controlling the **communication** between the components of the multimodal dialog system (SIP, MRCPv2, WebRTC, EMMA), see Sect. 13.4.2;
- standards describing the **dialog flow** and how **modalities** interact (SCXML, EMMA), see Sect. 13.5.

13.2 The HALEF Dialog System

The multimodal HALEF framework [21, 24, 30] is composed of the following distributed open-source modules (see Fig. 13.1 for a schematic overview):

- Telephony servers—Asterisk [28] and Freeswitch [16]—that are compatible with SIP (Session Initiation Protocol), PSTN (Public Switched Telephone Network) and WebRTC (Web Real-Time Communications) standards, and include support for voice and video communication.
- A voice browser—JVoiceXML [22]—that is compatible with VoiceXML 2.1, can process SIP traffic, via a voice browser interface called Zanzibar [20] and incorporates support for multiple grammar standards such as JSGF (Java Speech Grammar Format), ARPA (Advanced Research Projects Agency), and WFST (Weighted Finite State Transducer), which are described in Sect. 13.4.1.2.
- An MRCPv2 (Media Resource Control Protocol Version 2) speech server—which allows the voice browser to control media processing resources such as speech recorders, speech recognizers, or speech synthesizers over the network. It relies on other protocols such as SIP for session handling, RTP (Real-time Transport Protocol) for media streaming, and SDP (Session Description

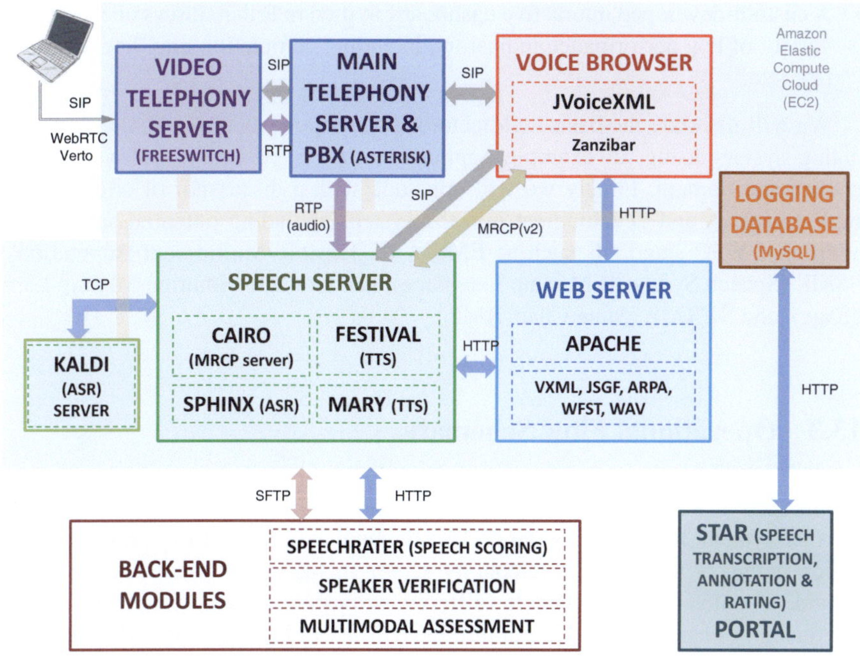

Fig. 13.1 System architecture of the HALEF spoken dialog system depicting the various modular open-source components as well as W3C standard protocols that are employed

Protocol) to allow the exchange of other capabilities such as supported codecs over the network. HALEF supports multiple speech recognizers (Sphinx [13], Kaldi [19]) and synthesizers (Mary [23], Festival [27]).

- A webserver—Apache Tomcat[15] that can host web applications that serve dynamic VoiceXML pages, web services, as well as media libraries containing grammars and audio files.
- OpenVXML, a voice application authoring suite that generates dynamic web applications that can be housed on the web server (also see Sect. 13.4.1.1).
- A MySQL[16] database server for storing call log information. All modules in HALEF connect to the database and write their log messages to it. We then post-process this information with stored procedures into easily accessible views.
- A custom-developed, open-source Speech Transcription, Annotation and Rating (STAR) portal that we implemented using PHP and the JavaScript framework jQuery. The portal allows one to analyze, listen to (or watch) full-call (video) recordings, transcribe them, rate them on a variety of dimensions such as caller experience and latency, and perform various semantic annotation tasks required to train automatic speech recognition and spoken language understanding modules.

[15]http://tomcat.apache.org/.

[16]https://www.mysql.com/.

- A custom-developed interactive dashboard written in R that allows one to view a variety of key performance indicators, including completion rate, latency, busy rate, etc.

We will illustrate the basic architecture and components of the HALEF spoken dialog system using an example application that is currently deployed in the educational domain. Finally we will conclude with a discussion of ongoing and future research and development into the system, including potential support for additional W3C standards such as EMMA (Extensible Multimodal Annotation), SSML (Speech Synthesis Markup Language), EmotionML (Emotion Markup Language), and SCXML (State Chart XML).

13.3 Operational Flow Schematic

In this section we describe how video and audio data flow to/from the multimodal HALEF system. In case of regular PSTN telephony, users call into a phone number which connects them to the telephony server in the cloud where they need to provide an extension to connect to (different extensions are associated with different dialog system instances that in turn have different task content). Alternatively, users can use softphones (or SIP phones) to connect directly to the IP address of the cloud-based telephony server using the extension. Even more convenient is the use of a web application to call directly out of a web browser application on either a computer, smartphone or tablet device. Here, the only information required by the user is the URL of the website containing the connection configuration (which includes the telephony server IP address and the extension). The Media Capture and Streams API[17] enables access to the computer's audio and video input devices via the web browser. WebRTC[18] is then used via a Javascript implementation to send video and audio to FreeSWITCH and receive audio back from FreeSWITCH. When the call comes in from the user, HALEF starts the dialog with an audio prompt that flows out of the HALEF system via Asterisk over SIP/RTP to FreeSWITCH. FreeSWITCH then sends the audio to the web browser via WebRTC. The user then gives a response to the system that flows through WebRTC to FreeSWITCH and then through SIP/RTP to Asterisk. During the teleconference, the user's video and audio interactions are continuously streamed and recorded.

Once the Asterisk server receives the call, it sends a notification to the voice browser to fetch the VXML code from the web server. The voice browser in turn identifies the resources that the speech server will need to prepare for this application. It then notifies the MRCPv2 server and starts sessions and channels for all required resources including the provisioning of speech recognition grammars.

[17]https://www.w3.org/TR/mediacapture-streams.

[18]http://www.w3.org/TR/webrtc/.

Finally, the speech server sends a SIP response back to the voice browser and Asterisk to confirm session initiation. Completion of this process successfully establishes a communication channel between the user and HALEF's components. Once the session is established, Asterisk streams audio via RTP to the speech server. When the caller starts speaking, the Sphinx engine's voice activity detector fires and identifies speech portions; then, the speech is sent to the ASR engine (HALEF supports both Kaldi and Sphinx) which starts the decoding process. When the voice activity detector finds that the caller has finished speaking, the recognition result is sent back to the voice browser, which processes it and sends this answer to the spoken language understanding module. The output of the natural language understanding module is subsequently sent to the dialog manager which evaluates and generates VXML code with the final response to be spoken out by the speech synthesizer (either Festival or Mary). The voice browser then interprets this VXML code and sends a synthesis request to the speech server with the response. The speech synthesizer synthesizes the response and passes the result back via RTP to Asterisk, which forwards the audio signal to the user. At the same time, Cairo sends a confirmation signal to the voice browser. After receiving this signal, the voice browser sends a cleanup request to close all open channels and resources. This ends the SIP session with Asterisk, which finally triggers Asterisk to send an end-of-call signal to the user.

There are other endpoints that are supported or likely can be supported by HALEF. An endpoint is defined as a device at the edge of the network (e.g., a telephone or a soft phone). Note that HALEF also natively supports audio-only dialogs with PSTN (public switched telephone network) or soft phone endpoints (that, for example, can use PSTN/SIP proxies such as ipKall).[19] We have successfully tested and used SIP clients for this purpose such as Peers[20] for PC and 3XC[21] for smartphones. We have also used SIP over WebRTC, and SIP/WebRTC clients such as sipml5,[22] jssip,[23] etc. to connect to HALEF directly through Asterisk as well as via webrtc2sip[24] to Asterisk.

13.4 Standards Used in HALEF

The following section examines in more detail how different specific industry standard specifications are synergistically combined within the HALEF multimodal dialog framework. Since HALEF is primarily a spoken dialog system, we first

[19]http://www.ipkall.com/.

[20]http://peers.sourceforge.net/.

[21]http://www.3cx.com/voip/sip-phone/.

[22]https://www.doubango.org/sipml5/.

[23]http://www.jssip.net/.

[24]http://webrtc2sip.org/.

examine the key voice standards used in its operation. We then describe the various communication standards used to transport voice and video data across different components of the dialog system.

13.4.1 Voice Standards

13.4.1.1 VoiceXML

The origins of VoiceXML[25] began in 1995 as an XML-based dialog design language intended to simplify the speech recognition application development process within an AT&T project called Phone Markup Language (PML). VoiceXML or VXML was designed for creating audio dialogs that feature synthesized speech, digitized audio, recognition of spoken and DTMF key input, recording of spoken input, telephony, and mixed initiative conversations. It was conceived to integrate the advantages of web-based development and content delivery into interactive voice response applications. The code listing below shows an example VXML page as used by HALEF. This example VXML page illustrates how various system parameters can be specified, such as the timeout value of 3 s specified in the `timeout` variable. Also, this example shows several of the components required for the interactive conversation, such as the system prompt (a prerecorded audio file, in this case) specified in the `<prompt>` element and the grammar file (see Sect. 13.4.1.2) that controls which user utterances can be recognized by the ASR system.

```
<vxml version="2.1">
<form id="InputRequestForm" scope="document">
 <field name="A_try_peanuts">
  <property name="bargein" value="true"/>
  <property name="timeout" value="3s"/>
  <property name="confidencelevel" value="0.5"/>
  <property name="sensitivity" value="0.5"/>
  <property name="speedvsaccuracy" value="0.5"/>
  <property name="completetimeout" value="3s"/>
  <property name="incompletetimeout" value="3s"/>
  <property name="maxspeechtimeout" value="10s"/>
  <property name="inputmodes" value="voice"/>
  <property name="com.telera.speechenabled" value="true"/>
  <prompt bargein="true" xml:lang="en-US">
  <audio
     src="/7703/-/resources/EPS_Builder_Voice/Default/peanuts_offer.wav"/>
  </prompt>
  <grammar mode="voice" type="application/srgs+xml"
     src="/7703/-/resources/EPS_Builder_Voice/Default/try_peanuts.
     gram"/>
 <filled>
```

[25] http://www.w3.org/TR/voicexml20/.

```
  <var name="lastresult" expr="'<lastresult>'"/>
  <submit
      next="/7703/-/next?Action_216121ee52ce43378ca2e014b92f71b4=
      success.filled"
      method="post" namelist="A_try_peanuts last result"/>
  </filled>
  <noinput></noinput>
 <nomatch></nomatch>
 <catch event="connection.disconnect.hangup"></catch>
 </field>
<catch event="externalmessage.cpa.machine"></catch>
<catch event="externalmessage.cpa.beep"></catch>
<catch event="externalmessage.cpa.machine"></catch>
</form>
<catch event="connection.disconnect.hangup"></catch>
</vxml>
```

However, developers of dialog applications who are not familiar with the VXML markup language may prefer to define dialog flows using a simpler, flowchart-based GUI instead of manual coding. Therefore we have integrated the OpenVXML toolkit into the HALEF framework. OpenVXML is an open-source software package[26] written in Java that allows designers to author dialog workflows using an easy-to-use graphical user interface, and is available as a plugin to the Eclipse Integrated Developer Environment.[27] OpenVXML allows designers to specify the dialog workflow as a flowchart, including details of specific grammar files to be used by the speech recognizer and text-to-speech prompts that need to be synthesized. In addition, they can insert "Script" blocks of Javascript code into the workflow that can be used to perform simple processing steps, such as natural language understanding on the outputs of the speech recognition. The entire workflow can be exported to a Web Archive (or WAR) application, which can then be deployed on a web server running Apache Tomcat.

Figure 13.2 shows a simple OpenVXML dialog flow where callers are required to accept or decline an offer of food in a pragmatically appropriate manner. This example can be compared to the example VXML page shown in the earlier code listing to illustrate the differences between designing a dialog directly using VXML or through the OpenVXML authoring tool. The VXML code therein corresponds to the first block in Fig. 13.2 in which a system prompt is played ("Would you like some of these chocolate covered peanuts? ...") By double-clicking on this block in the OpenVXML tool, the designer specifies the prompt that should be played or generated by the TTS engine (as indicated in the `<prompt>` element in the VXML page), the grammar that should be used to recognize the utterance by the ASR system (corresponding to the `<grammar>` element in the VXML page), as well as a variety of system parameters, such as the timeout variable. This GUI-based representation in OpenVXML is then translated into VXML pages at run-time so that it can be interpreted by the voice browser.

[26]https://github.com/OpenMethods/OpenVXML.

[27]www.eclipse.org.

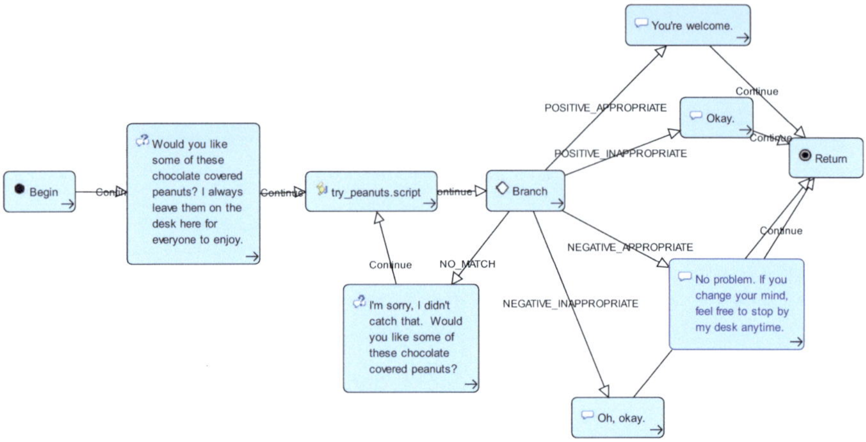

Fig. 13.2 Example design of a workplace pragmatics-oriented application targeted at non-native speakers of English where the caller has to accept or decline an offer of food (peanuts, in this case) in a pragmatically appropriate manner

The aforementioned item was designed to measure two primary constructs of English language proficiency: (1) task comprehension, i.e., correctly understanding the stimulus material and the questions being asked and (2) pragmatic appropriateness, i.e., the ability to provide a response that is appropriate to the task and the communicative context. The caller dials into the system and then proceeds to answer one or more questions, which can either be stored for later analysis (so no online recognition and natural language understanding is needed) or processed in the following manner: depending on the semantic class of the callers' answer to each question (as determined by the output of the speech recognizer and the natural language understanding module), they are redirected to the appropriate branch of the dialog tree and the conversation continues until all such questions are answered.

13.4.1.2 Voice Grammar and Language Model Standards

Grammars are used by speech recognizers to determine what a speech recognizer should listen for, and so describe the utterances a user may say. This section describes the standard grammar formats (JGSF, ARPA, WFST, and SRGS) in use by the spoken dialog community. Note that while currently HALEF only includes support for the first three, we plan to include support for this in the future.

1. *JGSF*:
   The JSpeech Grammar Format (JSGF[28]) is a platform- and vendor-independent textual representation of grammars for use in ASR. It adopts the style and

[28]JSGF (see http://www.w3.org/TR/jsgf/) is technically not a W3C standard. It is a member submission and is published as a W3C note.

conventions of the Java Programming Language in addition to use of traditional grammar notations. For example, the following JSGF grammar accepts one of two speech recognition outputs, “yes” or “no.”

```
#JSGF V1.0;
grammar yesno;
public <yesno> = yes | no;
```

2. *ARPA*:
   Although not a W3C recommendation, the Advanced Research Projects Agency (ARPA) format was one of the first popular ones that allowed specification of *statistical* grammars (also called language models or LMs) such as finite state automata (FSA) or statistical n-gram models. The language model is a list of possible word sequences. Each sequence listed has an associated statistically estimated language probability tagged to it. The following listing shows an example of a yes/no ARPA grammar.

```
This is an example ARPA-format language model file
\data\
ngram 1=4
ngram 2=4
ngram 3=4

\1-grams:
-0.7782 </s> -0.1761
-0.3010 <s> -0.5228
-0.7782 no -0.3978
-0.7782 yes 0.0000

\2-grams:
-0.1761 </s> <s> -0.0791
-0.3978 <s> no 0.1761
-0.3978 <s> yes -0.2217
-0.1761 no </s> 0.1761

\3-grams:
-0.3010 </s> <s> yes
-0.3010 <s> no </s>
-0.3010 <s> yes </s>
-0.3010 no </s> <s>

\end\
```

3. *WFST*:
   Speech and dialog system developers nowadays are increasingly moving to the Weighted Finite State Transducer (WFST) representation to write statistical grammars for their applications owing to its simplicity and power, even though it is not an official W3C recommendation. WFSTs are automata where each transition has an input label, an output label, and a weight. The weights can be used to represent the cost of taking a particular transition. The following shows an example of a WFST grammar (in text form) that accepts the words “yes” or “no.”

```
# arc format: src dest ilabel olabel [weight]
# final state format: state [weight]
# lines may occur in any order except initial state must be first line
# unspecified weights default to 0.0 (for the library-default Weight type)
0 1 yes yes 0.5
0 1 no no 1.5
1 2.0
EOF
```

4. *SRGS*:
   The Speech Recognition Grammar Specification (SRGS[29]) allows the grammar syntax to be written in one of two forms—an Augmented Backus-Naur Form (ABNF) or an Extensible Markup Language (XML) form—which are semantically mappable to allow transformations between themselves. Note that although the current version of HALEF does not include support for SRGS grammars, we plan to include this in the future. The following code snippet shows how a yes/no grammar can be defined in the ABNF format of SRGS.

```
#ABNF 1.0 UTF-8;
language en-US; //use the American English pronunciation dictionary.
mode voice; //the input for this grammar will be spoken words.
root $yesorno;
$yes = yes;
$no = no;
$yesorno = $yes | $no;
```

13.4.2 Communication Standards

WebRTC[30] or Web Real-Time Communication is a free, open W3C project that provides browsers and mobile applications with Real-Time Communications (RTC) capabilities via simple APIs. It defines a set of ECMAScript APIs in WebIDL to allow media to be sent to and received from another browser or device implementing the appropriate set of real-time protocols. As explained earlier, HALEF leverages the Verto protocol implemented in the Freeswitch video telephony server that is WebRTC-based to transmit video and audio data between the user and the dialog system.

The Media Resource Control Protocol Version 2 (MRCPv2) is a standard communication protocol for speech resources (such as speech recognition engines, speech synthesis engines, etc.) across VoIP networks which is designed to allow a client device to control media processing resources on the network.

[29]http://www.w3.org/TR/speech-grammar/.

[30]See http://www.w3.org/TR/webrtc/ and https://webrtc.org/.

13.5 Other Useful Standards for Multimodal Dialog Systems

There are several other useful standards that we are exploring for potential future integration into the HALEF framework. This section takes a closer look at some of these standards.

13.5.1 EMMA

The Extensible MultiModal Annotation (EMMA[31]) markup language is intended for use by systems that provide semantic interpretations for a variety of inputs, including but not necessarily limited to speech, natural language text, GUI, and ink input. The language is focused on annotating single inputs from users, which may be either from a single mode or a composite input combining information from multiple modes, as opposed to information that might have been collected over multiple turns of a dialog. The language provides a set of elements and attributes that are focused on enabling annotations on user inputs and interpretations of those inputs. EMMA would be a very useful standard to integrate into the HALEF framework given the focus on multimodal dialog, and hence this is one standard we are looking to include support for in HALEF going forward.

13.5.2 EmotionML

Emotion Markup Language or EmotionML,[32] as the name suggests, is "intended to be a standard specification for processing emotions in applications such as: (1) manual annotation of data; (2) automatic recognition of emotion-related states from user behavior; and (3) generation of emotion-related system behavior." Given the importance and ubiquity of emotions in dialog interactions and the subsequent requirement for automated analysis and processing of emotional state data, developing systems that are compatible with EmotionML would extend the accessibility and generalizability of those systems.

[31]http://www.w3.org/TR/emma.

[32]https://www.w3.org/TR/emotionml/.

13.5.3 SCXML

State Chart XML (SCXML[33]) is, according to the spec, "a general-purpose event-based state machine language that combines concepts from Call Control eXtensible Markup Language (CCXML) and Harel State Tables." CCXML[34] is "an event-based state machine language designed to support call control features in Voice Applications (including, but not limited to, VXML). The CCXML 1.0 specification defines both a state machine and event handing syntax and a standardized set of call control elements." Harel State Tables are a state machine notation that was developed by the mathematician David Harel [8]. They offer a clean and well-thought out semantics for sophisticated constructs such as parallel states. Although we do not require an additional state machine language as such in the current version of HALEF for smooth function, including support for SCXML in HALEF would lead to an expanded and more versatile dialog functionality, allowing one to specify dialog trees as generic state machines.

13.5.4 SSML

SSML, or Speech Synthesis Markup Language,[35] is an XML-based markup language that provides users with a standardized method for controlling different aspects of the speech output generated by a text-to-speech synthesizer. SSML allows one to alter prosody attributes such as rate, pitch, and volume. It also includes support for inserting pauses of any length, changing the speaking voice while reading, and controlling many other aspects of how the text is read by the synthetic voice.

13.6 Conclusions and Outlook

We have presented the current state of the art of the HALEF system—a fully open-source, modular, and standards-compliant spoken dialog system that can be interfaced with a number of potential back-end applications. We have illustrated the various open and W3C recommendations such as VoiceXML, WebRTC, and MRCPv2, among others, associated with different parts of the HALEF operational flow, demonstrating how these help in seamlessly assembling multiple components into a fully functional multimodal dialog system. The HALEF sourcecode is open-source and accessible online.[36]

---

[33]https://www.w3.org/TR/scxml/.

[34]https://www.w3.org/TR/ccxml/.

[35]https://www.w3.org/TR/speech-synthesis/.

[36]http://halef.org.

There remain many exciting directions for future research and development. For instance, the current HALEF implementation allows for audio and video input from the user and can synthesize output audio, but does not support full-fledged multimodal synthesis. In the future we would like to be able to incorporate support for video and emotion generation, as well as the control of avatars and simulations. Additionally, we would like to incorporate W3C recommendations such as EMMA and EmotionML into the HALEF architecture.

References

1. Baggia, P., Burnett, D., Marchand, R., & Matula, V. (2016, to appear). The role and importance of speech standards. In *Multimodal interaction with W3C standards: Towards natural user interfaces to everything*. Springer.
2. Baumann, T., Buß, O., & Schlangen, D. (2010). *Inprotk* in action: *Open-source software for building German-speaking incremental spoken dialogue systems*. Fachbereich Informatik: Hamburg.
3. Bohus, D., & Horvitz, E. (2010). Facilitating multiparty dialog with gaze, gesture, and speech. In *International Conference on Multimodal Interfaces and the Workshop on Machine Learning for Multimodal Interaction* (ICMI-MLMI'10), November 8–12, 2010, Beijing, China (p. 5). ACM.
4. Bohus, D., Raux, A., Harris, T., Eskenazi, M., & Rudnicky, A.: Olympus: An open-source framework for conversational spoken language interface research. In *Proceedings of the HLT-NAACL*, Rochester (2007).
5. Damnati, G., Béchet, F., & De Mori, R. (2007). Experiments on the France telecom 3000 voice agency corpus: Academic research on an industrial spoken dialog system. In *Proceedings of the Workshop on Bridging the Gap: Academic and Industrial Research in Dialog Technologies*, NAACL-HLT, Rochester, NY, April 2007 (pp. 48–55). Association for Computational Linguistics.
6. DeMara, R. F., Gonzalez, A. J., Jones, S., Johnson, A., Hung, V., Leon-Barth, C., et al. (2008). Towards interactive training with an avatar-based human-computer interface. In *The Interservice Industry Training, Simulation & Education Conference, ITSEC (December 2008)*. Citeseer.
7. Gorostiza, J. F., Barber, R., Khamis, A. M., Pacheco, M., Rivas, R., Corrales, A., et al. (2006). Multimodal human-robot interaction framework for a personal robot. In *The 15th IEEE International Symposium on Robot and Human Interactive Communication, 2006. ROMAN 2006* (pp. 39–44). Hatfield, UK: IEEE.
8. Harel, D., & Politi, M. (1998). *Modeling reactive systems with statecharts: The STATEMATE approach*. New York: McGraw-Hill, Inc.
9. Hartholt, A., Traum, D., Marsella, S.C., Shapiro, A., Stratou, G., Leuski, A., et al. (2013). All together now. In *Proceedings of the 13th International Conference on Intelligent Virtual Agents, IVA 2013*, Edinburgh, UK, August 29–31, 2013 (pp. 368–381). Berlin/Heidelberg: Springer.
10. Hastie, H. W., Johnston, M., & Ehlen, P. (2002). Context-sensitive help for multimodal dialogue. In *Proceedings of the 4th IEEE International Conference on Multimodal Interfaces* (p. 93). Washington, DC, USA, IEEE Computer Society.
11. Johnston, M., Bangalore, S., Vasireddy, G., Stent, A., Ehlen, P., Walker, M., et al. (2002). Match: An architecture for multimodal dialogue systems. In *Proceedings of the 40th Annual Meeting on Association for Computational Linguistics (ACL)*, Philadelphia, July 2002 (pp. 376–383).

12. Jurčíček, F., Dušek, O., Plátek, O., & Žilka, L. (2014). Alex: A statistical dialogue systems framework. In *Proceedings of the 17th International Conference on Text, Speech and Dialogue, TSD 2014*, Brno, Czech Republic, September 8–12, 2014 (pp. 587–594). Switzerland: Springer.
13. Lamere, P., Kwok, P., Gouvea, E., Raj, B., Singh, R., Walker, W., et al. (2003). The CMU SPHINX-4 speech recognition system. In *Proceedings of the ICASSP '03*, Hong Kong, China.
14. Lison, P. (2013). *Structured probabilistic modelling for dialogue management*. Ph.D. thesis, University of Oslo.
15. López-Cózar, R., Callejas, Z., Griol, D., & Quesada, J. F. (2015). Review of spoken dialogue systems. *Loquens, 1*(2), e012.
16. Minessale, A., & Schreiber, D. (2012). *FreeSWITCH Cookbook*. Packt Publishing Ltd.
17. Neßelrath, R., & Alexandersson, J. (2009). A 3D gesture recognition system for multimodal dialog systems. In *6th IJCAI Workshop on Knowledge and Reasoning in Practical Dialogue Systems* (pp. 46–51).
18. Pieraccini, R., & Huerta, J. (2005). Where do we go from here? Research and commercial spoken dialog systems. In *6th SIGdial Workshop on Discourse and Dialogue*.
19. Povey, D., Ghoshal, A., Boulianne, G., Burget, L., Glembek, O., Goel, N., et al. (2011). The Kaldi speech recognition toolkit. In *Proceedings of the ASRU*, HI, USA.
20. Prylipko, D., Schnelle-Walka, D., Lord, S., & Wendemuth, A. (2011). Zanzibar openIVR: An open-source framework for development of spoken dialog systems. In *Proceedings of the TSD*, Pilsen, Czech Republic.
21. Ramanarayanan, V., Suendermann-Oeft, D., Ivanov, A., & Evanini, K. (2015). A distributed cloud-based dialog system for conversational application development. In *16th Annual SIGdial Meeting on Discourse and Dialogue (SIGDIAL 2015)*, Prague, Czech Republic.
22. Schnelle-Walka, D., Radomski, S., & Mühlhäuser, M. (2013). JVoiceXML as a modality component in the W3C multimodal architecture. *Journal on Multimodal User Interfaces 7*(3), 183–194.
23. Schröder, M., & Trouvain, J. (2003). The German text-to-speech synthesis system MARY: A tool for research, development and teaching. *International Journal of Speech Technology, 6*(4), 365–377.
24. Suendermann-Oeft, D., Ramanarayanan, V., Teckenbrock, M., Neutatz, F., & Schmidt, D. (2015). HALEF: An open-source standard-compliant telephony-based modular spoken dialog system—A review and an outlook. In *Proceedings of the IWSDS Workshop 2015*, Busan, South Korea.
25. Swartout, W., Artstein, R., Forbell, E., Foutz, S., Lane, H.C., Lange, B., et al. (2013). Virtual humans for learning. *AI Magazine, 34*(4), 13–30.
26. Swartout, W., Traum, D., Artstein, R., Noren, D., Debevec, P., Bronnenkant, K., et al. (2010). Ada and grace: Toward realistic and engaging virtual museum guides. In *Proceedings of the 10th International Conference on Intelligent Virtual Agents, IVA 2010*, Philadelphia, PA, USA, September 20–22, 2010. Lecture Notes in Computer Science (pp. 286–300). Berlin/Heidelberg: Springer.
27. Taylor, P., Black, A., & Caley, R. (1998). The architecture of the festival speech synthesis system. In *Proceedings of the ESCA Workshop on Speech Synthesis*, Jenolan Caves.
28. van Meggelen, J., Smith, J., & Madsen, L. (2009). *Asterisk: The future of telephony*. Sebastopol: O'Reilly.
29. Yu, Z., Bohus, D., & Horvitz, E. (2015). Incremental coordination: Attention-centric speech production in a physically situated conversational agent. In *16th Annual Meeting of the Special Interest Group on Discourse and Dialogue* (p. 402).
30. Yu, Z., Ramanarayanan, V., Mundkowsky, R., Lange, P., Ivanov, A., Black, A.W., et al. (2016). Multimodal HALEF: An open-source modular web-based multimodal dialog framework. In *Proceedings of the IWSDS Workshop 2016*, Saariselka, Finland.

Chapter 14
A Case Study of Audio Alignment for Multimedia Language Learning: Applications of SRGS and EMMA in Colibro Publishing

Deborah A. Dahl and Brian Dooner

Abstract The synchronization of read-aloud audio and text in language learning is a powerful reinforcement for learners at all levels. In order to provide this kind of synchronized media experience, audio must be aligned with the text so that the correct audio plays while the related text is being presented or highlighted. One solution for aligning text and audio in this way is a manual process using an audio editor, but this is time-consuming, expensive, and error-prone. A much faster and less expensive alternative is automatic alignment through the use of speech recognition. Since the text and the matching audio are known ahead of time, the speech recognizer can perform this task with a very low error rate. Further enhancing accuracy is the fact that read-aloud stories are typically recorded with careful speech at a lower word-per-minute rate than is typical of conversational speech. In Colibro Publishing's approach, a Speech Recognition Grammar Specification grammar is generated from the text and provided to a speech recognizer, which then generates Extensible Multimodal Annotation output with the exact audio timestamps for the beginning and end points of each sentence. The alignment is then used in the interactive story production process so that the correct audio is played with highlighted text.

D.A. Dahl (✉)
Conversational Technologies, Plymouth Meeting, PA, USA
e-mail: dahl@conversational-technologies.com

B. Dooner
Colibro Publishing, Philadelphia, PA, USA

D.A. Dahl (ed.), *Multimodal Interaction with W3C Standards*,
DOI 10.1007/978-3-319-42816-1_14

14.1 Introduction

Whether motivated by business, travel, entertainment, or simply general interest, language learning is a large and growing industry. In the European Union alone 93 % of secondary students are studying at least one foreign language and over 50 % are studying more than one foreign language [1]. Knowing a second language, especially English, not only increases job options and opens up business opportunities, but bilingual people even perform better on many tests of cognitive ability [2].

Although many people study foreign languages in traditional classrooms, electronic and online tools are also becoming more widely available. This is especially true when traditional classroom learning options are limited by the inability to find a teacher for the language of interest, by restrictions on taking a class due to work or family commitments, or for various other reasons. In those cases, e-learning is sometimes the best option. Of course, e-learning can also supplement traditional classroom instruction.

One well-recognized approach to language learning is reading for enjoyment, also called Free Voluntary Reading [3, 4]. Not only is the learner motivated by the desire to learn the new language, but also by the enjoyment of reading intrinsically interesting material. However, reading can be frustrating if the material is above the student's level. Material that is too difficult interrupts the learner's focus on the reading material with too many trips to the dictionary and the grammar reference. Bilingual texts can help the learner avoid these kinds of digressions, because the meaning of the text in the learner's native language is readily available.

Moreover, electronic bilingual texts open up a very exciting opportunity to add multimedia to the bilingual reading experience. The learner cannot only see and compare the text in both languages, but he/she can also hear it, spoken in a perfect native accent. It is particularly helpful if the text and audio are synchronized so that the corresponding text is highlighted when the audio is played. In this way the text and speech continually reinforce each other as the learner moves through a story, which has been shown to increase comprehension [5]. This was our goal in developing the Colibro technology described in this chapter.

14.2 About Colibro Interactive, Bilingual Stories

Colibro publishes specialized, digital content and develops multimedia reading technology for the global language learning market. With an emphasis on English Language Learning or ELL, Colibro's "recreational language learning" approach enables students to thrive in a global, connected world with social intelligence and sensitivity to different cultures.

Instead of focusing on grammar and vocabulary like more traditional language learning solutions, Colibro's content and technology is about reading and listening to fun, engaging narratives presented in bite-size stories which are read aloud by a native speaker. Colibro employs leveled text that is structured to best benefit a second language learner at his or her current stage of second language acquisition.

Colibro enables "open learning," so that mastering a new language can happen anytime, anywhere. Colibro meets language learners where they are—at their level and when they want it—in a fast-paced, global, mobile world.

This application is currently available for iPads in the iTunes store.

14.3 Efficient Alignment of Text and Audio

To provide the user experience of simultaneous text and audio, the text and audio must be aligned during the process of preparing the story. There are several options for alignment that can be considered:

1. The reader (or voice talent) could be asked to read the story one sentence at a time, saving each sentence into a separate audio file. This would disrupt the reading process for the voice talent and result in unnatural prosody.
2. Manual alignment of text and audio can be done using standard audio editors such as Audacity, or specialized speech research editors. However, the manual process is slow and error-prone, and is impractical for large quantities of text.

Given that manual approaches are impractical, we can look at approaches to automatic alignment.

1. Simply searching for silences longer than some threshold in the audio and assuming that they are sentence boundaries might be considered. However, this strategy can result in errors both due to missing actual sentence boundaries and to incorrectly assigning sentence boundaries to sentence-internal pauses. Automatically quantifying what is meant by "silence" and "length" is also problematic, since these will differ between recordings, and even between different utterances in the same recording.
2. Forced alignment of text and audio with a phonetic speech recognizer is an approach commonly used in speech research, but this capability is complex to implement and is available only in research systems. In our case, a requirement for this product was to use a supported, commercially available, speech recognizer.

In this paper we describe a new approach to alignment for text and audio using the Extensible Multimodal Annotation (EMMA) [6–8] and Speech Recognition Grammar Specification (SRGS) [9] standards. In addition, EMMA metadata such as timestamps, human language of the input, confidence, and the location of the input speech signal have all been very helpful.

Although we do not discuss it in detail in this paper, we have also found that when there is complete control of the recording process, as in our product, the voice talent can be asked to ensure that there are distinct pauses between sentences, making it possible to use technique 3a as a supplement to the speech recognition-based process described here. Future research will quantify the incremental benefit of this technique.

14.4 User Experience

At this point it is useful to provide an example to show what the user experience is like for a reader of a Colibro book.

Figure 14.1 shows an example of a page from a bilingual English/Spanish Colibro story, "Dominic the Dragon." The English and Spanish texts are aligned, and the user can hear the English audio by pressing one of the "play" buttons. The user can either hear the audio for each sentence individually or for the entire story. This page shows how the story appears for an English-speaking learner of Spanish. In some cases we have also prepared reversible versions of the same story, so that it can be used, for example, by a Spanish-speaking learner of English.

Fig. 14.1 Page from a Colibro child's book

14.5 Aligning Text and Audio

This user experience requires alignment of the text and audio so that the matching sentence text can be highlighted when the audio plays. In Colibro, this process consists of the following steps:

1. Author text and record audio.
2. Segment text into sentences.
3. Create an SRGS grammar from the sentences where each grammar rule corresponds to one sentence of the text.
4. Recognize the audio using the grammar created from the sentences.
5. Output an EMMA document with timestamp information marking the beginnings and ends of the sentences.
6. Integrate the timing information into the story bundle or package for presentation to the user.

14.6 Steps in the Alignment Process

14.6.1 Text Segmentation into Sentences

The first step in the process is segmentation of the story text into sentences. The Stanford CoreNLP [10] sentence segmentation module is used for this. The result is a text file where each sentence appears on a new line. Note that our use of sentence segmentation in this way means that currently the alignment is only at the sentence level. The alternative possibility of word-level alignment would mean that there is a much greater amount of animation in the user interface as each word is highlighted. Since animation in a GUI interface is known to be distracting [11], we wished to minimize this. Full sentence alignment is also reasonable for the simple stories currently available in the Colibro library, because the sentences are short. As the Colibro library expands to include more advanced books, we will investigate alignment of units within sentences, balancing this against the desire to minimize distracting animations.

14.6.2 Grammar Creation

The next step is to create an SRGS grammar based on the sentence segmentation, where each sentence corresponds to an alternative rule within a `<one-of>` element that spans the entire story. Thus, each sentence in the story is treated as a possible speech recognition result. Representing the entire story as a single sequence of grammar rules is not flexible enough, since a problem with one sentence can result in a "nomatch" for the entire story. While a statistical ngram language model could also have been used for this purpose, we found in our initial technology assessment that using ngram language models did not lead to sufficiently accurate results. Since the text to be recognized is fixed and known, robustness to unexpected inputs, a major factor motivating the use of ngram models, was not a significant factor. Discrepancies between the written text and the audio will only occur if the reader makes an error.

The grammar creation process is a simple text reformatting step which could be done with a number of tools such as Perl scripts or, as in our case, with regular expressions. A grammar showing the result of the grammar creation step is shown in Fig. 14.2.

```
<?xml version="1.0" encoding="UTF-8" ?>
<!DOCTYPE grammar PUBLIC "-//W3C//DTD GRAMMAR 1.0//EN"
"http://www.w3.org/TR/speech-grammar/grammar.dtd">
<grammar xmlns="http://www.w3.org/2001/06/grammar" xml:lang="en-us"
xmlns:xsi="http://www.w3.org/2001/XMLSchema-instance"
xsi:schemaLocation="http://www.w3.org/2001/06/grammar http://www.w3.org/TR/speech-
grammar/grammar.xsd" version="1.0" tag-format="semantics/1.0" mode="voice"
root="grammar">
  <rule id="grammar" scope="public">
    <one-of>
      <item>Charlemagne A Really Big Dog</item>
      <item>Charlemagne (we call him Charlie) came home as a small puppy only eight
weeks old.</item>
      <item>When he was really young he had what looked like an old man's beard.</item>
      <item>Charlie liked to eat everything in sight.</item>
      <item>He not only ate his food, but he also ate wood, dirt, flowers, and even
worms!</item>
      <item>Soon he was growing faster.</item>
      <item>And faster.</item>
      <item>Faster than anyone had ever seen a dog grow.</item>
      <item>Charlie loved to play with lots of new small puppy friends.</item>
      <item>But they just could not believe how BIG he was for his age!</item>
      <item>Charlie is now almost a year old.</item>
      <item>He is a really big dog.</item>
      <item>His dream is to be a police dog, but  he is still too young!</item>
      <item>He just doesn't know that yet...</item>
      <item>So for now, he just keeps dreaming.</item>
      <item>He's big, he's just not all grown up yet!</item>
      <item>THE END</item>
    </one-of>
  </rule>
</grammar>
```

Fig. 14.2 SRGS grammar of an English story text

14.6.3 Alignment of Sentence Audio with Text

The next step in the alignment process is to recognize the audio corresponding to the story text, using the grammar created from the story sentences. We selected Microsoft Speech Server as the recognizer for this purpose, as it is free, supports SRGS, and covers 26 languages. Because it is server-based, it also fits well into our processing pipeline. However, its recognition result is in the form of a SAPI [12] object, and we needed a standard, vendor-independent text-based format for our results. We also needed to identify both the timing information and confidence values. Although we have no specific plans to use a different speech recognizer in the future, vendor-independence prepares us for this possibility.

The W3C EMMA specification [6–8] provides this detailed metadata, and as such is ideal for representing our recognition results. It was only necessary to convert the SAPI speech results to EMMA. This was accomplished by working with Chant, Inc. (www.chant.net) to design and implement an EMMA wrapper for SAPI. The fact that EMMA is a standard was an important advantage in our collaboration with Chant because it meant that the intended output format was well-documented. Thus, it was easy to define the required output by simply referring to the EMMA specification.

The resulting EMMA for the first two sentences of the grammar in Fig. 14.2 is shown in Fig. 14.3. The entire output is contained in an `<emma:sequence>` container, with each result in a separate `<emma:one-of>`. For the purposes of alignment, the key information is contained in the `<offset-to-start>` attribute (bold in Fig. 14.3) of each `<emma:one-of>` which indicates how long after the beginning of the audio file (the "`time-ref-anchor-point`") the input began. The "`duration`" attribute, along with the "`offset-to-start`" attribute, tells when the utterance ended.

In addition to the basic information for alignment, the EMMA result contains other valuable metadata which can be used for debugging. The attributes "`emma:lang`", "`emma:signal`", "`emma:process`" and "`emma:grammar-ref`" all can be used to confirm that the processing pipeline was correctly configured. For example, "`emma:lang`" can be used to confirm that the recognizer was set to use the right language. Similarly, "`emma:signal`" can be used to confirm that the correct audio file was used.

We added one piece of non-standard information in the `<emma:info>` element, the extension point of EMMA. This was used to record the time that the audio file was processed by the recognizer. An actual processing time attribute was not anticipated in the EMMA standard because the most common use case for EMMA is real time interaction, where processing occurs during or at least immediately after the time of the utterance. In contrast, our use case is offline speech recognition which may occur considerably after the original speech.[1]

---

[1] In fact, the speech recognition in our application could occur many years after the original speech. This might happen, for example, if we wanted to align an historic speech with its

```
<emma:emma version="1.0" xmlns="http://www.example.com/example"
xmlns:emma="http://www.w3.org/2003/04/emma"
xmlns:xsi="http://www.w3.org/2001/XMLSchema-instance"
xsi:schemaLocation="http://www.w3.org/2003/04/emma http://www.w3.org/TR/2009/REC-
emma-20090210/emma.xsd">
        <emma:info>
                <process-time>
                        04/08/14 19:06:19 GMT
                </process-time>
        </emma:info>
        <emma:sequence id="seq1">
                <emma:one-of confidence="0.998" disjunction-type="recognition" du-
ration="2380" end="1396983983278" function="transcription" grammar-
ref="charlemagneISO.xml" id="oneof1" lang="en-us" media-type="audio/x-wav" medi-
um="acoustic" mode="voice" offset-to-start="780" process="Microsoft Speech Recog-
nizer 8.0 for Windows (English - US)" signal="CharlieEngOriginal.wav" signal-
size="194040" start="1396983980898" time-ref-anchor-point="start" time-ref-
uri="1396983980898" tokens="Charlemagne A Really Big Dog" verbal="true">
                        <emma:interpretation confidence="0.998"
id="interpretation1">
                                <text>Charlemagne A Really Big Dog</text>
                        </emma:interpretation>
                </emma:one-of>
                <emma:one-of confidence="0.998" disjunction-type="recognition" du-
ration="6200" end="1396983990887" function="transcription" grammar-
ref="charlemagneISO.xml" id="oneof2" lang="en-us" media-type="audio/x-wav" medi-
um="acoustic" mode="voice" offset-to-start="4440" process="Microsoft Speech Recog-
nizer 8.0 for Windows (English - US)" signal="CharlieEngOriginal.wav" signal-
size="412776" start="1396983984687" time-ref-anchor-point="start" time-ref-
uri="1396983984687" tokens="Charlemagne (we call him Charlie) came home as a small
puppy only eight weeks old." verbal="true">
                        <emma:interpretation confidence="0.998"
id="interpretation2">
                                <text>Charlemagne (we call him Charlie) came
home as a small puppy only eight weeks old.</text>
                        </emma:interpretation>
                </emma:one-of>
```


Fig. 14.3 EMMA results

The attribute "`emma:confidence`" is also useful for debugging, since a low confidence result could result from a mismatch between the audio file and the text. Mismatches can occur, for example, if the voice talent misreads the text during the recording process, or if there were problems with audio quality. Similarly, an "`uninterpreted`" attribute means that the recognizer was unable to recognize any utterance in that position, as shown in Fig. 14.4.

transcription. In that case, the standard "`emma:start`" and "`emma:end`" timestamps would be very different from the processing time, since they refer to the start and end of speech.

```
<emma:interpretation
          function="transcription"
          grammar-ref="charlemagneISO.xml"
          id="interpretation3"
          lang="en-us"
          media-type="audio/x-wav"
          medium="acoustic"
          mode="voice"
          process="Microsoft Speech Recognizer 8.0 for Windows (English - US)"
          signal="CharlieEngOriginal.wav"
          uninterpreted="true"
          verbal="true">
</emma:interpretation>
```

Fig. 14.4 Uninterpreted result

14.7 The System in Use

The system has been used for over one year to align text and audio in Colibro stories. Since Colibro is a commercial system rather than a research system, we have not been quantifying errors in detail. However, we have found that the efficiency and accuracy of this system is sufficient for its commercial purposes. Nevertheless, a manual quality control process is in place so that any errors can be caught and corrected during the story production process. One possible error is the recognizer failing to recognize a sentence, usually due to quiet audio. This is easy to detect in the quality control step because there will be an interpretation with an "`uninterpreted='true'`" value in the EMMA output, as shown in Fig. 14.4.

We have also used this system with Spanish without modification, except to replace the speech recognizer. Part of an EMMA result from a Mexican Spanish recognizer is shown in Fig. 14.5. In general we have observed anecdotally that the recognizers for languages other than US English are slightly more likely to miss sentences, which may be due to the fact that US English recognizers themselves are more accurate. However, the performance of the Spanish recognizer nevertheless meets our requirements for accuracy in a commercial system.

14.8 Related Applications

Alignment of text and audio has many applications, in addition to supporting speech research, and speech recognition has often been used for this task. For example, speech recognition has been used for alignment in applications such as karaoke [13], audio indexing [14], and literacy [15].

```
<emma:one-of confidence="0.98296"
        disjunction-type="recognition"
        duration="4970"
        end="1397230615915"
        function="transcription"
        grammar-ref="charlieSpanish.xml"
        id="oneof5"
        lang="es-mx"
        media-type="audio/x-wav"
        medium="acoustic"
        mode="voice"
        offset-to-start="23730"
        process="Microsoft Server Speech Recognition Language - TELE (es-MX)"
        signal="charlieSpanish.wav"
        signal-size="204960"
        start="1397230610945"
        time-ref-anchor-point="start"
        time-ref-uri="1397230610945"
        tokens="A Carlitos le gustaba comer todo lo que estaba a la vista."
        verbal="true">
        <emma:interpretation
                confidence="0.98296"
                 id="interpretation3">
                <text>A Carlitos le gustaba comer todo lo que estaba a la vista.</text>
        </emma:interpretation>
</emma:one-of>
```

Fig. 14.5 Mexican Spanish result

14.9 Conclusions and Future Directions

The standards discussed in this paper, SRGS and EMMA, were originally developed to support interactive spoken and multimodal dialogues. However, they also support automatic alignment of text and audio very well. Because the standards are fully defined it has been easy to communicate with both vendors and developers about the precise formats required.

The fact that SRGS is a standard also means that this application can be more vendor-independent than if we were using a proprietary grammar format.

EMMA also works well for this use case. Nearly all of our requirements for alignment metadata were already met by the EMMA standard, especially the offset time for the start of the signal, and duration, because those attributes give us our alignment. However, the language of input, confidence, tokens of input, process, and whether the input was interpreted or not are also useful for debugging and identifying problems with audio files. The only feature that we added (through the standard "`info`" element) was the actual time that recognition was performed.

We have found this approach to be sufficiently accurate and efficient for our commercial product. Future work will provide a fuller and more quantitative evaluation of our approach.

References

1. Eurostat (2016). Foreign language learning statistics. European Union. http://ec.europa.eu/eurostat/statistics-explained/index.php/Foreign_language_learning_statistics. Accessed 18 Jan 2016.
2. Bhattacharjee, Y. (2012). Why bilinguals are smarter. *New York Times*, March 17.
3. Krashen, S. (1989). We acquire vocabulary and spelling by reading: Additional evidence for the input hypothesis. *Modern Language Journal, 73*, 393–407.
4. Krashen, S. (2007). *Free voluntary reading*. Santa Barbara, CA: ABC-CLIO, LLC.
5. Lomicka, L. L. (1998). To gloss or not to gloss: An investigation of reading comprehension online. *Language Learning and Technology, 1*(2), 41–50.
6. Johnston, M. (2016). Extensible multimodal annotation for intelligent interactive systems. In D. Dahl (Ed.), *Multimodal interaction with W3C standards: Towards natural user interfaces to everything*. New York, NY: Springer.
7. Johnston, M., Baggia, P., Burnett, D., Carter, J., Dahl, D. A., McCobb, G., et al. (2009). EMMA: Extensible MultiModal Annotation markup language. W3C. http://www.w3.org/TR/emma/. Accessed 9 Nov 2012.
8. Johnston, M., Dahl, D. A., Denny, T., & Kharidi, N. (2015). EMMA: Extensible MultiModal Annotation markup language Version 2.0. World Wide Web Consortium. http://www.w3.org/TR/emma20/. Accessed 16 Dec 2015.
9. Hunt, A., & McGlashan, S. (2004). W3C Speech Recognition Grammar Specification (SRGS). W3C. http://www.w3.org/TR/speech-grammar/. Accessed 9 Nov 2012.
10. Stanford Natural Language Processing Group (2014). Stanford CoreNLP. Stanford University. http://nlp.stanford.edu/software/corenlp.shtml.
11. Galitz, W. O. (2007). *The essential guide to user interface design: An introduction to GUI design principles and techniques* (3rd ed.). Indianapolis, IN: Wiley Publishing, Inc.
12. Microsoft (2007). Microsoft Speech API 5.3 (SAPI). http://msdn2.microsoft.com/en-us/library/ms723627.aspx.
13. Shenoy, A., Wu, Y., & Wang, Y. (2005). Singing voice detection for karaoke application. *Paper Presented at the Proceedings SPIE 5960, Visual Communications and Image Processing 2005*, Bellingham, WA, USA.
14. Wilcox, L. (1988). Annotation and segmentation for multimedia indexing and retrieval. In *System Sciences, Proceedings of the Thirty-First Hawaii International Conference on System Sciences* (Vol. 252), pp. 259–266. doi:10.1109/HICSS.1998.651708.
15. Lee, K., Hagen, A., Romanyshyn, N., Martin, S., & Pellom, B. (2004). Analysis and detection of reading miscues for interactive literacy tutors. *Paper Presented at the Proceedings of the 20th International Conference on Computational Linguistics*, Geneva, Switzerland.

Part IV
Future Directions

Chapter 15
Discovery and Registration: Finding and Integrating Components into Dynamic Systems

B. Helena Rodríguez and Jean-Claude Moissinac

Abstract One of the major gaps in the current HTML5 web platform is the lack of an interoperable means for a multimodal application to discover services and applications available in a given space and network, for example, in a smart house with a network of connected objects. To address this gap, the Multimodal Interaction Working Group has produced a draft specification based on distributed services, which aims to support the Discovery and Registration of multimodal components. In this approach, the components are described and virtualized in a Resources Manager communicating bidirectionally through dedicated events. To facilitate the fine-grained management of concurrent multimodal interactions, the Resources Manager registers the distributed components and provides to the Interaction Manager the means to control them. In this way, interoperable search, discovery, and selection of heterogeneous and dynamic features on the Web of Things can be performed by multimodal applications producing natural interaction and a semantically rich user experience.

15.1 Introduction

The Multimodal Architecture and Interfaces is a current Recommendation of the World Wide Consortium (W3C) [1] introducing a generic structure and a communication protocol to allow the components in a system using multiple interaction modalities for input and outputs, called a "*multimodal system*" to communicate with each other. It proposes a generic event-driven architecture and a general frame of

B.H. Rodríguez (✉)
W3C's MMI Working Group Editor for the Discovery and Registration Activity,
Paris, France
e-mail: b.helena@soixante-dix.com

J.-C. Moissinac
Institut Mines Télécom, Télécom ParisTech CNRS, LTCI 46,
Paris Cedex 13, France

D.A. Dahl (ed.), *Multimodal Interaction with W3C Standards*,
DOI 10.1007/978-3-319-42816-1_15

reference focused in the control of the flow of data messages [2]. Thus the documents about the multimodal system architecture published by the W3C's Multimodal Interaction Working Group are mainly the description of good practices to manage multimodal—distributed—interfaces.

This web-oriented frame of reference has been proposed before the current growing evolution of distributed approaches for systems "in the cloud." At the time, the multimodal systems were mostly produced in an ad hoc manner, as shown by a state-of-the-art overview from 2012 covering 100 relevant multimodal systems where it was observed that more than 97 % of the systems had little or no distribution and consequently, no discovery and registration support [3].

Thus, for historical reasons the W3C's MMI Architecture and Interfaces and its runtime framework [4] were not addressing:

- The component's discovery and registration to support the fusion (integration) and the fission (composition) mechanisms.
- The modality component's data model needed to build this registry.
- The modality component's description to facilitate the orchestration (and even the turn-taking) mechanism in a dynamical system.

These are the three issues covered by a new services-oriented proposal, which is now published as a W3C Working Draft [5].

The proposal is an extension of the MMI Architecture and Interfaces model, designed to support the automation of the semantic discovery, registration, and composition of multimodal services. It is also designed to fulfill the requirements of a high-level Quality of Service such as the accurate selection of components when these are not available anymore, do not meet the expected functionality or disrupt the context of use.

With these goals in mind, the Working Draft was structured in three parts:

1. A new addressing method needed for the component's announcement at bootstrapping.
2. An architectural extension in order to support the handling of the state of the multimodal system using a virtual component approach for registration.
3. Two new events for the messaging mechanism, to address the requirements of discovery and registration in distributed systems.

These three results will be completed by the creation of a common and interoperable vocabulary of generic features to allow the gross and generic discovery of modalities in large networks over a concrete networking layer.

In this chapter we present this work on open standardization as follows: first in Sect. 15.2 we give an overview of the problem, followed by a study of the related work, secondly in Sect. 15.3 we describe the work on Discovery and Registration with a case study for the Web of Things, and finally in Sect. 15.4 we give a conclusion and some perspectives.

15.2 Dynamic Multimodal Systems

Historically, multimodal systems were implemented in stable and well-known environments [3]. Their complexity demanded laboratory-like implementations and very few experiences were developed for real-time contexts or the distribution of components.

Today, this situation has evolved. Due to the evolution of mobile media systems and networks, the web developer's community is progressively confronted with the need of integrate multiple modalities in dynamic contexts. Besides, this challenge is expected to be huge in the years to come, when large-scale networks like the Internet of Things (IoT) will attain a state of maturity.

The increasing amount of user-produced and collected data will require a more dynamic software behavior with a more dynamic and adequate approach. It will be necessary to handle the user's technical environment in a context where the demand for energy supply is getting higher and higher.

15.2.1 The Internet of Things and Distribution

We are also facing the need to encourage and improve the efficiency in consumption by boosting the creation of systems compatible with the IoT including the smart-grid technologies.

In a context in where many of these technologies existed as silos (communication modules, mobile, enterprise software, specific single function hardware, home automation, etc.) now they are becoming integrated and taking shape in a new class of solutions enabled through information infusion from heterogeneous sources that needs to be coupled with decision systems.

For example, in Japan [6] (as in European countries) the distributed applications will play a very strategic role in the reduction of energy consumption, helping to evolve to an on-demand model. With this goal, the sustainable consumption in houses must be handled distantly, using data collected by multimodal applications for the IoT. These applications must interact in a distributed and coordinated manner collaborating in the energetic efficiency, home automation management, and user segmentation and profiling.

Today the response to these emerging needs primarily focuses on connecting things, however, the real value will be produced by giving sense to the "connected information" that these things collect and by being able to utilize it constructively in rich applications providing natural interaction. We should focus on end-to-end scenarios and applications; the user experience has to be given high priority combined with effectively managing accuracy and reliability of information flow.

Thus, we have multiple issues, concerning the multimodal user interaction with very heterogeneous features, devices (some of them with low resources), protocols, and messaging mechanisms to be synchronized in an interoperable way.

In this context, on one hand, modality discovery and selection for distributed applications become a new working horizon giving new challenges for multimodal systems, user-centric design, and web research. And, on the other hand, generic and interoperable approaches will be unavoidable, preferably using web technologies, but capable of going beyond the browser model.

Today, there is an enormous variety and quantity of devices interacting with each other and with services in the cloud: by 2019, according to the Cisco VNI Forecast of this year, 11.5 billion mobile devices will be connected to the internet from a total of 149 billions of connected objects that are expected and the vast majority will use some form of wireless for access [7]. Web technologies are expected to be at the center of the IoT, thanks to their universal adoption and huge scalability.

15.2.2 Current Approaches

Nevertheless the definition of a standardized programming model for objects beyond the page-browser mechanism has not been established yet, and the classical internet of documents or the internet of knowledge has being built with a series of architectural premises that could be inadequate and even a foundational obstacle to this new challenge.

To address it, device-centric technologies, proposals, and protocols are spreading all over the current discussion around the Web of Things, focused on the device connection through API's and assuming that the "infinite things problem" will be resolved by creating a "virtual image" of this reality on IoT systems, classifying families of products according to the device-vendor models. But this solution will just transfer a real-world problem to a virtualized one, with the concurrency of policies, architectures, platforms, protocols, and standards that such a transfer implies.

On one hand, browser vendors are advocating for browser-based solutions assuming that a model that works well for web pages (based on the document model) and web apps (mostly based on client–server models) in computers and mobiles, can be easily extended to any other kind of objects.

But, how can we model a precipitation sensor as a document? Is it really logical to communicate with a rice cooker as a "data resource"? How do we apply a client–server model to reflexive objects in the network, acting at the same time as server and clients of their own features? Are address registries of devices as stable as the address directories of web pages or web apps? How to express on, off, or stand-by states with web technologies? And what will be the environmental and energy price to these choices?

On the other hand, device vendors are advocating for energy efficient and lightweight protocols fine-tailored for constrained devices; and willing to provide a web gateway (using RESTFul interfaces) to allow communication between these devices and web applications. After nearly 20 years of research, some of the industrial consortia, led by energy providers and home appliance vendors, built a series of low-level protocols and technologies supported by national policies: KNX [8], ZigBee SEP 2.0 [9, 33], Z-Wave [10], Echonet [11], ETSI M2M [12], DLNA [13], UPnP [14], ZeroConf [15], etc.

As the list above shows, these concurrent protocols and technologies have to be evaluated and selected by a developer or a new device producer. If this panorama of exploded technologies continues, the situation that mobile developers endured during years will reappear in the web of objects: heterogeneous operating systems, closed SDK's, app distribution circuits, and developing models for an infinity of objects.

The W3C is working on a response to a real and urgent need of a vendor-agnostic model of components and communication, to encompass the diversity of proposals and technologies of the IoT.

Besides the Multimodal Interaction Working Group two other groups are addressing this need to reduce the effort of implementation for developers and app vendors groups. They are the Web of Things Interest Group [16] and the WebApps Working Group [17].

While these two groups work mainly in the connection to web browser API's and the low-level control of distributed devices, our effort in the MMI Working Group has been always focused to evolve web technologies from device-centric applications, to natural interaction experiences and user-centered models that will extend the definition of an application to seamlessly encompass multiple heterogeneous devices collaborating and sharing resources and computational capabilities, both locally and across the web.

As an illustration of the problem, in a multimodal system, devices may contain nested logical devices, as well as functional units, or services.

A functional categorization of devices is currently defined by the UPnP protocol with 59 standardized device templates and a generic template profile, the Basic Device. With the same spirit, the Echonet consortium defines a number of eight device groups, Zigbee classifies the devices with six profiles and KNX proposes more than 146 functional blocks for a non-defined classification of devices (this standard is service-oriented). In all cases, the specification defines explicitly the device's properties and access methods.

In contrast, more generic protocols like Z-wave use the Generic device approach and three abstract classes of devices, Zigbee SEP 2.0 defines only three device categories while BACnet uses 13 Function Groups and five service types.

Classifications of Devices are provided also by the Composite Capability Preference Profiles Specification or even with the User Agent Profile Specification extension of CC/PP [18] maintained by the Open Mobile Alliance (formerly the WAP Forum) with the Specification's Part 7: Digital Item Adaptation, in which Terminals and Terminals capabilities are described.

Finally according to the OCDE, the things of the IoT can be classified according to eight dimensions: mobility, size, complexity, dispersion, power supply, placement, connectivity, and animatedness [19].

The W3C's WebApps Working Group [17] is working on a set of heterogeneous deliverables going from the device object level to very specific features, including browser extensions, HTML5 extensions, and event networking issues: Vibration API, Battery Status API, HTML Media Capture, Proximity Events, Ambient Light Events, Media Capture and Streams, MediaStream Image Capture, Media Capture Depth Stream Extensions, Network Service Discovery (HTTP-based services advertised via common discovery protocols within the current network), Wake Lock API, Menu API, and the sensor API to come.

This example showing the WebApps description proposals illustrates the concurrency of concerns and approaches around the "Thing" indexing and registration problem.

On the other hand, the Web of Things Interest Group is working from an avatar perspective on the duplication of real-life objects through virtualized objects to be used by applications [16].

This involves the identification of use cases and requirements for direct access to sensors and actuators from the browser, gateways that bridge IoT technologies and the Web, service platforms at the network edge, like home hubs, and scalable cloud-based platforms. From a networking perspective this group is focused on creating adapters and gateways to allow low-level technologies used in M2M interaction.

This work can be made more extensible and less driven by the specific capabilities of today's mobile devices by binding it with the generic, device-independent multimodal Interfaces API.

It would also be very useful to integrate these proposals with the taxonomic efforts already made by consortia like Echonet during the last 20 years in a common and standardized vocabulary and a generic API.

15.2.3 Multimodal Interaction in the Web of Things

We can imagine that the horizon opened by the Web of Things is as exponential as the technical solutions currently available. This situation explains and supports the MMI Working Group generic approach and our proposal for a generic discovery and registration for dynamic Multimodal Components.

Multimodal systems are computer systems endowed with rich capabilities for human–machine interaction and able to interpret information from various communication modes.

According to [20] the three principal features of multimodal systems are

1. The fusion (integration) of different types of data.
2. Real-time processing and temporal constraints imposed on information processing.
3. The fission (composition) of restituted data: a process for realizing an abstract message through output on some combination of the available channels.

On these systems, the management of modalities is most of the time hard-coded, leaving aside the problem of a generic architecture and a dynamic system. This mode of implementation neglects the issues around extensibility and the need of discovery, monitoring, and coordination of modalities in real-time with context-aware solutions.

Consequently, developers manually compose multimodal applications in an ad hoc and proprietary way, with consequences like time-consuming maintenance and inadequate interoperability, which increases the cost burden of the industry stakeholders.

15.2.4 A Standard Approach to Multimodal Interaction

To address this lack of a generic approach, the MMI Architecture and Interfaces defines an architectural pattern for any system communicating with the user through different modalities simultaneously instantiated in the same interaction cycle. In this unique context of interaction the final user or the application automation can dynamically switch modalities.

This kind of bidirectional system combines inputs and outputs of multiple sensorial modes and modalities (e.g., voice, gesture, handwriting, biometrics capture, temperature sensing, etc.) and can be used to identify the meaning of the user's behavior or to compose intelligently a more adapted, relevant, and pertinent message.

The Multimodal Architecture and Interfaces specification uses the Model-View Controller (MVC) design pattern generalizing the View to the broader context of the multimodal dynamic presentation, where the information can be rendered in a combination of various modalities depending on their availability.

Thus, the MMI recommendation distinguishes (Fig. 15.1) three types of components: the Interaction Manager which is responsible for controlling the interaction flow, the Data Component which stores the application data, and the Modality Components, covering the forms of representing information in a known and recognizable rendered structure.

For example, acoustic data can be expressed as a musical sound modality (e.g., a human singing) or as a speech modality (e.g., a human talking).

The component representing the presentation layer in the MMI Architecture is, indeed, a Modality Component in Fig. 15.1. This is a logical entity that handles the input and output of different hardware devices (e.g., microphone, graphic tablet, and keyboard) or software services (e.g., motion detection, biometrics sensing).

The Modality Components are loosely coupled software modules that may be either co-resident on a device or distributed across a network. This aspect promotes low dependence between Modality Components, reducing the impact of changes and facilitating their reuse in distributed systems. As a result, these components have little or no knowledge of the functioning of any other modules and the

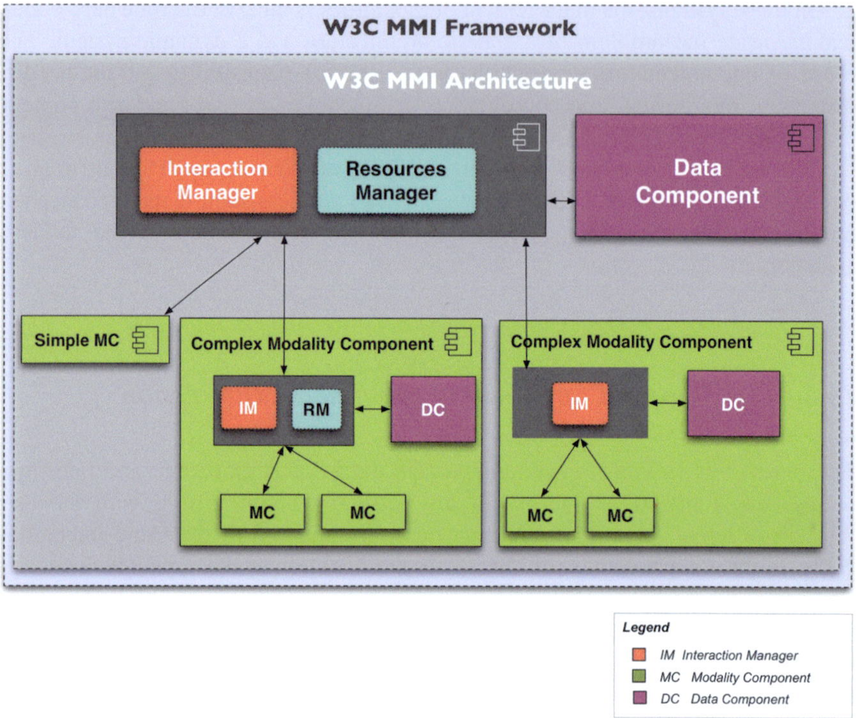

Fig. 15.1 The MMI architecture

communication between modules is done through the exchange of events following a protocol provided by the MMI architecture.

Despite this important feature, which provides an excellent support for distribution, the architecture was focused only on the interaction cycle, leaving aside the support of the lifecycle of Modality Components from a system perspective, a more dynamic application behavior, and the non-functional goals of some features to adapt the application to a particular space, device family and interaction type, using context-aware techniques.

15.2.5 A Survey of Multimodal Architectures

In a similar way, other multimodal systems also lacked a distributed approach, but provide us very interesting approaches that will enhance future extensions needed for the implementation of applications in the Web of Things. We carefully studied [3] this sample of 16 multimodal architectures that were selected from an analysis of a larger set of a 100 of multimodal implementations.

The selection criterion was the amount of information provided by the authors about the architectural facets of the implementation, its completeness and its responses for three needs: distribution, the description of modalities, and the use of semantic technologies to allow a generic selection of features.

Two groups were detected: a first group of emerging systems covering the discovery and registration criteria in functional topics like event and state handling, and a second group, covering more transversal topics like the use of generic models, a distributed architecture and the delegation of control.

15.2.6 Event Handling

The first recurrent topic was event handling. Six architectures tried to address the management of events, which is normal in the human computer interaction field because user interfaces are highly event-oriented. In OAA [9], triggers provide a general mechanism to express conditional requests using a blackboard style of communication. Each agent in the architecture can install triggers either locally, on itself, or remotely on its facilitator or on peer agents. These triggers are used for requesting that some action be taken when some set of conditions is met.

The installation of a trigger within an agent can be thought of as a representation of that agent's obligation to carry out the specified action whenever the specified condition holds true, while programming predefined events. The possible actions are not hard-coded: in OAA the action to execute may be any compound goal executable by the dynamic community of agents. When an unanticipated agent joins the community, no modifications to the existing code are required for an application to make use of it.

GALATEA [21] uses macro-commands while an Agent Manager that possesses a macro-command interpreter expands each received macro-command in a sequence of commands sending them sequentially to the designated modules in a one-to-many design. This capability of broadcasting is very useful in the Web of Things applications and is already used by M2M short range technologies like LonWorks [22], Modbus [23], KNX [24], ZigBee [9], or Z-wave [10].

With task control layers in OPENINTERFACE [25], the communication paradigm (event-based, remote procedure call, pipe, etc.) is implemented using adapters and connectors for two types of events: instantaneous events and persistent events. The mechanism implements the event dispatching following using predefined rules, as in OAA.

In MEDITOR [26], events are handled with three specialized managers: the input messages queue, the input messages generator, and the output messages generator. The temporal order is ensured and disambiguation is handled with a routing table and predefined rules. Hard-wired reactions are the tool in REA [27], for quick reactions to stimuli. These stimuli then produce a modification of the agent's behavior without much delay, working as predefined events.

In HEPHAISTK [28], events are handled by the Event Manager, which ensures the temporal order of events. The application not only is a client, but also is another input source, and consequently the Event Manager is designed also as a recognition agent, which communicates through a set of predefined messages.

In contrast to these six multimodal architectures, the MMI Architecture responds to the event management concern with the Interaction Life-Cycle Events, and the proposal of a dedicated component: the Interaction Manager. This solution provides a clear separation between the interaction control and the interaction content data.

Nevertheless, hardwired mechanisms or predefined events are not envisioned and the event protocol as currently defined only covers user interaction, leaving aside the announcements for registration or the information about the availability of components, which the system needs to handle.

The transport queue mechanism implemented in MEDITOR, GPAC [29], and HEPHAISTK could also be an important support for the fusion/fission of modalities. In consequence, these mechanisms were detected as possible extensions to the W3C's Architecture to provide some complementary resources to handle multimodal events in an interoperable way.

15.2.7 State Management

The second key topic, recovered from five of the studied architectures is state management. This feature is oriented to register the evolution of the interaction cycle and provides the information about any modification of the state of the system and its components.

State management is designed as a monitoring process in support of the decision layer (SMARTKOM [30], HEPHAISTK), as a display list manager in support of the fusion and fission mechanisms (DIRECTOR) or as a blackboard (OAA, HEPHAISTK) which is a central place where all data coming from the different sources are standardized, and other interested agents can use them at will. Finally, the states are handled by an object manager—for decoding and rendering purposes—(GPAC), and even as a routing table (MEDITOR).

Concerning state management the MMI Runtime Framework recommends a specific component to handle the multimodal session and the state of components; yet, it does not give details about the interfaces needed to use this component or about its role in the management of the interaction cycles.

As a result, the Discovery and Registration extension to the MMI Architecture and Interfaces is conceived to complete this generic description with specific details about the eventual implementation, behavior, and responsibilities of a resources manager, which is responsible for the management of the states of the system.

15.2.8 Models

In respect to the definition of models 12 architectures propose interesting approaches for the modeling of the entities that participate in the multimodal interaction. However, only SMARTKOM addresses the modeling task with a proposal coming from web semantic technologies.

In addition, depending on the entity, the models are more or less expressive or homogeneous, and consequently, usable. The modeling of the multimodal interaction (SMARTKOM, HEPHAISTK, and MEDITOR), the multimodal task (GALATEA, OPENINTERFACE, and SQUIDY [31]), the dialog interaction (REA, GALATEA, and SMARTKOM), and the devices (SMARTKOM) is more extensive, tested and advanced than the modeling of the user (REA), the application (OAA, SMARTKOM, ELOQUENCE, GPAC, and HEPHAISTK) or the environment and the context of usage (SMARTKOM).

This growing and common interest on models expressed in SMARTKOM as a foundational principle opened the way to reinforce the MMI Architecture and Interfaces recommendation by addressing this issue of data modeling with a semantic approach, to produce a multimodal vocabulary and a semantic description and taxonomy of modalities.

15.2.9 Distribution

Concerning the distribution issue, it is tackled with solutions like the remote installation of triggers (OAA), the distribution of the fusion/fission mechanisms into nodes and components that can even be external to the multimodal system, for example, in the *cloud* (OPENINTERFACE, SQUIDY), the management of inputs as "sensed" data with the use of broadcasted media containing behavior and interaction information in distributed streams (GPAC) and finally, the distribution of application services (SMARTKOM, HEPHAISTK).

The distribution is also reflected on service-oriented proposals of application services and service advertisement (OAA, SMARTKOM) and the networking services layer to manage the broadcasted input and output data of a rich application (GPAC).

The MMI Framework and Architecture addresses this topic with its distributed nature based on web standards. Nevertheless, there are few current implementations using the web services or a service-oriented approach for multimodal systems.

The current implementations are oriented to prototype mobile interfaces (Orange Labs), to provide a multimodal mobile client for health monitoring (Openstream), to test an authoring tool (Deutsche Telekom R&D), and to complete JVoiceXML, an open source platform for voice interpretation (TU Darmstadt). We believe that it is possible that interesting extensions arise from a fully SOA implementation of the MMI Framework and Architecture standard according to its distributed nature.

15.2.10 Interaction Management by a Client Application

The transversal topic is the delegation of the interaction management by a client application. It is present in the form of application agents (CICERO, OAA) or application services (SMARTKOM, HEPHAISTK).

The MMI Architecture and Interfaces recommendation does not deal with this subject because the application is meant to be the concrete implementation of the architecture. A delegation approach supposes that an external functional core can delegate the management of the interaction to a multimodal system built in accordance with the open standard, and providing multimodal functionalities to the client application installed on devices with low processing capabilities.

The MMI Architecture and Interfaces do not currently address this problem, even if it could be the type of requirement of a multimodal browser, a home gateway virtualization, or a Web of Things application. Our current work on the W3C's MMI Working Group addresses the possible extensions that such approach could bring and how the MMI Architecture standard can support this type of future implementation.

To sum up, the study of the related work allowed us to express the context and goals of the current recommendation for Discovery and Registration of Multimodal Components in order to facilitate the implementation of distributed applications for the Web of Things. In the following section we will describe the results coming from this analysis as a new W3C draft recommendation.

15.3 Discovery and Registration for Multimodal Systems

To the best of our knowledge, there is no standardized way to build a web multimodal application that can dynamically combine and control discovered components by querying a registry structured according to the modality information. At the same time—as we showed in Sect. 15.2—the multimodal research efforts also lack of this distributed perspective.

Based on this previous analysis the MMI Working Group decided to focus on three complementary extensions to the MMI Architecture and Interfaces to support distributed applications:

1. We completed the current addressing method in order to evolve from a client–server model to an anycast model.
2. We reinforce the management of the "*multimodal session*" with a dedicated component to handle the system's state and support the system's virtualization of components.
3. We extend the transport layer with two new events designed to complete and enhance the interaction Lifecycle Events.

15.3.1 Extending the Flow of Events to Support an Anycast Model

To inform the system about the changes in the state of the Modality Components an adaptive addressing mechanism is needed. We believe that the combination of push/pull mechanisms is crucial to extend the MMI Architecture to the Web of Things.

For example, in the case of the unavailability of a given Modality Component, it needs to communicate with the control layer. This situation is not necessarily related to the interaction context itself, but it can affect it, because the interaction cycle can be stopped or updated according to this change on the global state of the system.

In the current state of the Multimodal Architecture Specification [1], interaction events like Prepare or Start, must be triggered only by the Interaction Manager and sent to the Modality Components.

In result, a Modality Component cannot send messages to the Interaction Manager other than the message beginning the interaction cycle: the newContextRequest event. Any other event originated by an internal command or like in our example, by a change on the component's state cannot be raised.

Nevertheless, to start an interaction cycle the Modality Component needs to be already part of the system and to be registered. The registration process is part of a previous phase as Fig. 15.2 shows, when even the presence of the user is not mandatory and the communication must be bidirectional.

In Fig. 15.2 a Modality Component announces its presence to the network every 200 ms during a minute in order to join a multimodal system. After a while, and thanks to an event originated by a push mechanism for listening to the requests, the control layer receives the notification.

With the data provided by the notification the Resources Manager requests from the Data Component responsible of the registry an identification code for the Modality Component and verifies if the timeout does not need to change.

If the periodical request needs to change, the Resources Manager calculates the new value and builds a push request to inform the component. Finally, the Modality Component confirms that its state has changed.

This is an example of a bootstrapping process using HTTP and bidirectional communication. The pull mechanism is executed through the UpdateNotification while the push mechanism is executed using the checkUpdate event.

As Modality Components are reflexive objects in the network, acting at the same time as server and clients, they need to communicate and to receive messages as well. The flow of messages always initiated by the Interaction Manager is not sufficient to address use cases evolving in dynamic environments, like personal externalized interfaces, smart cars, home gateways, interactive spaces, or in-office assistance applications.

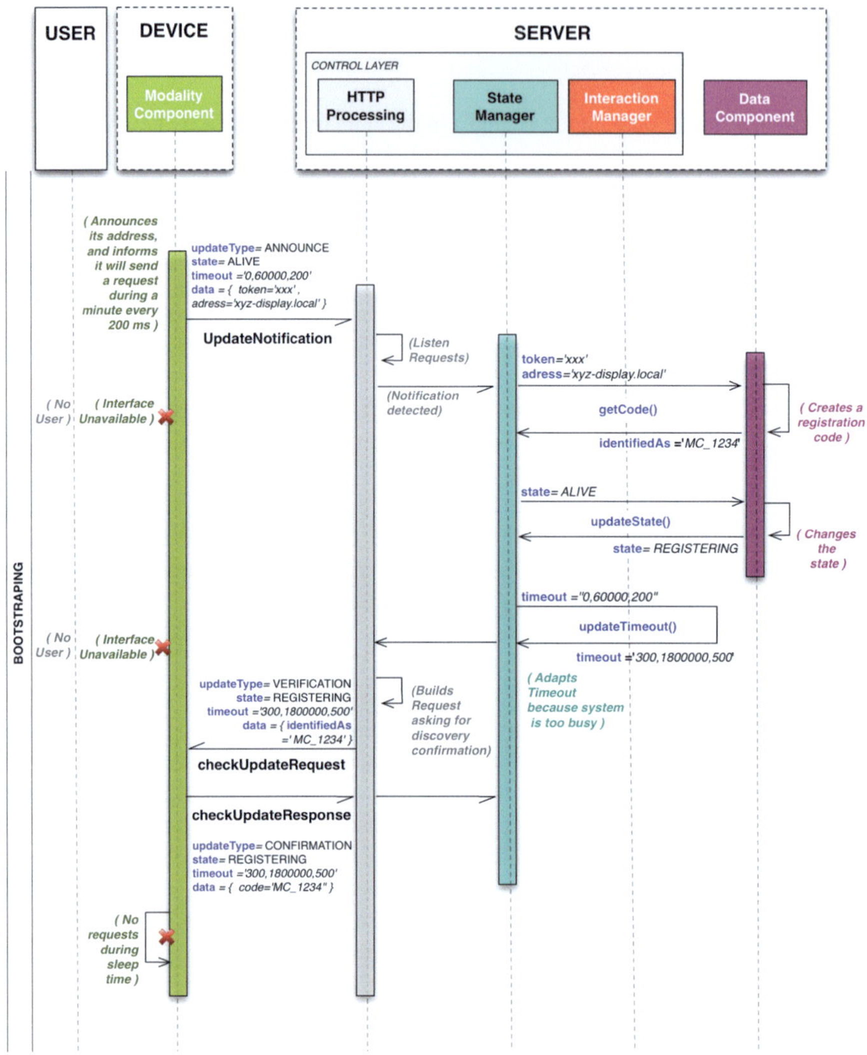

Fig. 15.2 The first phase of the registering process

In all these cases, Modality Components enter and quit the multimodal system dynamically, and they must declare their existence, availability, and capabilities to the system in some way.

To address this need, the bidirectional flow of messages supports a complete number of addressing methods and preserves a registry of the global state of the system.

One of the consequences of this new flow of messages is the capability to produce the advertisement of Modality Components. It allows the Multimodal System to reach correctness in the Modality Components retrieval and also affects the completeness in the Modality Component retrieval as required by our current use cases for discovery [32]. To return all matching instances corresponding to the user's needs, the request criteria must match some information previously registered before the interaction cycle starts, for example, at a minimum, its identifier.

For this reason, the MMI Architecture should provide a means for multimodal applications to announce the Modality Component's presence, address, and state. This was the first step to fulfill the distribution requirements: Modality Components must be distributed in a centralized, a hybrid, or a fully decentralized way [32].

For Discovery and Registration purposes the distribution of the Modality Components influences how many requests the Multimodal System can handle in a given time interval, and how efficiently it can execute these requests. Even if the MMI Architecture Specification is distribution-agnostic, with this bidirectional flow of messages, Modality Components can be located anywhere and communicate their state and their availability to a new dedicated component: the Resources Manager.

15.3.2 Defining an Architectural Module to Support Device Virtualization

The new flow of messages between the Modality Components and the control layer needed a mechanism tracing the relevant data about the session and the system state. This is the first of the responsibilities for a new module, the Resources Manager.

This manager is responsible for handling the evolution of the "*multimodal session*" and the modifications in any of the participants of the system that could affect its global state. It is also aware of the system's capabilities, the address and features of Modality Components, their availability, and their states.

As Fig. 15.3 shows, according to our recommendation, these states can be: Alive (the component is in On or Sleep mode), Loading (the component is awake and loading resources), Registering (the component is following the two step process to join the multimodal system), Available (the component is registered and capable to interact), Idle (the component is ready to receive an interaction), Busy Waiting (the component is not ready, waiting for the process executed on the system), Processing (the component is busy processing some data), Unregistered (the registration of the component is no more valid), or Unavailable state (some problem avoids the component to interact).

Thus, the Resources Manager is nested in the control layer of the multimodal system and keeps the control of the global state and resources of the system.

This newly extended control layer encompasses the handling of the multimodal interaction and the management of the resources of the multimodal system. In this way, with our extension, the architecture preserves its compliance with the MVC design pattern.

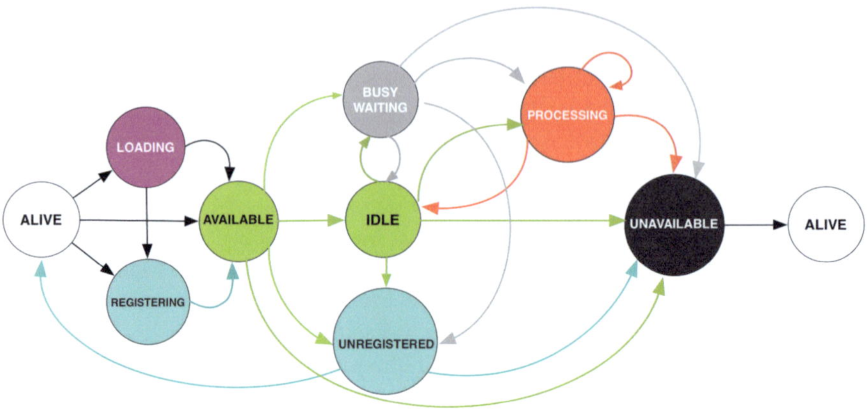

Fig. 15.3 States of the modality components

The data handled by the Resources Manager can be structured and stored in a virtualized manner. In this way, the Resources Manager can be calibrated for mediated discovery and federated registering [32].

The Resources Manager uses the scanning features provided by the underlying network, looking for components tagged in their descriptions with a specific group label. It can implement a discovery procedure to compose its register based on the information provided by the network layer.

For example, the Resources Manager can request to the ETSI M2M gateway technical information about a given device driver in order to register the virtualization as Modality Component of its features and operations.

First, it can request using HTTP the technical information about the current implementation of the gateway:

```
HTTP GET /m2m/applications/gateway_0/containers/descriptor/contentInstances/last/content
```

The Resources Manager must parse the response (which is a description of services including the network structure), to recover the technical metadata that will be useful to the Interaction Manager in order to command the given device. In this case, this is a device using the physical layer IEEE 802-15-4 and the device is designed for Home Automation using ZigBee 1.0.

```
<obj>
<str name="interworkingProxyID" val="Text for correlation purpose"/>
<list name="supportedTechnologies">
    <obj>
     <enum name="anStandard" val="ZigBee_1_0"/>
     <enum name="anProfile" val="ZigBee_HA"/>
     <enum name="anPhysical" val="IEEE_802_15_4_2003_2_4GHz"/>
    </obj>
</list>

<list name="networks"/>
   <ref href="/m2m/applications/hub_0/">
</list>
</obj>
```

Second, the Resources Manager can request more functional information if the technical metadata shows that the device is interesting enough to the current multimodal application. To take this decision a mapping between the descriptions provided by the networking layer and the multimodal vocabulary must be done (The Multimodal Working Group is currently working on this task).

```
HTTP GET /m2m/applications/hub_0/containers/descriptor/contentInstances/last/content
```

The response provides the information about the registered nodes on the target hub of nodes (called network in the ETSI M2M technology model).

```
<obj is="zigbee:NetworkDescriptor">
   <str name="networkID" val="Text for correlation purpose"/>
   <str name="extendedPanID" val"0x685B3C34"/>
   <list name="nodes" of="obix:ref m2m:NodeDescriptor">
      <ref href="/m2m/applications/node_0/containers/descriptor/contentInstances/last/content"/>
      <ref href="/m2m /applications/node_1/containers/descriptor/contentInstances/last/content"/>
   </list>
</obj>
```

Given that this list does not provide enough information, the Resources Manager must iterate over the nodes list in order to recover mode metadata and functional information to register the modalities and to compose the virtualized Modality Component.

The iteration must be done in the following way:

```
HTTP GET /m2m/applications/node_0/containers/descriptor/contentInstances/last/content
```

The response provides the information about the current services (called application in the ETSI M2M technology model):

```
<obj>
   <str name="nodeID" val="Text for correlation purpose"/>
   <str name="ieeeAddress" val="0x685B3C88"/>
   <enum name="type" val="endDevice"/>
  <list name="applications">
     <ref href="/m2m/applications/service_0/"/>
  </list>
</obj>
```

The Resources Manager must finally iterate over the services to acquire all the functional data needed to register the Modality Components on the given gateway:

```
HTTP GET /m2m/applications/service_0/containers/descriptor/contentInstances/last/content
```

In the case of the ETSI M2M technology, the response will contain all the information needed to register the Modality Components.

```
<obj is="zigBee:ApplicationDescriptor">
    <str name="applicationID" val="Text for correlation purpose"/>
    <str name="applicationProfileID" val="0x104"/>
    <str name="applicationDeviceID" val="0x0103"/>
    <int name="endpoint" val="1"/>
    <list name="interfaces" of="m2m:InterfaceDescriptor">
        <obj is="zigbee:OnOffLightCluster zigbee:InterfaceDescriptor">
            <str name="interfaceID" val="Text for correlation purpose"/>
            <str name="clusterID" val="0x0006"/>
            <enum name="clusterType" val="server"/>
            <ref name="onOffState"
             href="/m2m/application/service_0/containers/monitor_0/contentInstances/last/content"
             is="zigbee:Point"/>

            <op name="zclToggle" in="obix:nil" out="obix:nil"
              href="/ m2m /applications/service_0/appCommands/operation_0"/>
            <op name=" zclOn " in="obix:nil" out="obix:nil"
              href="/ m2m /applications/service_0/appCommands/operation_1"/>
            <op name=" zclOff " in="obix:nil" out="obix:nil"
              href="/ m2m /applications/service_0/appCommands/operation_1"/>

        </obj>
    </list>
</obj>
```

The description also allows the Resources Manager to communicate with the services or data points to monitor the state of the Modality Components through application dependent routines. For example, in our current use case, which is a controller for a set of lights in an automated house, the Resources Manager has now the information needed to monitor not only the availability of the lights but also it can monitor its state to keep track of the environmental conditions for the multi-modal system.

```
HTTP GET /m2m/applications/service_0/containers/descriptor/monitor_0/contentInstances/last/content
```

In response, the Resources Manager recovers the current state of the lighting service on the device.

```
<bool name=" onOffState " val="off" is="zigbee:Point"/>
```

If the discovered component is not tagged with a group label, or doesn't provide enough metadata, the Resources Manager can use some mechanism provided to allow subscriptions to a generic group.

In this case, the Modality Component should send a request using the new flow of messages and using one of the new discovery events to the Resources Manager, subscribing to the register of the generic group.

After recovering the information, not only the Resources Manager translates the Modality Component's messages into method calls on the Data Component, like the MVC pattern proposes, but also the Resources Manager broadcasts to the Modality Component the changes on the system's state or notifies it following a subscription mechanism. Upon reception of the notification, the Modality Component updates the user interface according to the information received.

As our use case shows, the implementation of the Resources Manager depends on the application design, the targeted technologies, and the available networks. Nevertheless, the procedure of registering of the metadata describing the Modality Components available and the storage of the functional data needed to communicate with them is generic in most cases.

In this generic approach, the Resources Manager supports the coordination between virtualized components and their communication through the control layer. This enables the synchronization of the input constraints across modalities and also enhances the resolution of input conflicts from distributed modalities before the actual communication with the devices. It is also the starting point to declare and process the advertised announcements and to keep them up to date. The Resources Manager is the core support for mediated and passive discovery and it can also be used to trigger active discovery using the push mechanism or to execute some of the tasks on fixed discovery [32].

The Resources Manager is the interface that can be requested to register the Modality Component's information. It handles all the communication between them and the registry. The flow of discovery queries transit through it, which dispatches the requests to the Data Component and notifies the Interaction Manager, if needed. These queries must be produced using the state handling events presented in the next section. To summarize, the Resources Manager delivers information about the state and resources of the multimodal system during and outside the interaction cycle.

15.3.3 Two Events for Discovery and Registration

With a new flow of messages and a new component handling the state of the system, a Modality Component can register its services for a specific period of time. This is the basis for the handling of the Modality Component's state. Every Modality Component can have a lifetime, which begins at discovery and ends in a date provided at registration. If the Modality Component does not re-register the service before its lifetime expires, the Modality Component's index is purged. This depends on the parameters given by the Application logic, the distribution of the Modality Components, or the context of interaction.

When the lifetime has no end, the Modality Component is part of the multimodal system indefinitely. In contrast, in more dynamic environments, a limited lifetime can be associated with it, and if it is not renewed before expiration, the Modality Component will be assumed to no longer be part of the multimodal system. Thus, by the use of this kind of registering, the multimodal system can implement a procedure to confirm its global state and update the "inventory" of the components that could eventually participate in the interaction cycle.

Therefore, registering involves some Modality Components' timeout information, which can be always exchanged between components and, in the case of a dynamic environment, can be updated from time to time. For this reason, a registration renewal mechanism is needed.

We proposed a registration mechanism based on the use of a timeout attribute and two new events: the checkUpdate Event and the UpdateNotification, used in conjunction with an automatic process that ensures periodical requests.

15.3.4 checkUpdate

The checkUpdate Event is provided (a) to verify if there are any changes in the system side; (b) to recover the eventual message; (c) to adapt the request timeout if needed, and (d) to trigger automatic notifications about the state of the Modality Component, if the automaticUpdate attribute in the response is true.

For example, in a multimodal system it could be necessary for the loading of some audio resources to execute a given task. In the discovery phase the audio Modality Component must send a CheckUpdateRequest indicating its state "LOADING" for a given action related to the discovery phase, like the "HANDSHAKE" process. This announcement can be addressed directly to the Resources Manager or it can be broadcasted to the underlying network.

```
<mmi:mmi xmlns:mmi=http://www.w3.org/2008/04/mmi- arch >
  <mmi:CheckUpdateRequest
        mmi:Source="URIForRM"
        mmi:Target="URIForMC"
        mmi:RequestID="request-1"
        mmi:State="LOADING"
        mmi:UpdateType="HANDSHAKE"
        mmi:AutomaticUpdate="true"
       <mmi:metadata
        src="URIForMetadata"
        info="{medium:{acoustic}, modality:{acoustic:SPEECH}}" />
       <mmi:Timeout sleep="0" validity="360000" interval="200"/>
  </mmi:CheckUpdate Request >
</mmi:mmi>
```

If the AutomaticUpdate attribute is defined to True, the timeout will define the pace for the periodic communication that will follow (in our example 200 ms with a messaging communication that must begin immediately).

```
<mmi:mmi xmlns:mmi=http://www.w3.org/2008/04/mmi- arch>
 <mmi:CheckUpdateRequest mmi:Source="URIForMC"
    mmi:Target="URIForRM"
    mmi:RequestID="request-1"
    mmi:State="REGISTERING"
    mmi:UpdateType="HANDSHAKE"
    mmi:AutomaticUpdate="true"
    mmi:data="{code=MC_1234}" >
    <mmi:Timeout sleep="100" validity="360000" interval="1000"/>
 </mmi:CheckUpdateRequest>
</mmi:mmi>
```

In this way, if a Resources Manager knows that it is useless or disruptive to have a message each 200 ms—for example, because it knows that its data is only updated each 2000 ms—then the Resources Manager can use the checkUpdate Event to change the timeout attribute to 1000 ms and to put the Modality Component on sleep mode for 100 ms—for example, because it needs some time for changing its state or for processing some other commands. (See the code below that illustrates this case.) These changes exchanged in the response enhance input/output synchronization in distributed environments.

15.3.5 UpdateNotification

On the other hand, the UpdateNotification is proposed (a) to periodically inform the Resources Manager about the state of the Modality Component; and (b) to help in the decision-making process (on the server side, for example). For the notification of failures, progress, or delays in distributed processing, the UpdateNotification event ensures periodical requests informing other components if any important change occurs in the Modality Component's state or in the system.

```
<mmi:mmi xmlns:mmi=http://www.w3.org/2008/04/mmi-arch>
  <mmi:UpdateNotification
     mmi:Source="URIForMC"
     mmi:Target="URIForRM"
     mmi:RequestID="request-1"
     mmi:UpdateType="REPORTING"
     mmi:State="BUSY WAITING">
  <mmi:Timeout sleep="100" validity="360000" interval="200"/>
  </mmi:UpdateNotification>
</mmi:mmi>
```

This can support, for example, grammar updates or image recognition updates for a subset of differential data (e.g., the face is the same but now there is a smile).

In the example above, the UpdateNotification event is used to inform the system after a sleep period, that the Modality Component is in Busy Waiting state, for example, in the case of a motion detector sensor that is not recognizing activity in a room.

The use of the timeout attribute helps in the management of the validity of the advertised data. If a Modality Component's communication is out-of-date, the system can infer that the data has the risk of being inaccurate or invalid.

Finally, the checkUpdate Event allows the recovery of small subsets of the information provided by the Interaction Manager, to maintain up to date the data in the Modality Components as in the Resources Manager.

To sum up, the MMI Architecture responds to its main goal: to ensure interoperability between heterogeneous systems coming from very different contexts and implementations. This is made on three subjects the addressing method needed for models different than the client–server model, the "*multimodal session*" component to handle the system's state, and the transport layer by completing and reinforcing the interaction Lifecycle Events.

15.4 Conclusion

The work on standardization produced by the MMI Working Group in the last 2 years is focused on distribution. Today, a generic first step needed to allow the component's discovery and registration is done. The multimodal system has now a means to announce or discover the Modality Component's capabilities and states through different addressing modes. Second, the modality component's data model needed as a building block for a multimodal registry is founded, starting with a common taxonomy of generic states (out of the scope of this document) and the construction of a generic classification system for devices and groups of modes and modalities. These premises of classification will facilitate the orchestration mechanism using the Modality Component's description. A mechanism that is now possible, thanks to the extension of the MMI Architecture's event model with two events specified for discovery and registration needs.

These three issues are covered by our results and now are available to the community of web developers as a Working Draft. The requirements extracted from the analysis of the state of the art and a series of use cases provided by the industry [32] were the basis to produce three pertinent extensions.

To handle multimodal events in an interoperable way, we extended the MMI Architecture by completing the current addressing method in order to evolve from a client–server model to an anycast model using bidirectional communication.

To ensure the handling of states we proposed to support the management of the "*multimodal session*" by a dedicated component. This component is responsible for building a registry of devices and services to be used on the multimodal system by using a virtualization procedure (out of the scope of this document, but available soon in a new deliverable).

To allow distribution, we proposed extending the transport layer with two new events, completing and reinforcing the MMI interaction Lifecycle Events. And finally, to support the delegation of control and to use generic models, the Resources Manager creates and stores the registry, based on a selection following a classification of generic multimodal properties and a generic model of states. Both are used also for keeping the registry up-to-date.

As our use case illustrates, the MMI's Modality Component is an abstraction flexible enough for any implementation of the IoT and networking model, while keeping an interoperable structure.

The MMI Architecture is built around the management of continuous media and their states not only as outputs (presentations) but also as inputs. This means that the architecture is fine-tuned to handle issues derived from very dynamic environments needing session control and recovering with all kinds of media and interaction modes.

In this paper we presented our current work on Discovery and Registration of Modality Components from a generic and interoperable technology that will allow us to face the infinity created by the Web of Things. From an extensive study of the state of the art, we produced a series of requirements and evaluation criteria that founds the results presented in Sect. 15.3. In a future activity, the W3C working group will produce an annotation vocabulary and the support of the semantic annotation in the "info" dedicated attribute on the new discovery and registration events. This vocabulary is a first step on the direction of a more expressive description of the interaction with Modality Components using ontologies and a more intelligent composition of semantic web services for multimodal applications with rich interaction features.

Acknowledgments The authors wish to thank the W3C MMI Working Group for its collaboration, with special thanks to Deborah Dahl and Kazuyuki Ashimura, and to Cyril Concolato, Jean Lefeuvre and Jean Claude Dufour for the very helpful insights and support during this work.

References

1. W3C's MMI-Arch. http://www.w3.org/TR/mmi-arch/. Accessed 1 Apr 2015.
2. Rodriguez, B. H. The W3C's multimodal architecture and interfaces. https://en.wikipedia.org/wiki/Multimodal_Architecture_and_Inter faces. Accessed 29 Sept 2015.
3. Rodriguez, B. H. (2013). *A SOA model, semantic and multimodal, and its support for the discovery and registration of assistance services*. Ph.D. thesis, Institut Mines-Télécom, Telecom ParisTech, Paris.
4. W3C's Multimodal Interaction Framework. http://www.w3.org/TR/mmiframework/. Accessed 1 Apr 2015.
5. Rodriguez, B. H., Dahl, D., Ashimura, K., Barnett, J., Tumuluri, R., & Kharidi, N. (Eds.) (2015). Discovery and registration of multimodal modality components: State handling. First Public Draft 11/06/2015. http://www.w3.org/TR/mmi-mc-discovery/. Accessed 29 Sept 2015.
6. International Symposium on Home Energy Management System—Joint Discussion with the W3C MMI WG, Keio University Shonan Fujisawa Research Institute. 25–26 Feb 2015. Organized by the Ministry of Industry, Echonet Consortium and Keio University. Participants: Mr. Sano, Director Ministry of Economy, Trade and Industry, Mr. Taniguchi, Executive Director ENNET Corporation, Mr. Kodama, Director ECHONET Consortium, Mr. Isshiki Director ECHONET HEMS Interoperability test center, Mr. Umejima, Senior Fellow ECHONET Consortium, Deputy Chair, JSCA, Mr. Murakami, Director ECHONET Consortium, Mr. Aida Leader Iene consortium, Sureswaran Ramadass, Chairman Asia Pacific Advanced Network, Richard Schomberg, IEC Chair Senior Vice President EDF France, Patrick Veron, Former Senior Vice President in Cisco Corporation, Ms Dahl chair W3C MMI Working Group, Ms RODRIGUEZ W3C Discovery and Registration Editor.
7. Cisco Visual Networking Index: Global Mobile Data Traffic Forecast Update 2014–2019 White Paper. http://www.cisco.com/c/en/us/solutions/collateral/service-provider/visual-networking-indexvni/white_paper_c11-520862.html. Accessed 4 Aug 2016
8. KNX Network Communications Protocol for Intelligent Buildings (EN 50090, ISO/IEC 14543). http://www.knx.org/knx-en/index.php.
9. Martin, D., Cheyer, A., & Moran, D. (1999). The open agent architecture: A framework for building distributed software systems. *Applied Artificial Intelligence, 13*(1–2), 91–128.
10. Z-Wave. http://www.z-wave.com.
11. ECHONET Energy Conservation and Homecare Network. http://www.echonet.gr.jp/.
12. ETSI Machine to Machine Communication. http://www.etsi.org/technologies-clusters/technologies/m2m.
13. Digital Living Network Alliance. http://www.dlna.org.
14. Universal Plug and Play. http://www.upnp.org.
15. Zero Configuration Networking. https://developer.apple.com/bonjour/index.html.
16. Web of Things Interest Group. https://www.w3.org/WoT/IG/.
17. WebApps Working Group. http://www.w3.org/2008/webapps/.
18. CC/PP. http://www.w3.org/TR/CCPP-struct-vocab/.
19. Machine-to-machine communications: Connecting billions of devices. OECD Digital Economy Papers, No. 192, OECD Publishing. doi:10.1787/5k9gsh2gp043-en. Accessed 25 April 2012.
20. Coutaz, J., Nigay, L., Salber, D., Blandford, A., May, J., &Young, R. (1995). Four easy pieces for assessing the usability of multimodal interaction: The CARE properties. In *Proceedings of the INTERACT'95*, Lillehammer.
21. T. Nitta et al. Activities of Interactive Speech Technology Consortium (ISTC) targeting open software development for MMI systems. Robot and Human Interactive Communication, 2004. ROMAN 2004. 13th IEEE International Workshop on, 2004, pp. 165-170. doi: 10.1109/ROMAN.2004.1374749

22. LonWorks: ISO CEI 14908-1/4 or ASI CEA 709.1. ISO/IEC 14908, Parts 1, 2, 3, and 4 available online: http://downloads.echelon.com/support/documentation/manuals/general/078-0183-01B_Intro_to_LonWorks_Rev_2.pdf Accessed 4 Aug 2016
23. http://modbus.org/docs/Modbus_Application_Protocol_V1_1b3.pdf. Accessed 1 Jan 2016.
24. KNX: ISO/IEC 14 543-3 (2006). http://www.iso.org/iso/catalogue_detail.htm?csnumber=59865 Accessed 4 Aug 2016.
25. Serrano, M., Nigay, L., Lawson, J.-Y., Ramsay, L., & Denef, S. (2008). The open-Interface framework: A tool for multimodal interaction, In: *Proceedings of the CHI'08 Extended Abstracts on Human Factors in Computing Systems—CHI EA08*, ACM, New York, NY, USA, pp. 3501–3506.
26. Bellik, Y. (1995). *Interfaces Multimodales: Concepts, Modèles et Architectures*. Ph.D. thesis, University Paris-South 11, Orsay.
27. Cassell, J. (2001). Embodied conversational agents: Representation and intelligence in user interfaces. *AI Magazine, 22*(4), 67–83.
28. Dumas, B., Lalanne, D., & Ingol, R. (2008). Démonstration: Hephais TK, une boîte à outils pour le prototypage d'interfaces multimodales.
29. Le Feuvre, J., et al. (2011). *Experimenting with multimedia advances using GPAC*. Scottsdale, AZ: ACM Multimedia.
30. Herzog, G., & Reithinger, N. (2006). The SmartKom architecture: a framework for multimodal dialogue systems. In W. Wahlster (Ed.), SmartKom: foundations of multimodal dialogue systems. Heidelberg: Springer. doi:10.1007/3-540-36678-4_4.
31. König, W. A., Rädle, R., & Reiterer, H. (2009). Squidy: A zoomable design environment for natural user in-terfaces. In *Proceedings of the 27th International Conference Extended Abstracts on Human Factors in Computing Systems (CHI EA'09)*, ACM, New York, NY, USA, pp. 4561–4566. doi:10.1145/1520340.1520700.
32. Rodriguez, B. H., Dahl, D., Tumuluri, R., Wiechno, P., & Ashimura, K. (2012). Registration & discovery of multimodal modality components in multimodal systems: Use cases and requirements. W3C Working Group Note 5 July. http://www.w3.org/TR/mmi-discovery/. Accessed 1.
33. The Smart Energy Profile 2. http://www.zigbee.org/zigbee-fordevelopers/applicationstandards/zigbeesmartenergy/. Accessed 4 Aug 2016

Chapter 16
Multimodal Interactivity in Foreign Language Testing

Teresa Magal-Royo and Jesús García Laborda

Abstract Multimodality in interactive digital environments for second language acquisition testing has been begun to be used only very lately. Some of the multimodality concepts have been recently developed in various research projects. Research into the automation of online university entrance exams has been prompted by the need to efficiently manage online tests and handle the task of exam marking semi-automatically. At the same time, we have addressed the use of multi-platform and/or multi-browser applications to handle the technical and functional validation of accessibility and so enable universal access. The application of multi-modularity to methods of navigation during the examination is possible in mobile devices that enable simultaneous interaction when students input data. This chapter aims to demonstrate the technical feasibility of the implementation process for an examination in regard to the technological and formal variables of navigation in the exam. These variables are currently being handled in mobile devices and the result is a further advancement in the process of Computer Aided Language Learning, CALL. This paper presents the multimodal approach has been taken into account in the development of an online prototype which validates the functional and technical assumptions adapted to university entrance exams for foreign languages.

16.1 Introduction

Language learning is crucial in today's world. Foreign languages are used for a variety of purposes from leisure such as visiting different countries to professional purposes (i.e., for international business). That is why measuring the students' competence has become an issue of major importance over the years. Students in many national and international contexts and for a number of purposes such as university or high school graduation are learning foreign languages. Like teaching

T. Magal-Royo (✉)
Universitat Politécnica de Valencia, Valencia, Spain
e-mail: tmagal@degi.upv.es

J. García Laborda
Universidad de Alcalá de Henares, Alcalá de Henares, Madrid, Spain

D.A. Dahl (ed.), *Multimodal Interaction with W3C Standards*,
DOI 10.1007/978-3-319-42816-1_16

methodologies, language teaching has also changed over time [1–3]. Likewise, the techniques to deliver tests have seen dramatic changes. Obviously, over time most delivery has been done through pen and paper for written tests and person-to-person. It would not be until the 1980s when the first tests in BASIC or FORTRAN were delivered. These consisted of just simple fill-in the gaps or multiple choice exercises. In the 1990s it was considered that this kind of exam failed to obtain real evidence of language competence across the skills. After all, language is a lot more than choosing from a set of responses. At the time, computer-based language testing was just seen as a convenient and fast way to deliver traditional exams [4]. However, since the middle 1970s, after the oil crisis, the number of people commuting or travelling abroad increased dramatically evidencing a great need to develop speaking communicative skills which, in turn, also needed to be measured. Thus, speaking became prevalent (almost in excess) in language testing. In the 1990s computer-based speaking tests (when done!) were limited to the deployment of audio files which students responded either on basic recording computer applications but more often in cassette recorders. It would not be until the very end of that decade when the Educational Testing Service began to implement computer speaking tasks [5, 6]. In 2004, the new IB TOEFL™ showed the importance of using a variety of task types. However, computer tests fall short, even today, since they have not conveniently solved problems such as ubiquitous testing, naturalness of test delivery, and more issues which are critical for the future of communicative language testing. As a consequence, the importance of multimodality is self-evident in the future of language testing.

Multimodal technology increases the validity of a test by integrating the different kinds of modes to assess the student's competence [3], for instance, in the aforementioned inclusion of listening, speaking, reading, and writing tasks. In this way, the student is required to show competence in many more ways than with traditional paper and pencil exams. As will be seen, for instance, the integration of images in tasks such as descriptions triggers the use of vocabulary and even grammatical structures which in turn respond to communicative demands. In short, there is dramatic shift from written texts as a primary source of assessment and communication representation to a more realistic situation. This can also be observed in the way in which students address their study and exams. While in many contexts they prepare most of their work through hypertexting on computers, their exams are still delivered in a traditional manner especially in the classroom where the production of computer-based assessments is more costly. Today's students are able to use video, graphics, and apply strategies of interaction with the text in iPADS or tablet PCs even in test situations which were unthinkable just a few years ago. This is also connected to their way of integrating semiotic symbols to improve their navigability and capacity to achieve adequately in language tasks [7].

In Spain, traditionally the foreign language section of the University Entrance Examination only includes three types of tasks: reading comprehension, writing, and grammar composition. The implementation of a computer-based exam online has been suggested for a long time [8]. In fact, The Spanish Ministry of Education has also undertaken some trials in the use of computers for the foreign language exam in the 4th level of primary which was to be implemented next year but has been postponed "sine die."

This chapter addresses a proposal presenting a review of current trends in computer language testing that lead to the need to use multimodality in language testing or in language acquisition.

Then, we define the task framework of interactive modality including its definition in language testing online and finally we address the OPENPAU™ project as an example of application of multimodality in language testing.

16.2 State of the Art: From Computers to Multimodality in Foreign Language Testing

Since the beginning of the new millennium, there has been an increasing interest in implementing computer-based language tests for a number of reasons: efficiency, equivalence of paper and pencil tests, innovation [9], economy [4], cited by [10], transference of results, security in test delivery and rating, and contextual validity. By contextual validity Suvorov and Hegelheimer refer to the enhanced response triggered by the use of enriched input especially visual enhanced tasks (such as the use of audiovideo digital format). They consider nine attributes and categories in describing the framework of computer-assisted language testing (see Table 16.1):

Obviously, although they are all interconnected, for the purpose of this paper we will focus just on media density, response type, and task type. As mentioned above, one of the most significant aspects of computer-based language testing is the inclusion of different types of media, especially videos and podcasts. As Suvorov and Hegelheimer mention:

> The use of multimedia, which may incorporate audio, images, videos, animation, and graphics, has gained much attention among researchers because it is believed to have the potential for enhancing the authenticity of language tasks (Suvorov and Hegelheimer [10], p. 597).

In relation to the type of item in a test or exam, not all of them may be adequate for the use of an enriched visuals task. For instance, grammar items such as matching, fill-in the gaps, and others may not provide an adequate ground or functionality for such inclusion. Besides, not all the students may be eager to use technology in their tests as not all the students would be happy to use technology-oriented language instruction [11].

The factors mentioned so far and the rapid advances in use and development of technology have led to an increase of testing platforms. A number of significant exams have gone online in the last years [2] which are mostly commercial [9, 12] due to the high costs of design, implementation, and validation. Among the most significant projects, the best known is Test of English as a Foreign Language Internet-Based Test, TOEFL iBT™ [1] but other computer-based tests are currently

Table 16.1 Framework for the description of-assisted language tests

#	Attribute	Categories
1	Directionality	Linear, adaptive, and semiadaptive testing
2	Delivery format	Computer-based and Web-based testing
3	Media density	Single medium and multimedia
4	Target skill	Single language skill and integrated skills
5	Scoring mechanism	Human-based, exact answer matching, and analysis-based scoring
6	Stakes	Low stakes, medium stakes, and high stakes
7	Purpose	Curriculum-related (achievement, admission, diagnosis, placement, progress) and non-curriculum-related (proficiency and screening)
8	Response type	Selected response and constructed response
9	Task type	Selective (e.g., multiple choice), productive (e.g., short answer, cloze task, written and oral narratives), and interactive (e.g., matching, drag and drop)

Source: Suvorov and Hegelheimer [10], p. 596

used to assume significant decisions such as BULATS™ Online, Pearson Test of English (PTE) Academic™, Versant™, English Test BEST Plus™ Computer-Adaptive Version, IELTS™, the Cambridge Board of Examinations suite (p.e. Cambridge First, Cambridge Advanced, PET™, and their versions for schools).

In Spain there have been just a few projects aiming at the design of a computer-based test for high stakes [13] (see Table 16.2).

Finally, we will add that there is serious research for the development of the *Diploma de Español como Lengua Extranjera*[1] *(DELE)* but so far no formal implementation in testing conditions has been done.

All these projects lead to the importance of the introduction and use of multimodality in three directions: (1) in the test implementation as related to task design and delivery; (2) navigability; and (3) ubiquity.

16.3 Interactive Multimodality Oriented to Language Testing

For language testing using ubiquitous devices, multimodality is a process in which devices and users are able to make a joint interaction, whether auditory, visual, tactile, or gesture, from anywhere, at any one time. Through multimodality the user can determine the mode or modes of interaction he wants to use to access

[1] http://cvc.cervantes.es/ensenanza/dele/default.htm. Accessed 22/03/2016.

Table 16.2 Projects developed related to computer-based test for high stakes in Spain

Project and year	Funding institution	Results
HIEO (2005)	Generalitat Valenciana	Design of a computer-based testing tool
SELECTOR (2007)	Generalitat Valenciana	It suggests construct, analyze institutional needs, improve PLEVALEX tool analyzes student reactions
PAULEX-PAUER (2007)	Ministry of Education	Defines the construct, addresses the institutional needs, studies teacher attitudes, develops the PAUER computer platform, analyzes students' reactions, students develop and test the first mobile application, suggests evaluating commercial applications. Uses mobile phones for the test.
OPENPAU (2012–2014)	Ministry of Education	It proposes a new construct focused on the implementation by the universities. Analyzes materials and proposes a test and new ways of delivering the PAU. Use of mobile phones and tablet PC for the test.

Source: García Laborda

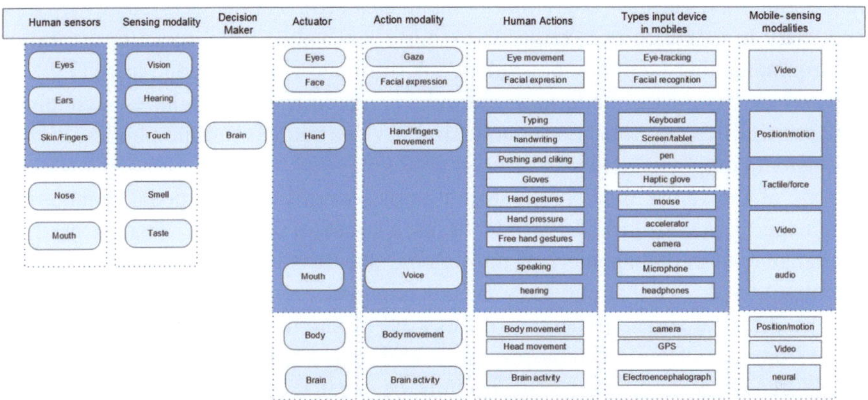

Fig. 16.1 General types of modalities in human sensing and actions task for language learning testing though mobile devices. *Source*: Adapted from Sharma et al. [14]

information via different types of input such as keyboard, mouse, pen, touch screen, voice, etc., especially developed a specific interface [14] (see Fig. 16.1).

We consider that the most effective option for a language test online in a ubiquitous device communication would be sequential multimodality applied to the navigability process of the test because it needs to be sequenced in order to solve the language tasks without inferences or potential biases originated by the computer skills of the person taking the test. That means that the use of computers or other devices should not make a difference according to the platform-use skills but according to language competence. In the end, a language test should measure the language abilities only (face to face validity). This is crucial to assist the student to learn how to use the application despite the mode of interaction that the testing device can provide. One important fact is that the testee should be able to implement the test tasks without any technical interference due to interface design. For instance, if a testee is to read a text for reading to assess his reading comprehension, the interface and navigation should help him to see and read clearly the text (as many times as desired by the test administrators, sometimes one and others more) as well as to confirm or modify the answers as many times as necessary within the time assigned for that task.

16.3.1 Programming Languages Adapted to Multimodality

One of the fundamental premises of accessibility is that accessibility is device-independence when entering or displaying data on the Web [15]. In fact, the challenge for most applications is to provide accessible contents on a growing number of the different devices with different screen sizes, different methods of interaction, and various programming languages [16]. That is the reason why it is necessary to establish user-centered usability and validation tests of the learning environments that facilitate the student's performance in a language test [17].

Over the last years, the term multimodal has been revitalized due to technological possibilities that most communication devices such as mobile phones, digital television, computers, and else offer.

One of the future more promising uses of Extensible Multimodal Annotation Language (EMMA) [18–20] would be as a protocol language. That is used to exchange data in management of multimodal interaction language testing systems in ubiquitous device that allows communication between the different components of a multimodal system. The aim of this kind of language protocol is to integrate input from users from different resources, and shapes the data to be processed in a single representation. That will in turn be treated by advanced information processing components and data processed in an examination language menu [21].

16.4 Multimodality in Language Tests

At present, there are numerous online computer-based applications for foreign language learning. These platforms allow, on the one hand, the management and control of digital audiovisual contents and, on the other hand, the creation of customized training itineraries that students can follow to acquire the language. They also offer the possibility of designing and using specific tests that allow the user to monitor his progress periodically through the revision of his speaking, writing, listening, and speaking performance at different competence benchmarks (especially the European Common European Framework for the Competence of the languages, in the US the ACTFL guidelines, etc.).

In the field of computer assisted language learning (CALL), standardized objective tests are continuously under revision especially to determine their effect in the classroom through the washback effect (the effect of testing in how classes are taught). This research has been empowered in the last 10 years because the data available on washback from computer-based exams is very limited. Data obtained on usability and ergonomics is also very limited [22] and sometimes does not clearly demonstrate whether enriched contexts have a beneficial effect on the testee's performance. That is why research projects like OPENPAU™ [13], PAULEX Universitas™ [22] and HIEO have addressed these two aspects as well as the students' and teachers' attitudes. All these projects created their own platforms for application in different devices and revisited the effect of technology on the students' reactions and platform design (see Table 16.2).

What has been especially attractive in this research on design and usability is the students' adaptation to online testing and to new devices. This is especially significant in the evolution of mobile devices. The use of ubiquitous devices would have a great impact on the online University Entrance Examination in Spain, because as observed in the field research, many high schools do not have an adequate wifi connection, computers do not always have an proper maintenance, schools also use different software, etc. In light of these facts, the use of mobile devices makes it possible to have the same equipment for all the students, with the same functionalities, software, etc., leading to fair conditions across schools and educational districts [23, 24].

The technological advances of ubiquitous devices and methods of interaction for accessibility are the starting point to analyze the potentiality that could come from the use of multimodal environments that favor communication and data transmission from the online language tests [25–29].

One of the most significant benefits that multimodality offers in language testing in our design in the OPENPAU™ application is to establish multiple access levels to the information (the language test) adapted to each testee through, at least, three simultaneous communication channels (i.e., keyboard, tactile, and voice), according to the type of test task whether oral or written [30–32].

The functional adaptation to the type of task has been developed in an interactive multimodal prototype application taking into account the technological conditions of mobile devices for user interactivity [33, 34].

The following sections will the conceptualization and factors implicated in the development of an interactive multimodal application for language testing alongside with the observations on the use of the OPENPAU™ platform where the user can choose from three modes of interactive navigation to take the exams.

Today, most devices use the conventional pattern data input (keyboard–mouse) and data output (screen–printer) [35]. This also the case for most online language tests along with the use of adequate media that is required in the specific conditions of a language exam (such as linguistic saliency, focus on meaning, and adequacy of tasks to the media) [36, 37]. Besides, research arising from the use of different methods of interaction [38], have suggested ways improve foreign language skills but there is not any specific project or research evaluating significantly the use of multimodality in exams. Out research goes a step forward, to our knowledge this is the first time that research seeking information on how multimodality is applied in language tests has been done.

16.5 Methodology for the Development of a Multimodal Application for Language Testing

To define a language testing online application we defined tasks, goals, and technical issues as the starting point for the development of the final prototype (see Fig. 16.2). Although the use of voice recognition can seem promising in this application and its use can be suggested for rating, there are a number of problematic issues involved in such use. The difficulty of establishing automatic rating parameters is not minor because it requires the creation of a recognition database that can discriminate among native accents without determined models (i.e., international students do not have a close-to-British or close-to-American accent but a combination of many) making recognition difficult. The other key issue is about how to add and deduct points in the scores (whether grammar, pronunciation, lexical content, speech cohesion, etc.). Thus, we believed that rating should be carried by traditional human raters and speech recognition should have its major value for navigation. We focused on:

- Validating an environment of communication with potential test users.
- Applying the concepts of technical accessibility in the creation of a communicative interface for the final user.
- Creating a multimodal navigation interface for testing language skills based on a series of tasks oriented for use in a ubiquitous device and in a web environment.

The designed interface used three general navigation and data input systems (touch screen, keyboard, and voice) through which the user can interact sequentially.

The internal programming of the application allows in its version of voice recognition, management actions, and processes on the navigation interface during the exam through the recognition of specific words like next, out, yes, no, etc., When the users first enters the test application (see Fig. 16.3), they can indicate the type of navigation interaction in the different screens created for each task (Figs. 16.4 and 16.5). This helps the users define the type of communication with the application they want to use.

As mentioned above, the task features of the test dictated the selection of the three communication systems. This led to establishing a coherent method of navigation for the different sections of the test (reading, writing, speaking, and listening) as well as a unidirectional method to input the student's production data that permitted secure storage and retrieval by the raters and administrators for an

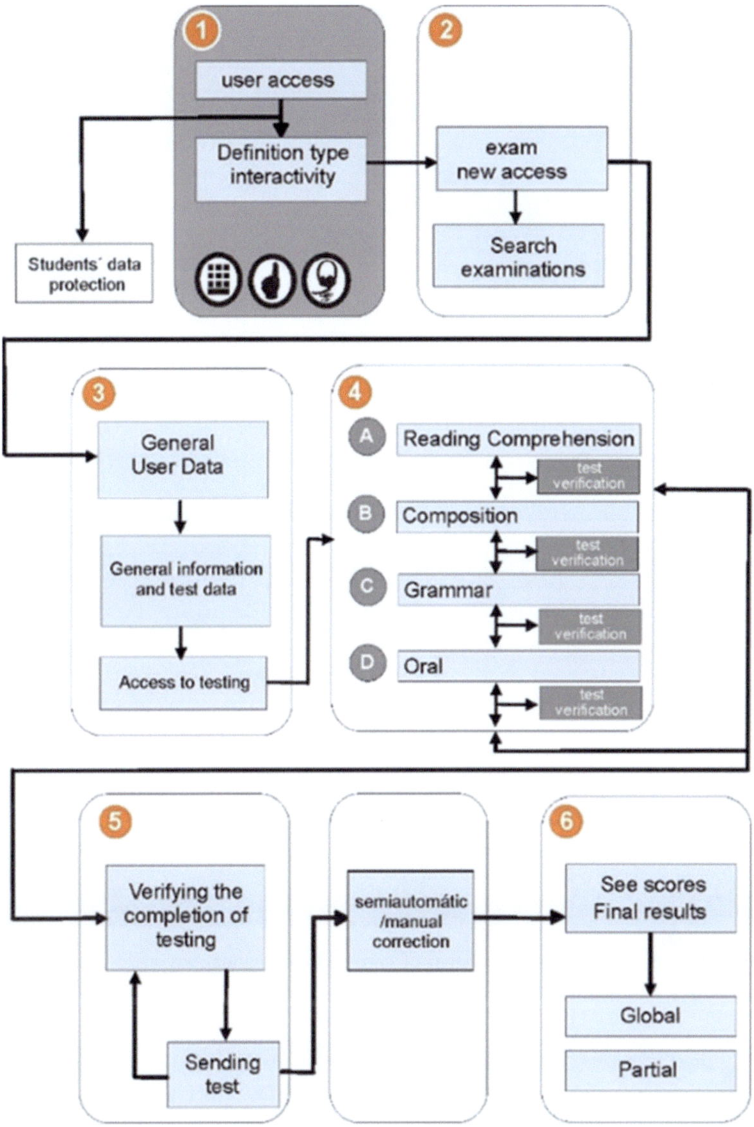

Fig. 16.2 Task definition for language learning testing through mobile devices

efficient automatized or human rating, the preparation of accurate reports, and the adequate and fast delivery of the final results to the student (see Fig. 16.6).

For instance, in the case of reading comprehension tasks, we used a system of consecutive linear interfaces where the person taking the test could first read the test and then respond to the questions in independent interfaces. We believed that this would be better than the traditional text deployment together with the questions

Fig. 16.3 OPENPAU™ student testing

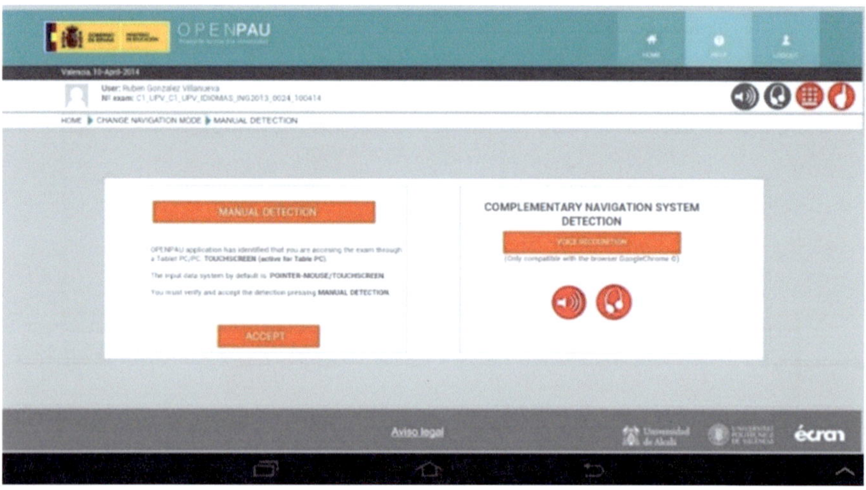

Fig. 16.4 Access to the detection system of conventional manual navigation (tactile and keyboard) and access to the multimodal detection system (voice recognition for navigation)

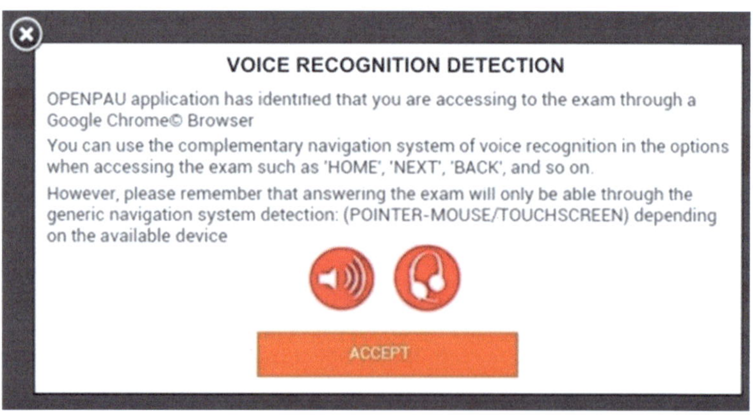

Fig. 16.5 Voice recognition detection

Fig. 16.6 OPENPAU™ multimodal prototype

which may be operational in large screens but not as much in tablet PCs or other mobile devices.

The platform permits the student to go forward and backwards within the official time limit assigned to each task. At the end of each section of the exam, the student has to save the answers. At the end of the exam the student has to do a final verification and validate the test before it is submitted (see Fig. 16.6).

The final verification interface makes it possible to verify whether all the sections have been completed and if the testee is ready to finish. As indicated in the instructions of the test, when the test taker validates the test, he grants permission to the platform to close the application and to the administrators to proceed to storage and the rating process.

The system had triple evaluation through experts and designers, through test takers whose interest was the test itself and users who looked at the design and presentation. The results indicate that while test takers tend to consider the online test difficult for younger students, they consider it adequate for 12th graders (18 years old) [39]. On the other hand, they consider the platform adequate and highly usable [40]. Limitations in multi-platform have been detected in strategies of interaction device-testee in cases such as enlargement of test, shadowing, and others that favor the use of tablet PC versus that tradition desktop PC. That means, IPad takers would have some advantages in front of traditional desktop users. Voice recognition could be used in both cases in a similar manner and with similar features.

As a summary of the process of creation of the user-oriented final interface (both for students and teachers), the importance of selecting the type of initial navigation test can be observed. An early decision in the mode of browsing increases the user's potential for the test. Thus an adequate decision of this type does not only favor the navigation but also enhances his potential for language test completion by reducing potential problems due to navigation.

16.6 Conclusions and Further Research

The results of the project research so far have served to implement and validate a future a real and effective multimodal prototype which has been designed by experts of different fields and expertise in language testing in cooperation with the target students (see Fig. 16.7).

Future research will have to look at the students' perceptions, teachers' and administrators' attitudes towards the use of multimodal applications oriented to second language acquisition and to create a new way to develop a test and exams online.

We will include design validation through questionnaires and field annotations in a multimodal prototype applied to the entire exam, not only navigation, but in the introduction and development of the same test and which is compatible with standards established by the W3C Multimodal Interaction Working Group [41].

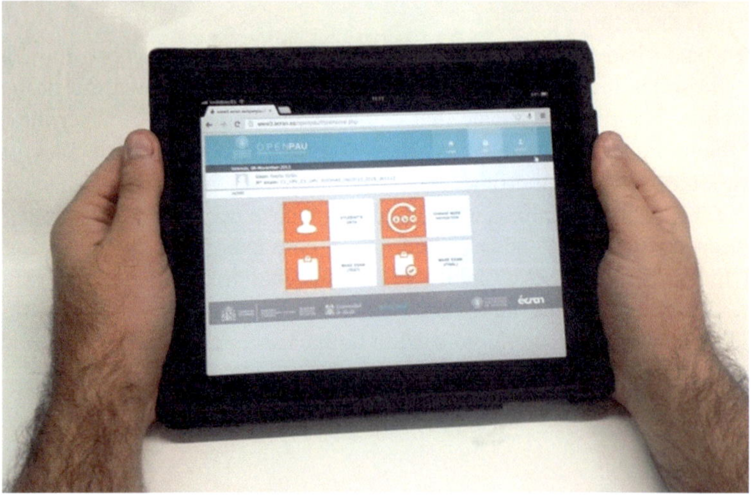

Fig. 16.7 Multimodal prototype as seen in the research Tablet PC

Acknowledgements The authors would like to thank the Ministry of Economy and Competitiveness for funding the research project (with co-financing by ERDF) within the framework of the National R + D + I (2011–2014) "Guidance, proposals and teaching for English section in the entrance examination to the University" (Reference FFI2011-22442). The researchers would also thank the Spanish Ministry of Education, Culture, and Sports because without the grant for the Senior Researchers Mobility this paper would probably have not been possible.

References

1. Chapelle, C. A., Enright, M. K., & Jamieson, J. M. (2007). *Building a validity argument for the test of English as a foreign language™*. New York, NY: Routledge.
2. García Laborda, J. (2007). On the net: Introducing standardized EFL/ESL exams. *Language Learning and Technology, 11*(2), 3–9.
3. Weir, C. (2005). *Language testing and validation: An evidence-based approach.* New York, NY: Palgrave Macmillan.
4. Garrett, N. (1991). Technology in the service of language learning: Trends and issues. *Modern Language Journal, 75*, 74–101.
5. Roever, C. (2001). Web-based language testing. *Language Learning and Technology, 5*(2), 84–94.
6. Stricker, L. J., Wilder, G. S., & Rock, D. A. (2004). Attitudes about the computer-based test of English as a foreign language. *Computers in Human Behavior, 20*, 37–54.
7. García Laborda, J., Magal-Royo, T., & Enriquez Carrasco, E. V. (2010). Teachers' trialing procedures for Computer Assisted Language Testing Implementation. *Eurasian Journal of Educational Research, 39*, 161–174.
8. García Laborda, J. (2010). Necesitan las universidades españolas una prueba de acceso informatizada? El caso de la definición del constructo y la previsión del efecto en la enseñanza para idiomas extranjeros. *Revista de orientación y Psicopadagogía, 21*(1), 71–80.
9. Chapelle, C. A., & Douglas, D. (2006). *Assessing language through computer technology.* New York, NY: CUP.
10. Suvorov, R., & Hegelheimer, V. (2013). Computer-assisted language testing. In A. J. Kunnan (Ed.), *Companion to language assessment* (pp. 593–613). Malden, MA: Wiley-Blackwell.
11. Karabulut, A., LeVelle, K., Li, J., & Suvorov, R. (2012). Technology for French learning: A mismatch between expectations and reality. *CALICO Journal, 29*(2), 341–366.
12. Chalhoub-Deville, M. (2010). Technology in standardized language assessments. In R. Kaplan (Ed.), *Oxford handbook of applied linguistics* (2nd ed.). Oxford: Oxford University Press.
13. García Laborda, J., Magal Royo, T., Litzler, M. F., & Giménez López, J. L. (2014). Mobile phones for a University Entrance Examination language test in Spain. *Educational Technology and Society, 17*(2), 17–30.
14. Sharma, R., Pavlovi, V. I., & Huang, T. S. (1998). Toward multimodal human–computer interface. *Proceedings of the IEEE, 86*(5), 853–869.
15. World Wide Web Consortium, W3C (2003). Device independence principles. http://www.w3.org/TR/2003/NOTE-di-princ-20030901. Accessed 12 Dec 2015.
16. Larson, J. A. (2010). Standard languages for developing multimodal applications. www.larson-tech.com/Writings/multimodal.pdf. Accessed 12 Dec 2015.
17. Zander, T. O., Kothe, C., Jatzev, S., & Gaertner, M. (2010). Enhancing human–computer interaction with input from active and passive brain–computer interfaces. *Human-Computer Interaction Series, 0*(3), 181–199. doi:10.1007/978-1-84996-272-8_11. Accessed 12 Dec 2015.

18. World Wide Web Consortium, W3C (2009). EMMA: Extensible MultiModal Annotation markup language. http://www.w3.org/TR/2009/REC-emma-20090210. Accessed 12 Dec 2015.
19. World Wide Web Consortium, W3C (2009). Guía Breve de Interacción Multimodal. Oficina Española del W3C. http://www.w3c.es/divulgacion/guiasbreves/Multimodalidad. Accessed 12 Dec 2015.
20. World Wide Web Consortium, W3C (2009). Multimodal architecture and interfaces. http://www.w3.org/TR/2009/WD-mmi-arch-20091201. Accessed 12 Dec 2015.
21. Oviatt, S., DeAngeli, A., & Kuhn K. (1997). Integration and synchronization of input modes during multimodal human–computer interaction. In *Proceedings of Conference Human Factors in Computing Systems, CHI'97*, Atlanta, pp. 415–422.
22. García Laborda, J., Magal-Royo, T., Da Rocha Siqueira, J. M., & Fernández Alvarez, M. (2010). Ergonomics factors in English as a foreign language testing: The case of PLEVALEX. *Computers and Education, 54*(2), 384–391.
23. Giménez-López, J. L., Magal-Royo, T., Garde Calvo, F., & Prefasi Gomar, S. (2009). The adaptation of contents for the creation of foreign language learning exams for mobile devices. IMCL2009. *International Journal of Interactive Mobile Technologies, 3*, Special Issue ISSN: 1865–7923.
24. Giménez-López, J. L., Magal-Royo, T., García Laborda, J., & Garde Calvo, F. (2010). Methods of adapting digital content for the learning process via mobile devices. *Procedia—Social and Behavioral Sciences, 1*, 2673–2677.
25. Fuster-Duran, A. (1996). Perception of conflicting audio-visual speech: An examination across Spanish and German. In D. G. Stork & M. E. Hennecke (Eds.), *Speech reading by humans and machines: Models, systems and applications* (pp. 135–143). New York, NY: Springer.
26. Magal-Royo, T., García Laborda, J., Peris-Fajarnes, G., & Spachtholz, P. (2007). Visual learning through guided iconography in wireless scenarios. In *Proceedings of ECEL 2007. 6th European Conference on e-Learning*, pp. 415–418.
27. Magal-Royo, T., Peris-Fajarnes, G., Tortajada Montañana, I., & Defez García, B. (2007). Evaluation methods on usability of m-learning environments. *International Journal of Interactive Mobile Technologies, 1*(1), ISSN: 1865–7923.
28. Magal-Royo, T., Gimenez-Lopez, J. L., & Pairy, B. (2011). Multimodal applications for foreign language teaching. In *14th International Conference on Interactive Collaborative Learning*, ICL, Piestany, Slovakia, pp. 145–148.
29. Magal-Royo, T., Laborda, G. J., & Gimenez-Lopez, J. L. (2011). Accessible multimodal interaction for language learning on mobile devices. In Q. Y. Zhou (Ed.) International Conference on Applied Social Science (ICASS 2011) Changsha, China, March 19–20.
30. Campillo, L., & Lanos, L. (2010). Tecnologías del habla y análisis de la voz. Aplicaciones en la enseñanza de la lengua. Revista Dialogo de la lengua. http://www.dialogodelalengua.com/articulo/pdf/2/1_campillos_DL_2010.pdf. Accessed 12 Dec 2015.
31. Lingle, V., & Deshpande, A. (2010). Online multimodal interaction for speech interpretation. *International Journal of Computer Applications, 1*(19), 81–85.
32. Magal-Royo, T., & Giménez Lopez, J. L. (2012). Multimodal interactivity in the foreign language section of the Spanish university admission examination. *Revista Educacion, 357*, 163–176.
33. Chittaro, L. (2010). Distinctive aspects of mobile interaction and their implications for the design of multimodal interfaces. *Journal on Multimodal User Interfaces, 3*(3), 157–165. SpringerLink (Ed.).
34. König, W. A., Rädle, R., & Reiterer, H. (2010). Interactive design of multimodal user interfaces. Reducing technical and visual complexity. *Journal on Multimodal User Interfaces, 3*(3), 197–213. doi:10.1007/s12193-010-0044-2. SpringerLink Ed.
35. Tan D., & Nijholt, A. (2010).Brain–computer interfaces and human–computer interaction. *Brain-Computer Interfaces. Human-Computer Interaction Series, 0*(1), 3–19. doi:10.1007/978-1-84996-272-8_1. Accessed 12 Dec 2015.

36. Alseid, M., & Rigas, D. (2008). An empirical Investigation into the use of multimodal E-learning interfaces. In S. Pinder (Ed.), *Advances in human-computer interaction* (Vol. 5, pp. 85–100). USA: Hindawi Publishing Corporation.
37. MCgee-lennon, M., Nigay, L., & Gray, P. (2010). The challenges of engineering multimodal interaction. *Journal on Multimodal User Interfaces, 3*(3), 155–156. SpringerLink Ed. www.springerlink.com/index/y55w472191907011.pdf. Accessed 12 Dec 2015.
38. Alwan, A., Vijian, B., Black, M., Casey, L., Gerosa, M., Heritage, M., et al. (2007). A system for technology based assessment of language and literacy in young children: The role of multiple information sources. In *IEEE 9th Workshop on Multimedia Signal Processing*, pp. 26–30. http://diana.icsl.ucla.edu/Tball/publications/tball_mmsp07.pdf. Accessed 12 Dec 2015.
39. García Laborda, J., & Litzler, M. F. (in press). Students' opinions about ubiquitous delivery of standardized English exams. *Porta Linguarum.*
40. García Laborda, J., Magal-Royo, T., & Bakieva, M. (2016). Looking to the future of language assessment: Tablet PCs usability in language testing. *Journal of Universal Computer Science, 22*(1).
41. Duarte, C. (2008). *Design and evaluation of adaptive multimodal systems.* Doctoral theses, Departamento de Informática, Universidad de Lisboa. http://hdl.handle.net/10455/3123. Accessed 12 Dec 2015.

Chapter 17
Multi-Device Applications Using the Multimodal Architecture

Nuno Almeida, Samuel Silva, António Teixeira, and Diogo Vieira

Abstract Nowadays, users have access to a multitude of devices at their homes, workplaces or that they can carry around. Each of these devices, given its features (e.g., interaction modalities, screen size), might be more suitable for particular users, tasks, and contexts. While having one application installed in several devices might be common, they mostly work isolated, not exploring the possibilities of several devices working together to provide a more versatile and richer interaction scenario. Adopting a multimodal interaction (MMI) architecture based on the W3C recommendations, beyond the advantages to the design and development of MMI, provides, we argue, an elegant approach to tackle multi-device interaction scenarios. In this regard, this chapter conveys our views and research outcomes addressing this subject, presenting concrete application examples.

17.1 Introduction

Mobile devices, smart TVs, media centers, and game boxes are widespread and used on a daily basis by millions around the world. These devices are important means for providing users with applications and services that support many aspects of their personal and professional life. Multimodal interaction (MMI) is a common feature for many of them. Each of these devices presents a set of characteristics (e.g., mobile nature and available input and output modalities) that make them more suitable for particular contexts, and being able to seamlessly change from one device to another while performing a task, commuting among environments (e.g., from a smartphone on the street to a tablet at home), following the concept of migratory interfaces [1], would provide a desirable degree of flexibility and user adaptation.

In addition, multiple devices, if used together, can provide new means for interacting with an application, whether providing complementary features to a

N. Almeida • S. Silva (✉) • A. Teixeira • D. Vieira
DETI—Department of Electronics, Telecommunications and Informatics, IEETA—Institute of Electronics and Informatics of Aveiro, University of Aveiro, Campus Universitário de Santiago, Aveiro, Portugal
e-mail: sss@ua.pt

D.A. Dahl (ed.), *Multimodal Interaction with W3C Standards*,
DOI 10.1007/978-3-319-42816-1_17

single user or supporting collaboration among users. The output modalities should also be able to provide feedback in multiple ways, adapting to the current device ecosystem, with the information made available in one device potentially adding to the contents provided by other accessible devices.

Therefore, the coexistence of multiple devices and applications provides (and demands) new interaction possibilities, posing challenges regarding how different devices can be used simultaneously to access a specific application, taking the most out of each device features (e.g., screen size) and sharing input and output modalities.

Designing and developing a single application to run on multiple devices, taking advantage of each device characteristics, and harnessing the power of multiple devices simultaneously, presents a number of challenges. These concern decisions on where the application logic will be instantiated and how the different modalities are managed and used to attain a seamless and fluid interaction experience. The main goal is that the experience with the different devices and modalities blends to the point that what is in focus is the interaction with the ecosystem for which each of the devices can be seen as a nested or complex modality component [2]. For multimodality, we aim at providing a versatile experience through a set of modalities, used simultaneously or not, to widen the data bandwidth between users and systems. With multi-device we aim to serve the same purposes at a higher level.

While working on the design and development of multimodal applications, in a variety of scenarios provided, for example, by projects AAL4All,[1] Paelife,[2] Smartphones for Seniors,[3] and Marie Curie IAPP project IRIS[4] [3], we envisaged several contexts where multiple devices are available and can be explored to enhance interaction, and in which several users may interact simultaneously over the same contents making use of their personal devices.

The support for multi-device interaction is the focus of several works presented in the literature. From the early multi-device proposals, such as those by Rekimoto [4], to recent works, e.g., Diehl [5], the design approaches vary considerably and despite interesting results having been attained, serving multimodal multi-device purposes, most approaches propose their own architectures with a potential negative impact on their dissemination and compatibility. Proposing approaches that build on existing standards should be, in our opinion, and to the extent that those standards serve the targeted use cases, our first option.

In our work context, we approach multi-device support starting from the multimodality perspective and we argue that the MMI architecture, adopted for the different projects, and proposed based on the W3C recommendations, provides enough flexibility to integrate all the features that would enable multi-device support. This is also emphasized by our view, conveyed above, that a multi-device

---

[1] http://www.aal4all.org/.

[2] http://www.paelife.eu/.

[3] http://www.smartphones4seniors.org/.

[4] http://iris-interaction.eu.

setting can be viewed as a set of complex (or nested) modality components (the devices) used to interact with the macro application made available to users on more than one device. This approach would exhibit two advantages: (1) an approach to multi-device mostly performed at the architecture level, rather than at application level, potentially enabling easier multi-device support to the set of applications running over the MMI architecture; and (2) a standards-based approach, serving the goals of a loosely coupled architecture.

This chapter presents our views and proposals to support multi-device applications based on the W3C MMI architecture. After a brief contextualization of the architectural and technical background at stake for our work, in Sect. 17.2, we describe two different multi-device approaches in Sect. 17.3. Then, Sect. 17.4 is devoted to two application examples extracted from our ongoing work on ambient assisted living and interactive visualization. While the first example shows how multi-device support can be easily added to an existing multimodal application, the second is a first step in exploring multi-device visualization. Finally, Sect. 17.5 presents some conclusions and discusses routes for further development.

17.2 Background

Our work on multimodal applications spans over a wide range of applications covering services in ambient assisted living, such as personal life assistants [6, 7] and telerehabilitation [8], medication adherence [9], and a recent evolution of other lines of research into the MMI domain concerning autism spectrum disorders [10], and interactive visualization [11]. Our work also includes research on interaction modalities, such as speech [12, 13] and gaze [14], in an effort to provide what we call generic modalities [11–13]. In many of these works there is a rising opportunity and need of addressing multimodal multi-device interaction and this section provides a very brief account of one of the most recent scenarios we are considering for requirement elicitation, in line with the use cases proposed, for example, in [15], and summarizes relevant aspects of our previous developments regarding the adopted MMI architecture and its role on multi-device support.

17.2.1 Research Context: Multimodal Multi-Device Scenarios

One of the scenarios motivating our interest in further developing and exploring multi-device MMI, and a representative example of our research context, is provided by the Marie Curie IAPP project IRIS [3]. One of the main goals of IRIS is to provide a natural interaction communication platform accessible and adapted for all users, particularly for people with speech impairments and elderly in indoor scenarios. The particular scenario under consideration, a household, where a family lives (parents, two children, one diagnosed with autism spectrum disorder, and a grandmother), and where different devices exist around the house, and are owned by the different family members, is the perfect setting for evolving multi-device MMI. In our view, communication can go beyond the exchange of messages through these media and profit from the dynamic multi-device ecosystem, where similar contents (e.g., a photo blog from the latest weekend activities or the family agenda) can be viewed in different manners, adapted to the device and user preferences, and supporting a collaborative interaction effort.

17.2.2 Multimodal Interaction Architecture

Our multimodal architecture [6, 12, 16–18], initially developed for AAL [6, 19], adopts the W3C recommendations and, as depicted in Fig. 17.1, can briefly be described as being composed of four main components [2, 17]: interaction manager (IM), data component, modality components, and the runtime framework.

The runtime framework provides the infrastructure to run the entire system, start components, and manage communication. Modality components enable the different inputs and outputs. The data component stores data related to the application and only the IM has access to it. The IM is a central component as it manages all received events from the modality components. It is a state chart configured through the State Chart extensible Markup Language (SCXML) [20].

The communication between components (or modalities) and the IM is performed by exchanging life cycle events, which contain Extensible MultiModal Annotation (EMMA) markup language [21], used to describe the events generated by modalities.

The implementation of the IM uses the Apache Commons SCXML[5] to manage the state machine defining the application logic. We extended the use of the SCXML to parse multimodal architecture life cycle events and trigger them into the state machine. The extension also includes the generation of the life cycle events to be transmitted to the modalities.

17.3 Multi-Device Support Using the W3C MMI Architecture

Our work on making multi-device interaction possible started from a set of simple ideas we considered as grounds for our first experiments: (1) have a unique application, running in the various devices, to reduce development cost, integrating

[5] http://commons.apache.org/proper/commons-scxml/.

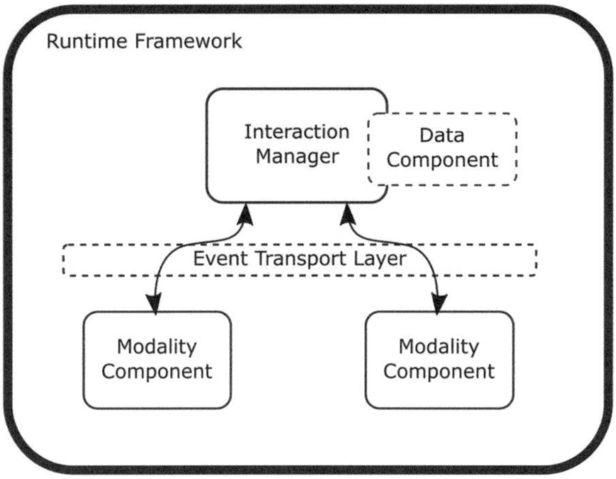

Fig. 17.1 Main components of the multimodal architecture as recommended by the W3C (adapted from https://www.w3.org/TR/mmi-arch)

means to adapt to the device, user, context, and to the existence and status of other devices running the same application; (2) make information from each of the existing device modalities available to the applications running in other devices; and (3) rely on the W3C MMI architecture standards and on our implementation of the MMI architecture to accomplish all the required features.

In our proposal, the IM module assumes the most importance. By using the IM as a pseudo-modality for applications running in other devices we were able to create a loosely coupled and extensible architecture that supports multiple modalities and the distribution of modalities across multiple devices, such as PCs, tablets, and smartphones. The architecture provides a flexible approach, allowing changing or adding modalities to the system without the other components being aware.

Two variants of the solution were proposed and prototyped, one considering an IM residing in each device, and other deploying the IM as a service, in the cloud, enabling the existence of a central IM per application, as detailed in the following sections.

17.3.1 Per Device Interaction Manager

Our first approach to multi-device applications considers that each device must run one IM. Figure 17.2 shows an illustration of a scenario with two devices, each running the application with a GUI modality, one IM, and any additional modalities, which can be different for each device. In this multi-device scenario, one IM behaves as a modality to the other. Following this approach, it is possible to disconnect the two devices and work with each device separately.

To enable each IM to discover the other, we use an UPnP server allowing all IMs to register their address. Each IM periodically sends broadcast requests to find UPnP servers with the service "MMI Discovery," registers its address on it, and

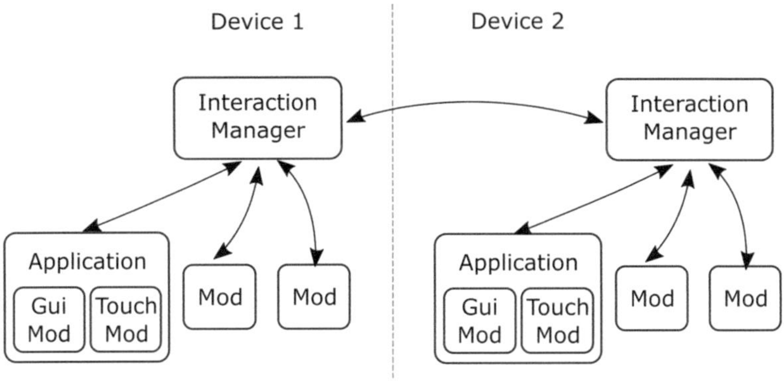

Fig. 17.2 Overview of the architecture supporting multi-device MMI considering an Interaction Manager for each device

obtains a list of existing IMs. From this point on, both IMs know of the other's existence and where to send the messages. Besides this process of discovery, all the communication between the IMs and the modalities is accomplished using the HTTP GET/POST protocol encapsulating the MMI life cycle events.

Only the IMs can exchange messages between devices. If a new event occurs in one modality of Device 1, this modality only sends the message to the local IM, which in turn, if it has information regarding the other device, sends the message to the other IM (Device 2). The IM for Device 2 processes that message as if it was sent by one of its modalities. Figure 17.3 presents an example of the messages between IMs and modalities after discovery.

Besides the simple registration of available devices, described above, a modality providing information regarding proximity among devices can be used in the multi-device application context as a trigger between single and multi-device use. As a proof of concept, and considering the scenario of a living room with a main unit (Home computer + TV) and a portable unit (tablet), we developed a proximity modality using wireless RSSI that computes the approximate distance between a tablet and the access point (placed near the main unit). Although this measurement is not accurate, it serves the purpose of identifying if the user is near or far away. Additionally, when the user is in front of the main unit, a Kinect is used to compute the distance between the devices. Proximity data is sent to the local IM that informs the other(s). With this information, the IM decides whether to use both devices for interaction. Figure 17.4 illustrates a scenario where the user is in different locations, enabling or not the use of the multi-device mode.

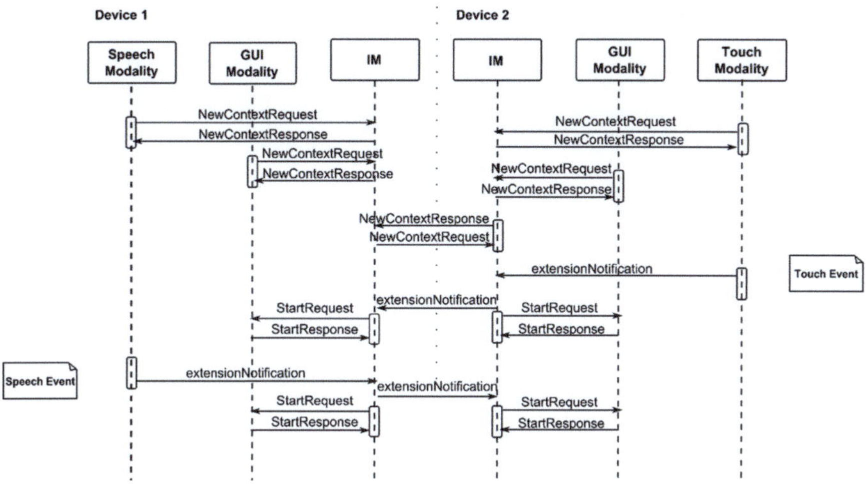

Fig. 17.3 Illustrative example of the communication between components for an MMI multi-device scenario. The touch and speech events, issued on one of the devices, are propagated to the other device through the Interaction Managers

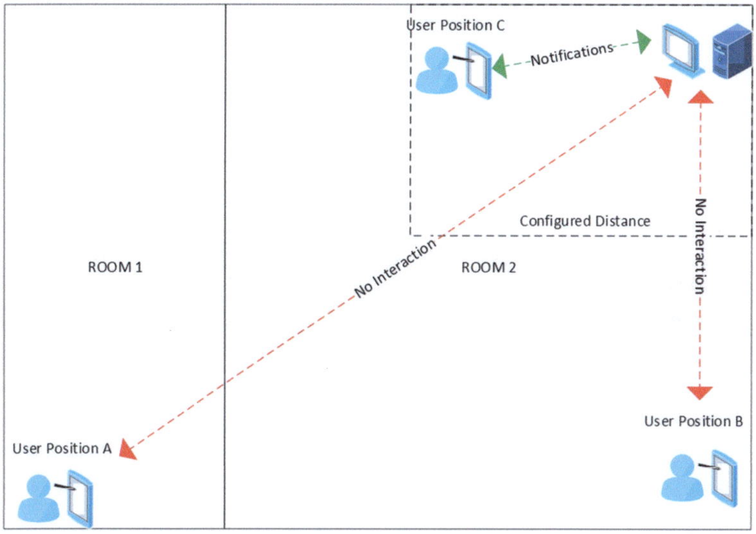

Fig. 17.4 Proximity modality in use illustrated for a fixed and a mobile device. The devices only enter multi-device mode when they are positioned near each other

17.3.2 Cloud-Based Interaction Manager

The approach described above allowed, to some extent, the use of multiple devices to improve the access to a particular application, profiting from all the available

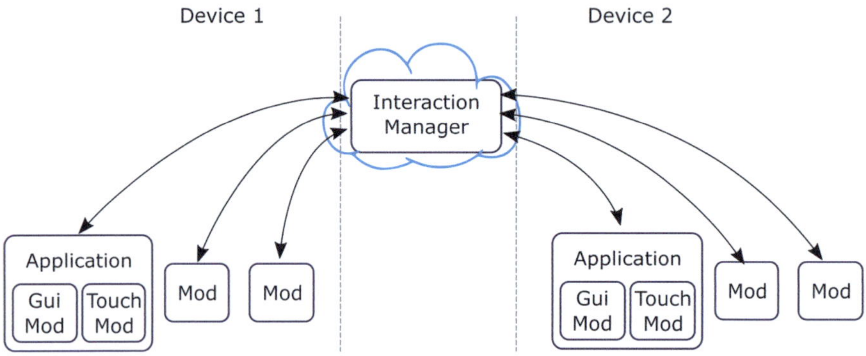

Fig. 17.5 Overall architecture for multi-device MMI supports using a single Interaction Manager located in the cloud

modalities and enabling a richer and more versatile output modality. Nonetheless, regarding the technical aspects of deploying these applications, a few issues arise, particularly for Microsoft Windows platforms. When making applications available through the Microsoft App Store, one of the limitations is that those applications, when installed, cannot communicate to internal services running on the device. Therefore, communication with the service providing the IM was compromised, which lead us to move into a solution with the IM located externally, in the cloud. This change in how our MMI architecture was implemented yielded a more generic approach to multimodal multi-device interaction.

In this novel proposal to creating multi-device applications, only one central IM was considered, located in the cloud, and capable of managing multiple devices and multiple clients. To enable multiple clients in the central IM, each modality registers in the IM with a unique identifier. Figure 17.5 presents the overview of the target architecture.

This approach is more generic and can encompass a larger number of devices with less complexity than the first approach without a central IM. Furthermore, despite the overall differences between the two approaches, the way the modalities and IM communicate is the same, i.e., the same life cycle events, containing the same EMMA markup. So, applications adopting the first approach to multi-device support can be easily migrated to this more versatile solution.

17.4 Application Examples

The following sections illustrate how the described features have been used in the context of two different applications, each using one of the alternative approaches described earlier. The first example builds on our work for ambient assisted living. It enables the use of a personal life assistant taking advantage of multiple devices to provide additional display space for accessing information by considering devices in the user's proximity. The second example concerns data and information visualization and is aimed at being an experimental platform to explore multimodal multi-device interactive visualization. Instead of the usual setting of custom applications to support collaborative multimodal interactive visualization and analysis, we argue that such features can be supported at the architecture level, enabling a more generalized use of such features among everyday applications.

17.4.1 AAL Device-Rich Scenarios: A Multi-Device Personal Life Assistant

This first example illustrates how an existing multimodal application, a personal life assistant [6] providing a set of modules such as news, weather, and a messaging hub, was evolved to support multi-device features.

The requirements for this application include not only the ones initially considered for the AALFred assistant [6], regarding the support for MMI including speech, touch, and gestures, but also new requirements were determined to create a multi-device experience. The application should be capable of running independently, connect among devices running it, change between autonomous or joint use, based on proximity, and allow showing the same or alternative content in each of the devices. Furthermore, the interface should be as similar as possible in both units, to minimize the need for additional learning.

In the first stage, only the news module was addressed, as a proof of concept for the multi-device features, updating the single device news module developed for the AALFred assistant. We considered, as the typical multi-device scenario, a static main unit (fixed computer) connected to a television and a mobile unit (tablet or smartphone), each working independently but simultaneously interoperable. In this multi-device scenario interaction can potentially be performed in three different ways: (a) through the main unit; (b) through the mobile unit; and (c) through both the main and mobile units.

Interaction through the main unit means that interaction is performed using only the modalities made available by the fixed-position device. In our prototype there are two main interaction modalities: voice (body), gestures and graphical output. In the interaction through the mobile unit, the user will interact by only using the modalities available on the tablet, particularly touch and graphical output. These two ways of interacting with the application are the typical single device scenario, although one should note that it is actually the same application running on each of the devices and not separate custom versions.

Interaction considering the two units takes advantage of the interaction capabilities of both devices to improve the usability of the system and implement new features. For example, when detecting that the user is within the range of the main unit, the application can allow using the main screen to visualize content while using the tablet as a controller.

In the news module there are three main information components that users can access: a list of the available news, an image illustration for the news content, and the news text. Considering these components, there are several content combinations possible when in the presence of two devices. For example, and without loss of generality, for a large TV set and a tablet, three combinations were considered, as depicted in Fig. 17.6: (1) TV and tablet showing the news content; (2) TV showing the image illustration in full screen and the tablet showing the complete news content; and (3) TV showing the whole news content and the tablet showing the list of news, serving as a news navigation device. If one of the applications is set to display only the news list and it starts working alone (the other device is not near) the application automatically reverts to working in single device mode.

This example depicts our first experiment with multi-device support. One of the first aspects that should be noted is that it consisted in adding multi-device capabilities to an existing multimodal application and, despite the changes to the IM, only minimal adjustments were required to the application, mostly concerning the extension of the output modality to support the different modes. In this example, even though, from the technical perspective, it would be possible to do so, we did not address the use of different input modalities connected to both devices, and the main unit was mostly used for its output capabilities. Although this example is for a particular scenario of two devices, this approach can be considered for any number of devices. Nonetheless, as previously mentioned, it would not serve the purpose of making the application available in the Windows App Store.

17.4.2 Collaborative Data and Information Visualization

The use of natural and MMI in Visualization, still a rather unexplored field of research [22], e.g., based on speech and touch, might bring advantages at different levels. The use of multiple interaction modalities can help bridge the gap between visualizations and certain audiences, by providing, for example, alternatives for the less technologically savvy, improving the visibility of certain aspects of the data, or by ensuring a richer communication channel between the user and the application. By supporting a multitude of interaction options, a system can also favor a more versatile scenario when it comes to the analysis of the data, enabling an active search for insight that otherwise might not have been foreseen by the interface designer [22]. In this regard, it is important to explore and understand the strengths and weaknesses of multimodality when used in the context of Interactive

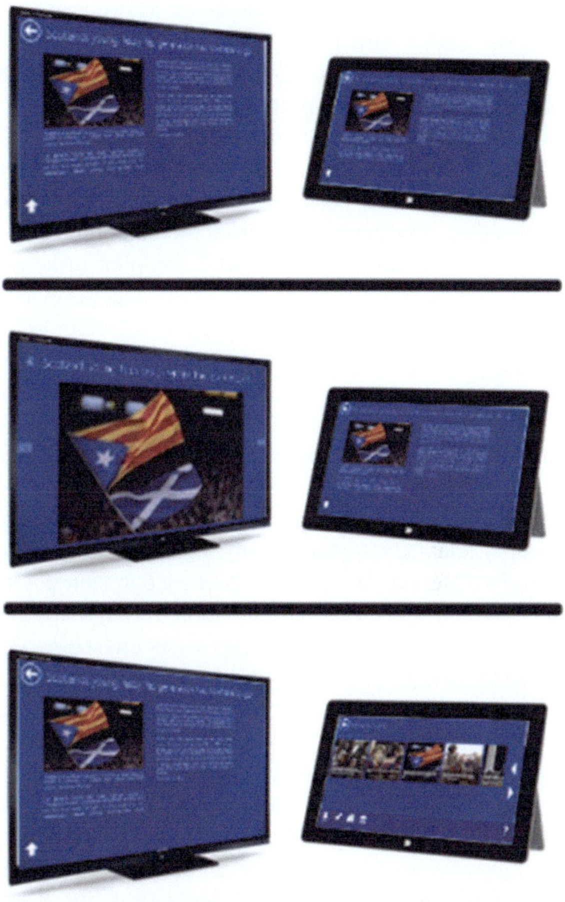

Fig. 17.6 Multiple ways to present the information available for a news content in a multi-device scenario including a TV and a tablet. From *top* to *bottom*: (**a**) both devices show the same content; (**b**) the TV shows a large image and the tablet the complete news contents; and (**c**) the TV shows the complete news contents and the tablet shows a news navigation menu

Visualization [23], exploring the potential advantages deriving from a richer interaction scenario, allowing adaptability to different contexts [24], and a wider communication bandwidth between the user and the application [25]. Furthermore, deriving from the wide range of devices available (smart TVs, tablet, smartphones, etc.), it is also relevant to explore how these multiple devices might be used to support visualization [26], whether individually, providing views adapted to the device characteristics [24], or simultaneously, providing multiple (complementary) views of the same dataset [27], fostering a richer interaction experience, or as the grounds for collaborative work [28].

To this purpose, we started the development of a prototype application that should allow for exploring the different challenges of collaborative multimodal interactive visualization [11] and how the MMI architecture could serve its requirements.

The application context that served as grounds for the prototype was inherited from our work on the proposal of evaluation frameworks for dynamic multimodal systems. Dynamic Evaluation as a Service (DynEaaS) [29] is a platform that supports the evaluation of complex multimodal distributed systems. The platform collects all data concerning the users' interaction with a system. The collected data is organized in a hierarchical form, according to the application components. Some insights of the users' performance with the application can be extracted by analyzing this data. Considering the amount and complexity of the resulting data, particularly in evaluation scenarios with complex tasks and several participants, it is important to create visualizations of the data allowing experts to interact, explore, and discuss the data.

To guide the design of the prototype application, we settled on a basic context scenario. In a meeting room, equipped with a TV connected to a computer, three experts meet to discuss some results of a previous system evaluation session. Each expert has a device capable of running the visualization application (each supporting multiple input and output modalities). One of the users is interacting with the TV, another user has a smartphone, and the other a tablet. The visualization modality adapts the default view to the screen size of each device. Also, users can choose a different visualization to be used in their device only. The outcomes of any user interaction over the visualization, through any of the available input modalities, are reflected in what the other users are seeing in their devices.

In the described context, the initial requirements for our prototype included: (1) Visualizations using different data representations, i.e., showing the same data but in a different way; (2) Multiple devices, adapting visualization to the screen size; and (3) Collaborative interactive visualization in a multi-device scenario.

The visualization system adopts the multimodal framework and, in this approach, only one IM is used (see Fig. 17.7) and is responsible for managing all the life cycle events coming from all devices. At this stage, we managed all the visualization modes as part of a single modality, the touch modality is part of the application, and the remaining modalities are allowed to connect to the framework.

One additional modality was created that took into consideration a smartphone-specific capability, using the accelerometer to detect the smartphone motion. The user can rotate the smartphone 90° to the right or left to navigate through data.

Following on the environment used in previous works, the application supports devices with Microsoft Windows, either desktop, tablets, and smartphones. To implement the visualization the D3js framework [30] was used as it naturally provides a large set of data and information representations.

For this first instantiation of the prototype, we opted for four different data representation alternatives, as depicted in Fig. 17.8: the sunburst visualization with breadcrumb and tooltips, the treemap, the treeview, and the timeline with tooltips. The consideration of the treeview, in particular, was meant to offer a compact representation that could be used in a device with a small display such as a small smartphone.

At its current stage, the prototype already illustrates some of the basic features we deem important as a proof of concept for a multimodal multi-device interactive

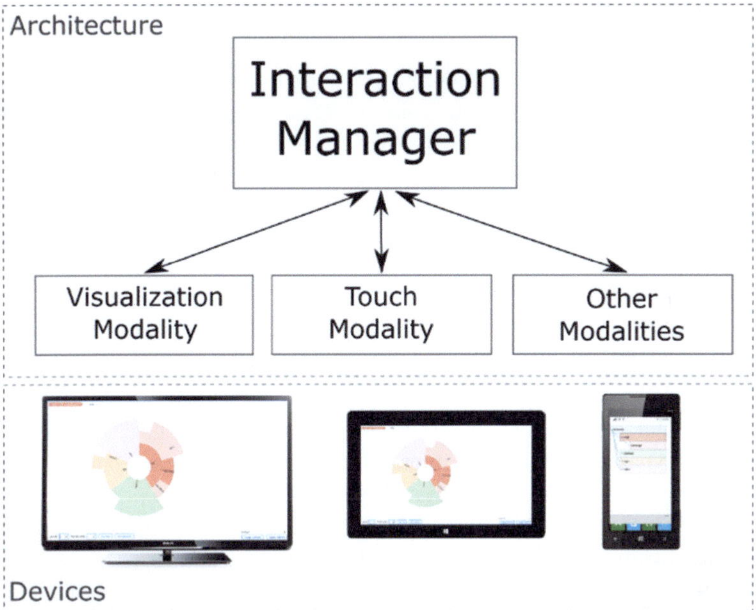

Fig. 17.7 Overview of the multi-device Visualization application architecture. Multiple devices, allowing multimodal interaction with different representations of the considered information, are connected to a central Interaction Manager located in the cloud

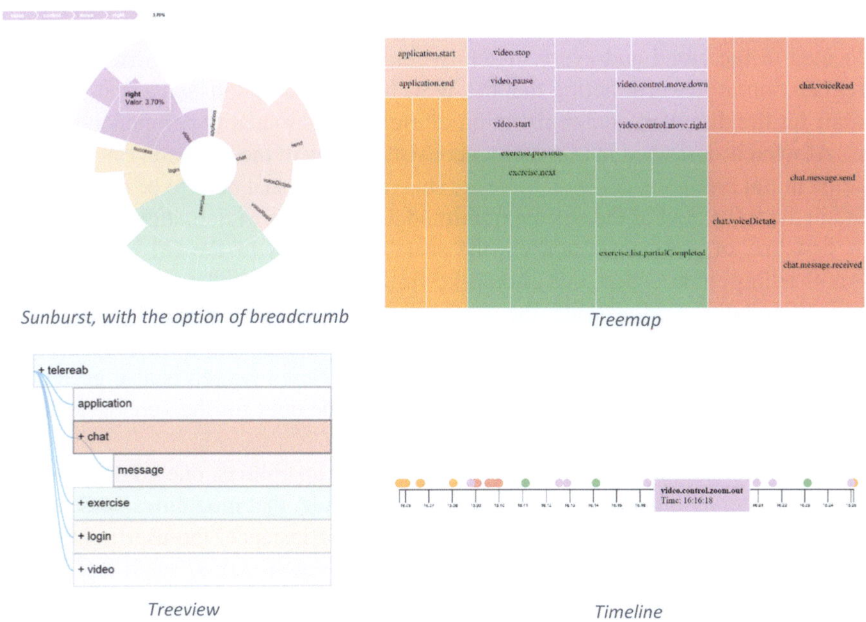

Fig. 17.8 Different data representations supported by the multi-device visualization application

visualization tool. By using the architecture to support the main features regarding the coordination between applications and the propagation of interaction, we place a complex aspect of such systems outside the application, yielding easier application development. In fact, from our point of view, this can be a first step towards a more general approach to multi-device support, where any application running over the multimodal framework can support, by default, multi-device features. One of the innovative features of using this second approach to multi-device is the deployment of the IM in the cloud, instead of an IM instance in each device.

17.5 Conclusions

Supporting interaction based on multiple devices is, we argue, fundamental to tackling the dynamic, device-rich interaction scenarios that have become so common nowadays. Supporting this feature at architecture level, as proposed, provides a simple and elegant approach that moves most of the need to support such features from the application into the architecture. In our view, such approach is critical to enabling widespread consideration of multi-device interaction not only in very specific applications, but also as a general feature available through the MMI architecture to all applications.

In our proposal we consider an approach to multi-device support where the IM is responsible for registering the available modalities. As to the discovery of modalities, in our first approach, we assume each device has an IM and finds other IMs in the network through an uPnP server; in a second approach, we consider a central IM located in the cloud and every modality connects to it. Regarding discovery and registration, it is worth noting that the W3C has recently published a first working draft for the discovery and registration of multimodal modality components [31].

As what is described in this chapter consists mainly of the adopted principles and some initial proofs-of-concept, there are several aspects that need to be addressed to attain the full extent of the desired multimodal multi-device interaction capabilities. A first line of work must address the scalability issues of the IM in the cloud. Another important aspect that should be further explored is the expansion of the output modality. For example, the graphical output modality should become more complex and autonomous to adapt to different devices and layouts in line with what we proposed for a complex speech modality [12]. New research is also needed to provide information regarding proximity between devices for the version with the IM in the cloud.

Finally, to get the most out of the multi-device capabilities, one could envisage complex interaction patterns considering, for example, that modalities made available by different devices can be used together. For instance, the user points to a large screen, equipped with a Kinect, and says "Show me this," with the speech recognized by her smartphone. Or several users may be in a room, analyzing the animation of a dynamic dataset, but given various computational and display characteristics of their personal devices, different representations are used for each (e.g., high resolution, graphical, text, and mixed). For both these examples, the timing for the different events assumes high relevance, and the importance of synchronization among the different devices and modalities is made clear. Addressing this challenge would largely benefit the evolution of multimodal multi-device interaction.

Acknowledgements The work presented in this chapter has been partially funded by IEETA Research Unit funding (Incentivo/EEI/UI0127/2014), Marie Curie IAPP project IRIS (ref. 610986, FP7-PEOPLE-2013-IAPP), project PaeLife (AAL-08-1-2001-0001), and QREN projects Smart Phones for Seniors (S4S), AAL4ALL and EMIF—European Medical Information Framework (EU FP7), co-funded by COMPETE and FEDER.

The authors thank all W3C MMI recommendations contributors for their insightful and inspiring approaches to MMI.

References

1. Ghiani, G., Polet, J., Antila, V., & Mäntyjärvi, J. (2015). Evaluating context-aware user interface migration in multi-device environments. *Journal of Ambient Intelligence and Humanized Computing, 6*(2), 259–277.
2. Dahl, D. A. (2013). The W3C multimodal architecture and interfaces standard. *Journal on Multimodal User Interfaces, 7*(3), 171–182.
3. Freitas, J., Candeias, S., Dias, M. S., Lleida, E., Ortega, A., Teixeira, A., et al. (2014). The IRIS project: A liaison between industry and academia towards natural multimodal communication. In *Proceedings of Iberspeech, Las Palmas de Gran Canaria, Spain*, pp. 338–347.
4. Rekimoto, J. (1998). A multiple device approach for supporting whiteboard-based interactions. In *Proceedings of the Conference on Human Factors in Computing Systems (CHI'98), Los Angeles, CA*, pp. 344–351.
5. Diehl, J., & Borchers, J. O. (2014). *Supporting multi-device iteraction in the wild by exposing application state*. (PhD thesis, No. RWTH-CONV-144030). Aachen:Fachgruppe Informatik.
6. Teixeira, A., Hämäläinen, A., Avelar, J., Almeida, N., Németh, G., Fegyó, T., et al. (2013). Speech-centric multimodal interaction for easy-to-access online services: A personal life assistant for the elderly. In *Proceedings DSAI 2013, Procedia Computer Science, Vigo, Spain*, pp. 389–397.
7. Hämäläinen, A., Teixeira, A., Almeida, N., Meinedo, H., Fegyó, T., & Dias, M. S. (2015). Multilingual speech recognition for the elderly: the AALFred personal life assistant. *Procedia Computer Science, 67*, 283–292.
8. Teixeira, A. J. S., Pereira, C., Oliveira e Silva, M., Alvarelhão, J., Silva, A., Cerqueira, M., et al. (2013). New telerehabilitation services for the elderly. In I. M. Miranda & M. M. Cruz-Cunha (Eds.), *Handbook of research on ICTs for healthcare and social services: Developments and applications*. Hershey, PA: IGI Global.
9. Ferreira, F., Almeida, N., Rosa, A. F., Oliveira, A., Casimiro, J., Silva, S., et al. (2013). Elderly centered design for Interaction—the case of the S4S Medication Assistant. In *5th International Conference on Software Development and Technologies for Enhancing Accessibility and Fighting Info-exclusion, DSAI, Vigo, Spain.*
10. Leal, A., Teixeira, A., & Silva, S. (2016). On the creation of a persona to support the development of technologies for children with autism spectrum disorder. In *Proc. HCI International LNCS 9739, Toronto, Canada*, 213–223. doi: 10.1007/978-3-319-40238-3_21

11. Almeida, N., Silva, S., Santos, B. S., & Teixeira, A. (2016). Interactive, multi-device visualization supported by a multimodal interaction framework: Proof of concept. In *Proc. HCI International. LNCS 9754, Toronto, Canada*, 279–289. doi: 10.1007/978-3-319-39943-0_27
12. Almeida, N., Silva, S., & Teixeira, A. (2014). Design and development of speech interaction: a methodology. In *Proc. HCI International, LNCS 8511, Crete, Greece*, 370–381.
13. Teixeira, A., Francisco, P., Almeida, N., Pereira, C., & Silva, S. (2014). Services to support use and development of speech input for multilingual multimodal applications for mobile scenarios. In *The Ninth International Conference on Internet and Web Applications and Services (ICIW 2014), Track WSSA—Web Services-based Systems and Applications, Paris, France*.
14. Vieira, D., Freitas, J. D., Acartürk, C., Teixeira, A., Sousa, L., Silva, S., Candeias, S., and Sales Dias, M. (2015). "Read that article": Exploring synergies between gaze and speech interaction. In *Proc. 17th International ACM SIGACCESS Conference on Computers & Accessibility (ASSETS '15). ACM, New York, NY, USA*, 341–342. doi: 10.1145/2700648.2811369
15. Wiechno, P., Dahl, D., Ashimura, K., & Tumuluri, R. (2012). Registration & discovery of multimodal modality components in multimodal systems: Use cases and requirements. [Online]. https://www.w3.org/TR/mmi-discovery/. Accessed 1 Jan 2016.
16. Almeida, N., Silva, S., & Teixeira, A. J. S. (2014). Multimodal multi-device application supported by an SCXML state chart machine. In *Workshop on Engineering Interactive Systems with SCXML, The sixth ACM SIGCHI Symposium on Computing Systems, Toronto, Canada*.
17. Almeida N., & Teixeira A. (2013). Enhanced interaction for the elderly supported by the W3C Multimodal Architecture. In *Proc. 5a Conf. Nacional sobre Interacção, Vila Real, Portugal*.
18. Teixeira, A., Almeida, N., Pereira, C., e Silva, M. O., & Pereira, J. C. (2013). Serviços de Suporte à Interação Multimodal. In A. Teixeira, A. Queirós, & N. Rocha (Eds.), *Laboratório Vivo de Usabilidade* (pp. 151–165). ARC Publishing.
19. Teixeira, A., Almeida, N., Pereira, C., e Silva, M. O., Vieira, D., & Silva, S. (2016). Applications in ambient assisted living. In D. Dahl (Ed.), *Multimodal Interaction with W3C Standards*. Springer.
20. Barnett, J., Akolkar, R., Auburn, R. J., Bodell, M., Burnett, D. C., Carter, J., et al. (2015), State chart XML (SCXML): State machine notation for control abstraction. W3C Recommendation. https://www.w3.org/TR/scxml/. Accessed 29 Jul 2016.
21. Baggia, P., Burnett, D. C., Carter, J., Dahl, D. A., McCobb, G., & Raggett, D. (2009). EMMA: Extensible multimodal annotation markup language. https://www.w3.org/TR/emma/. Accessed 1 Jan 2016.
22. Lee, B., Isenberg, P., Riche, N. H., & Carpendale, S. (2012). Beyond mouse and keyboard: Expanding design considerations for information visualization interactions. *IEEE Transactions on Visualization and Computer Graphics, 18*(12), 2689–2698.
23. Ward, M. O., Grinstein, G., & Keim, D. (2010). *Interactive data visualization: Foundations, techniques, and applications*. Natick, MA: CRC Press.
24. Roberts, J. C., Ritsos, P. D., Badam, S. K., Brodbeck, D., Kennedy, J., & Elmqvist, N. (2014). Visualization beyond the desktop—the next big thing. *IEEE Computer Graphics and Applications, 34*(6), 26–34.
25. Jaimes, A., & Sebe, N. (2007). Multimodal human-computer interaction: A survey. *Computer Vision and Image Understanding, 108*(1–2), 116–134.
26. Schmidt, B. (2014). *Facilitating data exploration in casual mobile settings with multi-device interaction*. Universitat Stuttgart, Holzgartenstr. 16, 70174 Stuttgart.
27. Chung, H., North, C., Self, J. Z., Chu, S., & Quek, F. (2014). VisPorter: Facilitating information sharing for collaborative sensemaking on multiple displays. *Personal and Ubiquitous Computing, 18*(5), 1169–1186.
28. Isenberg, P., Elmqvist, N., Scholtz, J., Cernea, D., Ma, K.-L., & Hagen, H. (2011). Collaborative visualization: Definition, challenges, and research agenda. *Information Visualization, 10*(4), 310–326.
29. Pereira, C., Almeida, N., Martins, A. I., Silva, S., Rosa, A. F., Oliveira e Silva, M., & Teixeira, A. (2015). Evaluation of complex distributed multimodal applications: evaluating a

telerehabilitation system when it really matters. In *Proc. HCI International, LNCS 9194, Los Angeles, CA, USA*, 146–157, doi:10.1007/978-3-319-20913-5_14
30. Bostock, M., Ogievetsky, V., & Heer, J. (2011). D^3: Data-driven documents. *IEEE Transactions on Visualization and Computer Graphics, 17*(12), 2301–2309.
31. Barnett, J., Dahl, D., Tumuluri, R., Kharidi, N., & Ashimura, K. (2016). Discovery and registration of multimodal modality components: State handling. [Online]. https://www.w3.org/TR/mmi-mc-discovery/. Accessed 15 Mar 2016.

Chapter 18
Multimodal Interaction Description Language Based on Data Modeling

Masahiro Araki

Abstract A methodology for developing a multimodal interactive system is required to be implemented as the number of input and output devices is increasing. Previous MMI (MultiModal Interaction) description languages used a state transition model for defining interaction, which has low maintainability and extensibility. In this paper, we design and implement a new MMI description language named MrailsScript from the software engineering point of view. MrailsScript consists of the data model definition that can be inherited from existing semantic web class. By expanding the existing web application development framework, our Mrails framework can automatically generate prototype code of MMI applications that are based on the MVC (Model-View-Controller) model referring to the annotations of task type and dialogue initiative. We also develop a helper application for the MMI system, MrailsBuilder, which assists in the coding of MrailsScript and contents management using semantic web standards, such as OWL, RDFS, and RDF.

18.1 Introduction

There has been continuing interest in the development methodology of multimodal dialogue systems (MDS). In recent years, statistical methods such as [1, 2] are attracting a lot of attention as a data-driven (i.e., corpus-driven) approach that can reduce the dependence on the troublesome handcrafted coding of dialogue management rules. Although statistical methods can also be applied to other components of MDS, such as the semi-automatic construction of semantic interpreters and response generators, the overall development process, including the connections with background applications and domain adaptation of speech/language components, has yet to be covered.

M. Araki (✉)
Department of Information Science, Kyoto Institute of Technology, Kyoto, Japan
e-mail: araki@kit.ac.jp

D.A. Dahl (ed.), *Multimodal Interaction with W3C Standards*,
DOI 10.1007/978-3-319-42816-1_18

There has been another "data-driven" approach, such as [3, 4], which covers the development process of all the components of dialogue systems and aims toward rapid construction of the entire system. These approaches start with a data model definition (hence, can be regarded as data-modeling driven approaches) and add some rules and templates that are used as task dependent information in the dialogue systems. As a data model definition, Kogure and Nakagawa [4] used a relational database (RDB) schema and Heinroth et al. [3] used OWL, which is an ontology definition language for the semantic web [5]. Although these data modeling schemes are familiar to the developers of web applications, the additional definitions of rules and templates for MDS are troublesome for ordinary web developers because such MDS related rules require some knowledge of linguistics and speech applications.

In this paper, we propose a new data-modeling driven approach for rapid development of MDS, which is based on collaboratively constructed semantic resources (CSRs). We show the automatic generation mechanism of code and data for simple MDSs. In addition, we compare the proposed approach with the ordinary data-modeling driven approach based on RDB. By using CSRs and the Rails framework of web application development, a large portion of troublesome definitions of rules and templates for MDS can be eliminated.

18.2 Data Modeling Driven Approach Based on CSRs

In this section, we explain our data-modeling driven approach and describe its usage of CSRs.

We proposed a data-modeling driven framework for rapid prototyping of multimodal dialogue systems [6, 7]. The overall architecture of our framework is shown in Fig. 18.1.

We define a data model definition language called MrailsScript. The framework, which is called Mrails, automatically generates the model, view, and controller (MVC) code from the data model definition written in MrailsScript. Mrails is implemented as wrapper program of Grails,[1] which is one of the popular web application development frameworks based on the MVC model. These frameworks realize rapid prototyping based on the "convention over configuration" concept and the extensibility of the prototype is brought by clearly separated MVC files. We add the use of existing ontology in defining a data model and the functionality of task-oriented interaction capability by the task type annotation and initiative type annotation.

We designed a class library which is based on class hierarchy and attribute definitions of existing semantic web ontology, Schema.org.[2] This class library is

[1] https://grails.org/

[2] https://schema.org/

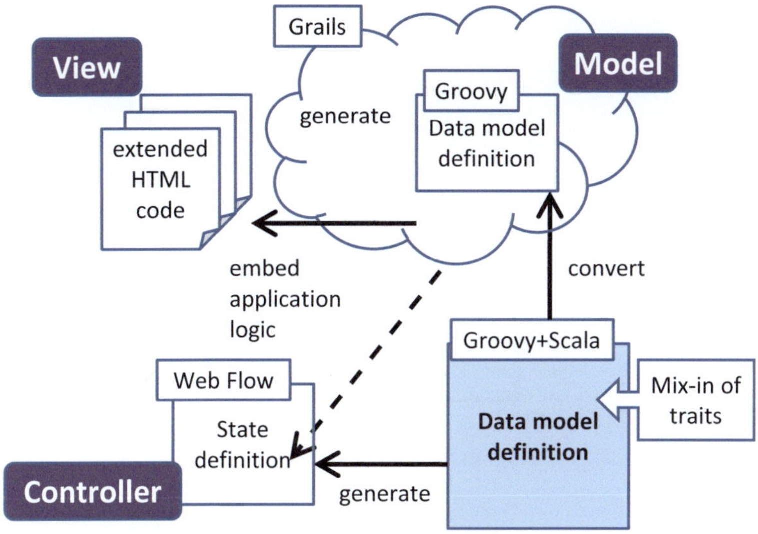

Fig. 18.1 Overview of the data-modeling driven MDS development framework

used as a base class of application-specific class definitions. An example of the class definitions is shown in Fig. 18.2.

The MyBook class inherits the Book class of the Schema.org ontology, which is the default ontology of MrailsScript. From this data model definition and attached annotation information, indicating the task type (Database search) and the initiative (system initiative) of the generated MMI system, the interpreter of Mrails automatically generates necessary view codes (such as create, list, and edit the instance), controller code (according to task type annotations such as slot-filling type, DB-search type, and explanation type), and model code, which accesses the backend database.

In Mrails, multimodal capabilities, such as speech input/output, are realized by adding JavaScript speech functions to the generated HTML5 view code, as shown in Fig. 18.3.

HTML5 contains the functionality of handling various types of modalities. Therefore, the developer can easily add modalities to the template of each view code.

18.3 Development Support: MrailsBuilder

We developed an MDS development support system called MrailsBuilder [8]. The overall concept of MrailsBuilder is shown in Fig. 18.4.

The development process of the MMI system in MrailsBuilder begins with the data model definition. As the base class of MrailsScript, CSRs cover popular

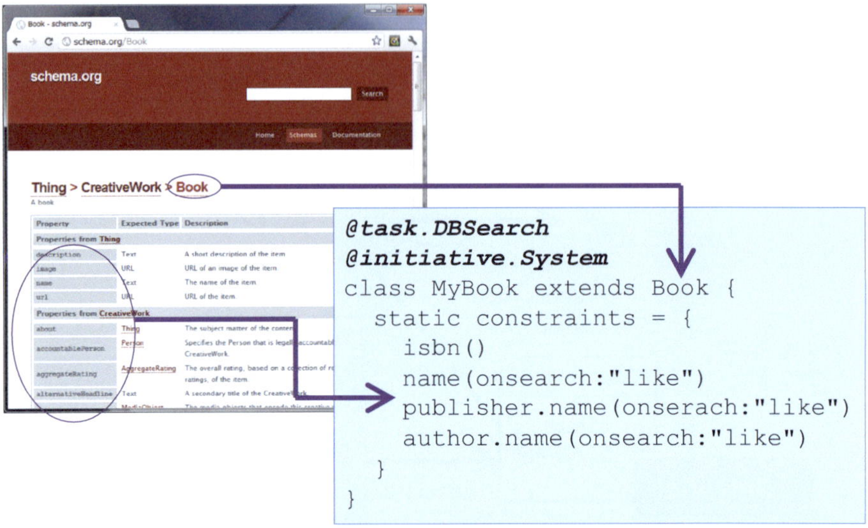

Fig. 18.2 An example of class definitions extending an existing class library

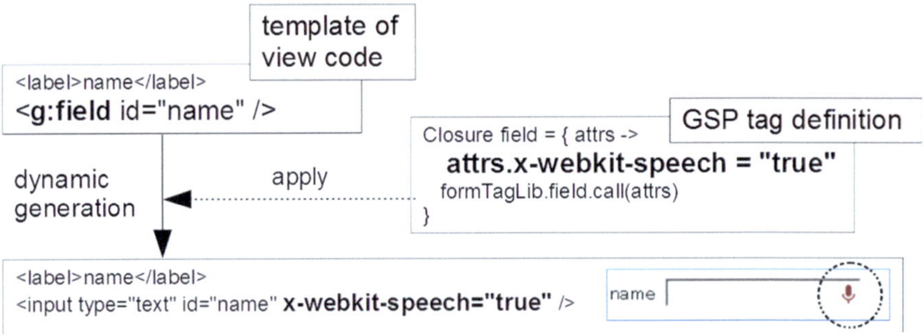

Fig. 18.3 Adding speech modality to generated view code

content for web applications. Therefore, it is a good practice for the developers to consult the existing data schema, which describes the target class in detail, to select appropriate properties dealt with by the target system. MrailsBuilder supports the data model definition process, which resembles the coding support of the popular IDE, as shown in Fig. 18.5.

The value of each property must be a literal. However, it is not easy to distinguish between a class and a literal based solely on the name of the property. MrailsBuilder supports this distinction by color highlighting (blue highlighting indicates a literal). Additionally, hovering the cursor over the target property will give a description of the target in a pop-up window. The contents of the pop-up window explanation are dynamically generated using the `rdf:description` property of each concept.

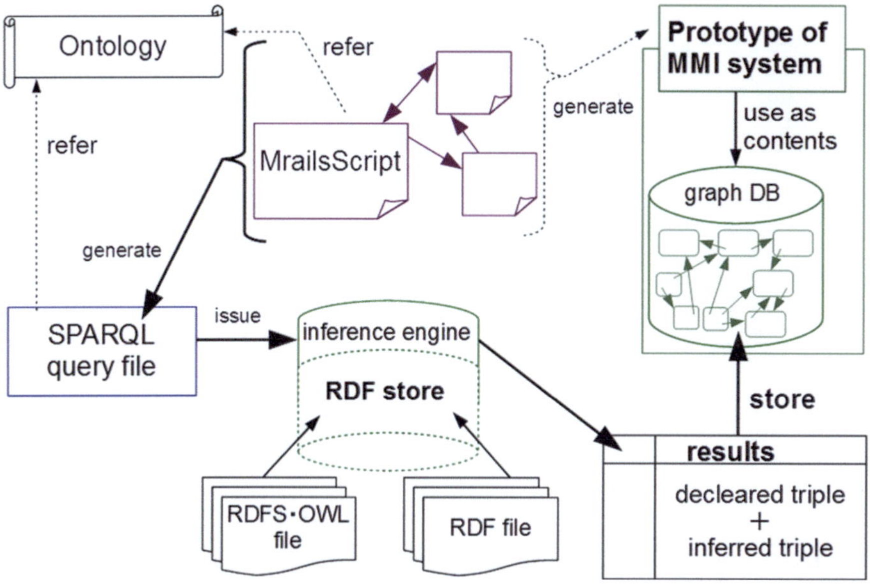

Fig. 18.4 The overall concept of MrailsBuilder

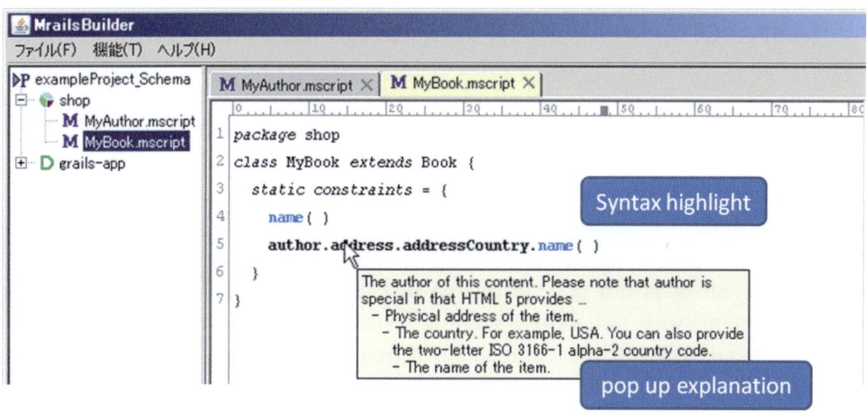

Fig. 18.5 Code highlighting and pop-up hint

Figure 18.6 shows a screenshot of the MrailsBuilder that demonstrates the selection process of properties by consulting the class property description.

Based on this data definition, MrailsBuilder automatically generates SPARQL queries [9] that retrieve the contents of open semantic web resources, such as DBpedia.[3] The retrieved contents are stored in the graph database engine.

[3] http://wiki.dbpedia.org/

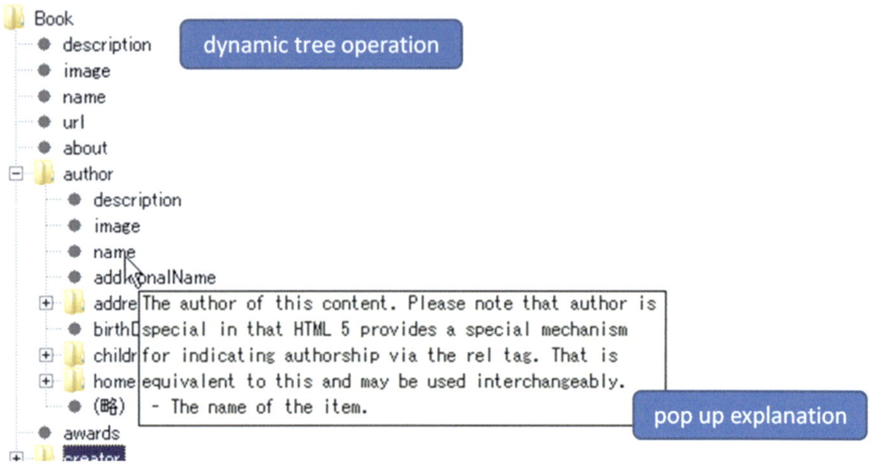

Fig. 18.6 Tree representation of class information

18.4 Example and Qualitative Evaluation

As an example application, we developed a book search MMI system using MrailsBuilder. The data model is defined in Fig. 18.7.

The SPARQL query is automatically generated and the contents of the target system can be retrieved from DBpedia. Using the Grails framework, we can easily generate MVC code. The templates of view code are written independent of the modalities. The speech modality specified in [10] is added at the stage of working code generation from these templates by consulting the Groovy Server Pages (GSP) tag definition, as shown in Fig. 18.3. By adding speech functionality to the template of the view code, we can implement the prototype book search MMI system shown in Fig. 18.8.

This system does not depend on the task or domain. To create multimedia archive systems, such as a motion learning system for traditional skills, the developer only needs to create a data model definition by following the already prepared schema of multimedia data [11].

In addition, this system does not depend on the language. The semantic web ontology itself is not dependent on the language. Although the contents are described in an individual language, this system can specify the language of the retrieved results by indicating the `lang` attribute [12].

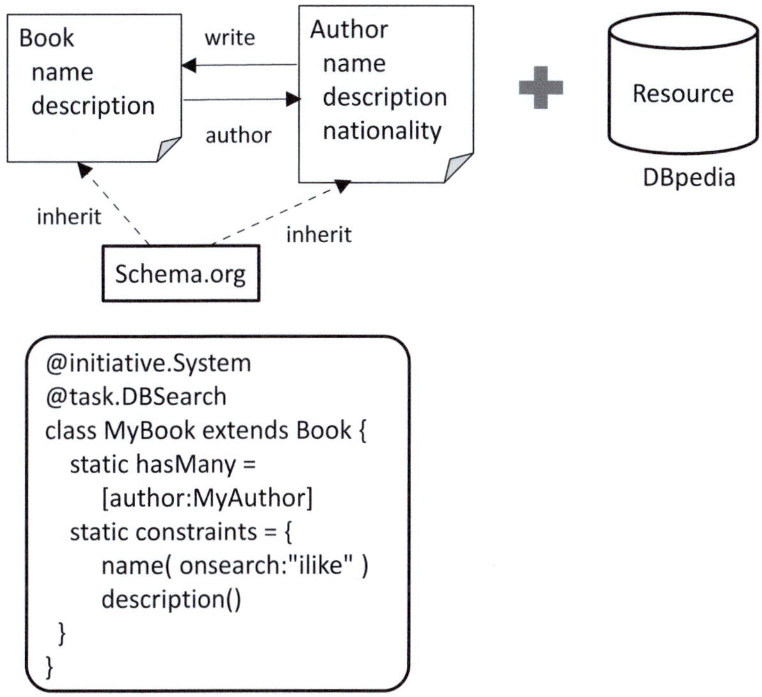

Fig. 18.7 Data model of the experimental system

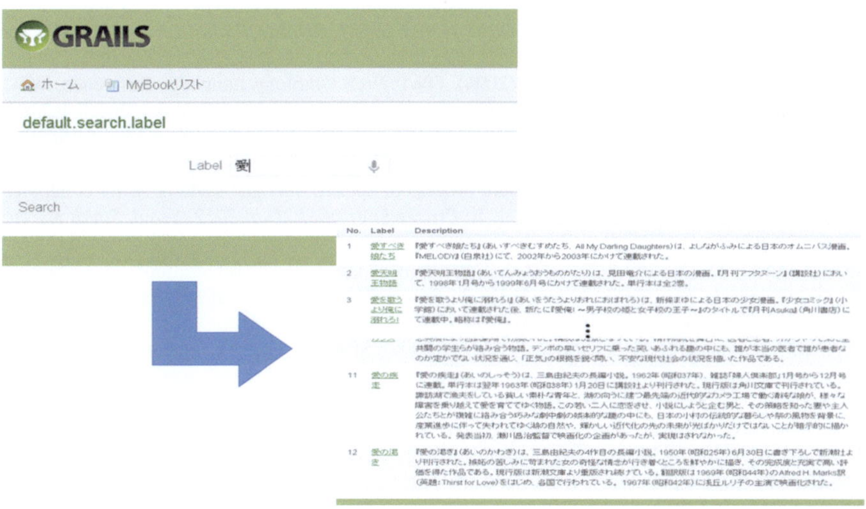

Fig. 18.8 Prototype book search system

18.5 Conclusion

In the present paper, we introduced MrailsScript, a multimodal dialogue system description language, Mrails, a multimodal interactive system development environment and MrailsBuilder, a helper application for MDS development. These are based on semantic web standards, such as OWL, RDFS, and RDF. Therefore, this development set utilizes CSRs as contents of the MDS systems. The HTML5 view codes are generated from the templates. It makes easy for developers to add multimodal functionality by customizing these templates.

In the future, we intend to implement a multi-domain dialogue system using this development environment and evaluate the efficiency of the development process.

Acknowledgment The collaborator of this research is Mr. Daisuke Takegoshi.

References

1. Hori, C., Ohtake, K., Misu, T., Kashioka, H., & Nakamura, S. (2009). Statistical dialog management applied to WFST-based dialog systems. In *Proceedings of ICASSP 2009, Taipei, Taiwan*, pp. 4793–4796.
2. Williams, J. D., & Young, S. (2007). Partially observable Markov decision processes for spoken dialog systems. *Computer Speech and Language, 21*(2), 393–422.
3. Heinroth, T., Denich, D., & Bertrand, G. (2009). Ontology-based spoken dialogue modeling. In *Proceedings of the IWSDS 2009, Irsee, Germany*
4. Kogure, S., & Nakagawa, S. (2001). A development tool for spoken dialogue systems and its evaluation. In *Proceedings of TSD2001 (LNAI 2166), Zelezna Ruda, Czech Republic*, pp. 373–380.
5. Dean, M., & Schreiber, G. (Eds.). (2004). OWL Web Ontology Language Reference, W3C Recommendation. 10 February 2004. https://www.w3.org/TR/2004/REC-owl-ref-20040210/.
6. Araki, M., & Mizukami, Y. (2011). Development of a data-driven framework for multimodal interactive systems. In *Proceedings of IWSDS 2011, Granada, Spain*, pp. 91–101.
7. Araki, M. (2012a). Rapid development process of spoken dialogue systems using collaboratively constructed semantic resources. In *Proceedings of SIGDial 2012, Seoul, South Korea*, pp. 70–73.
8. Takegoshi, D., & Araki, M. (2014). Development environment for multimodal interactive system based on ontological knowledge. In *Proceedings of IIAI AAI 2014, Kitakyushu, Japan*, pp. 785–788. doi:10.1109/IIAI-AAI.2014.158.
9. Prud'hommeaux, E., & Seaborne, A. (Eds.). (2008). SPARQL Query Language for RDF. W3C Recommendation 15 January 2008. https://www.w3.org/TR/rdf-sparql-query/.
10. Shires, G., & Wennborg, H. (Eds.). (2012). Web Speech API Specification. https://dvcs.w3.org/hg/speech-api/raw-file/tip/speechapi.html.
11. Araki, M. (2012b). Multimodal motion learning system for traditional arts. In *Proceedings of AHFE2012, San Francisco, USA*, pp. 5274–5281.
12. Araki, M., & Takegoshi, D. (2012). A rapid development framework for multilingual spoken dialogue systems. In *Proceedings of COMPSAC 2012, Izmir, Turkey*, pp. 195–202.

Chapter 19
Multimodal Fusion and Fission within the W3C MMI Architectural Pattern

Dirk Schnelle-Walka, Carlos Duarte, and Stefan Radomski

Abstract The current W3C recommendation for multimodal interfaces provides a standard for the message exchange and overall structure of modality components in multimodal applications. However, the details for multimodal fusion to combine inputs coming from modality components and for multimodal fission to prepare multimodal presentations are left unspecified. This chapter provides a first analysis of possible integrations for several approaches for fusion and fission and their implications with regard to the standard.

19.1 Introduction

With the advent of the W3C Multimodal Architecture and Interfaces recommendation a first promising candidate to standardize multimodal systems is available [2]. However, the actual approach on how input from multiple sources is fused into a coherent meaning (multimodal fusion) as well as state-of-the-art concepts on how to deliver information using more than a single available modality (multimodal fission) is addressed only superficially in the respective standards. As Schnelle-Walka et al. pointed out in [24], the W3C already suggests several markup languages to cope with the issues in a general multimodal architecture (see Fig. 19.1).

They argue that the W3C MMI architecture is reduced to a specification of the relationship of interaction managers and modality components communicating via

D. Schnelle-Walka (✉)
Harman International, Connected Car Division, Stuttgart, Germany
e-mail: dirk.schnelle-walka@harmn.com

C. Duarte
LaSIGE, Faculdade de Ciências, Universidade de Lisboa, Lisboa, Portugal
e-mail: caduarte@fc.ul.pt

S. Radomski
Telecooperation Lab, Technische Universität Darmstadt, Darmstadt, Germany
e-mail: radomski@tk.informatik.tu-darmstadt.de

D.A. Dahl (ed.), *Multimodal Interaction with W3C Standards*,
DOI 10.1007/978-3-319-42816-1_19

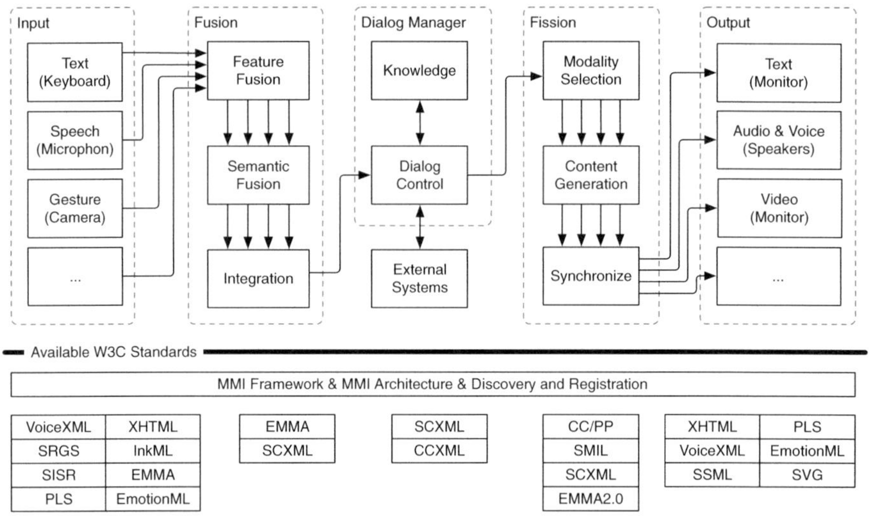

Fig. 19.1 High-level architecture of a multimodal dialog system and available W3C standards [24]

life-cycle events. In contrast, the earlier W3C MMI Framework made an attempt to consider the specifics of multimodal fusion and multimodal fission as well. The aim of the earlier approach was closer to what has been the state of the art for multimodal fusion and multimodal fission. In this chapter, we will, based on actual applications in the smart home context (e.g., [23]), elaborate on different approaches to integrate today's proposals, as detailed in the following sections, into the W3C MMI architecture. It is our hope that this will stimulate respective discussions, eventually resulting in respective standards as well.

19.2 Multimodal Fusion

Multimodal fusion has been the subject of research for more than two decades. This resulted in diverse solutions consisting of different levels, architectures, and algorithms. Sharma et al. [26] proposed one of the earliest classification schemes with three levels for fusion of incoming data: sensor-level (or data-level) fusion, feature-level fusion, and decision-level fusion. Later, Sanderson and Paliwal [22] defined another set of terms, with similar meanings: pre-mapping, midst-mapping, and post-mapping fusion. The proposed levels essentially differ at which time information combination takes place.

- **Pre-mapping data-level fusion** deals with raw data coming from recognizers. This represents the richest form of information available from a quantitative perspective. Given that the signal is directly processed, there is no loss of information. On the other hand, this makes it very susceptible to noise and failures. As a consequence of the heavy processing required by this type of fusion, it is better suited for situations where multiple streams of a single modality are involved.

- **Pre-mapping feature-level fusion** is oriented for closely coupled or time-synchronized modalities, e.g., speech and lips movement recognition. In feature-level fusion, features are extracted from sensor-collected data. If the features are commensurate they can be combined. Compared to data-level fusion, feature-level fusion has better noise interference management, but is more susceptible to data loss.
- **Midst-mapping fusion** processes a group of streams concurrently, while the mapping of the sensor-data/features space to the decisions/opinions space takes place. Given its nature, this type of fusion is also oriented for closely coupled modalities.
- **Decision-level fusion** allows multimodal systems to make effective use of loosely coupled modalities, which makes it one of the more popular forms of fusion. With this architecture, the fusion engine does not need to deal with issues of noise and failure, given that it receives information that has already been processed. This means that the engine is responsible for deriving the semantic meaning from the combination of partial semantic information coming from each input mode. Each preprocessed information constitutes a concrete decision that was made by one or more recognizers.
- **Opinion-level fusion** (also known as score-level fusion) is very similar to decision-level fusion. Both operate after the mapping of data/feature space into decision/opinion space. In fact, some authors [9] consider the former a subset of the latter. However, other authors [22] prefer to distinguish between them, given that in opinion-level fusion a group of experts (i.e., recognizers) provide opinions instead of hard decisions. Opinions can be combined through, e.g. weighted summation or weighted product approaches, before using a classification criterion in order to reach a final decision.

In addition to the classification through their architecture of fusion type, fusion engines can also be classified by whether they are adaptive or non-adaptive [16]. The main concept of adaptive, or quality fusion, is to assign different weights to different modalities. This allows to imprint adaptive features to a fusion system, by setting the reliability and discrimination of experts through time according to the state of the environment, signal quality, knowledge regarding users, or application logic. Several options have been proposed for implementing the different approaches. In particular for decision-level fusion, Dumas et al. [5] consider the following ones:

- **Frame-based fusion** using data structures known as frames for meaning representation of data originating from various sources, modeling objects as attribute-value pairs.
- **Unification-based fusion** based on merging, recursively, attribute-value structures to obtain a logical meaning representation.

- **Symbolical/statistical fusion** an evolution of standard unification-based approaches, adding statistical processing techniques.

We will elaborate on these approaches in the following sections. As an example consider a user standing in the living room of a smart home. He/She points to a shutter and utters *"close."* In this case, we have two modalities: speech (M_S) and gesture (M_G). For each fusion type we will consider a proper recognition result, as well as the case of the pointing gesture being falsely interpreted as a move up gesture to open the shutter in addition to pointing to the wanted shutter.

For now, we expect cooperative users. This means that the user does not provide contradicting inputs. For instance, he/she will not point to a shutter and ask the system by voice to turn it off.

19.2.1 Frame-Based Fusion

The concept of frame-based fusion was introduced by Vo and Wood [8]. An overview of the concept is illustrated in Fig. 19.2.

Modality processors are responsible for capturing input per modality and extract knowledge in form of semantic frames. Frames are usually related information slots occurring in unison. Each modality M_i may fill in the slots S_k with a value V_k as it can be retrieved in a dialog turn and assign it a score p_k. Hence in feature fusion the information provided per modality results in the set

$$I_{M_i} = \{(S_1(V_1), p_1), (S_2(V_2), p_2), \ldots, (S_n(V_n), p_n)\} \tag{19.1}$$

In a subsequent step of semantic fusion, a multimodal interpreter combines the input of the various modality processors as a union of values per slot S_k.

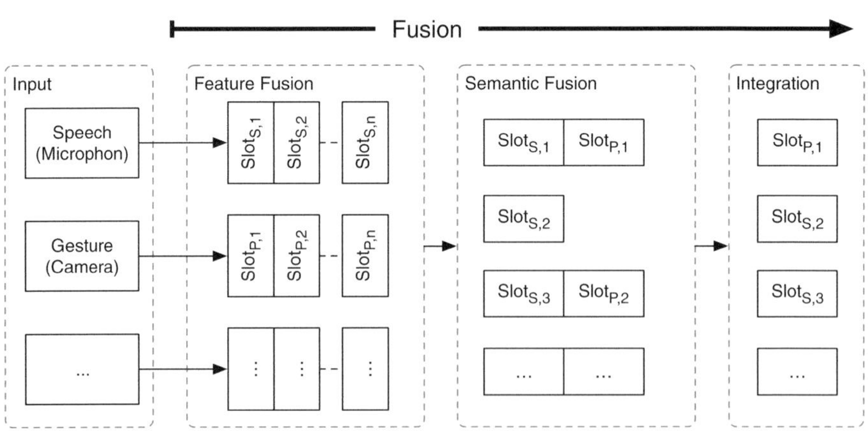

Fig. 19.2 Frame-based multimodal fusion

$$I_M = \bigcup_i I_{M_i} = \bigcup_{S_k} \{(S_k(V_{M_1}), p_{M_1}), (S_k(V_{M_2}), p_{M_2}), \ldots, (S_k(V_{M_l}), p_{M_l})\} \quad (19.2)$$

Next, the values of the slots that have the same value for a slot are added. Hence, the information I_S provided per slot

$$I_S = \bigcup_l (\{(S_k(V_l), \sum_l p(l)) | V_l = V_k\}) \quad (19.3)$$

Finally, in an integration step the values with the maximum score are selected per slot as the best hypothesis.

$$\hat{I}_{M_i} = \{(S_1(V_1), \max(p_1)), (S_2(V_2), \max(p_2)), \ldots, (S_n(V_n), \max(p_n))\} \quad (19.4)$$

In our example the output of feature fusion is

$$\begin{aligned} I_{M_S} &= \{(S_{\text{action}}(\text{“close”}), 0.9)\} \\ I_{M_G} &= \{(S_{\text{action}}(\text{“close”}), 0.2), \quad (S_{\text{object}}(\text{“shutter”}).0.8)\} \end{aligned}$$

This leads in the semantic fusion to

$$\begin{aligned} S_{\text{action}} &= \{(S_{\text{action}}(\text{“close”}), 0.9), (S_{\text{action}}(\text{“close”}), 0.2)\} \\ S_{\text{object}} &= \{S_{\text{object}}(\text{“shutter”}).0.8)\} \end{aligned}$$

The result of integrating the results is

$$\begin{aligned} S_{\text{action}} &= \{(S_{\text{action}}(\text{“close”}), 1.1)\} \\ S_{\text{object}} &= \{S_{\text{object}}(\text{“shutter”}).0.8)\} \end{aligned}$$

The example with the false interpretation of an open command will lead to

$$\begin{aligned} I_{M_S} &= \{(S_{\text{action}}(\text{“close”}), 0.9)\} \\ I_{M_G} &= \{(S_{\text{action}}(\text{“open”}), 0.2), \quad (S_{\text{object}}(\text{“shutter”}).0.8)\} \end{aligned}$$

For semantic fusion this leads to

$$\begin{aligned} S_{\text{action}} &= \{(S_{\text{action}}(\text{“close”}), 0.9), (S_{\text{action}}(\text{“}open\text{”}), 0.2)\} \\ S_{\text{object}} &= \{S_{\text{object}}(\text{“shutter”}).0.8)\} \end{aligned}$$

and integrates to

$$\begin{aligned} S_{\text{action}} &= \{(S_{\text{action}}(\text{“close”}), 0.9)\} \\ S_{\text{object}} &= \{S_{\text{object}}(\text{“shutter”}).0.8)\} \end{aligned}$$

As a result, the shutter will still be closed as with the correct recognition.

19.2.2 Unification-Based Fusion

The concept of unification-based fusion was introduced by Johnston et al. [11]. An overview of the functionality is shown in Fig. 19.3.

They based their concept on Prolog to unify data structures. Similar to frame-based fusion, this concept relies on slot-value pairs to produce the common set of pairs as a result of the fusion. Hence, the *feature fusion* produces a list of value candidates per slot.

$$I_{M_i} = \{S_1(V_1), S_2(V_2), \ldots, S_n(V_n))\} \tag{19.5}$$

These are considered to express fragments of an intended meaning. Now, unification $\oplus$ is used to arrive at a predefined command pattern C_k to express.

$$C_k = \bigoplus_i I_{M_i} \tag{19.6}$$

In case there are insufficient interpretations to fulfill any command, this leads to no result such that $C_k = \emptyset$.

If the same value has different or conflicting values, unification will result in a set of results suited to resolve the command pattern.

$$C = \{C_k | C_k \neq \emptyset\} \tag{19.7}$$

In our example the output of feature fusion is

$$\begin{aligned} I_{M_S} &= \{S_{\text{action}}(\text{"close"}), \quad S_{\text{object}}()\} \\ I_{M_G} &= \{S_{\text{action}}(), \quad S_{\text{object}}(\text{"shutter"})\} \end{aligned}$$

The expected command C will receive an action and an object. This leads to

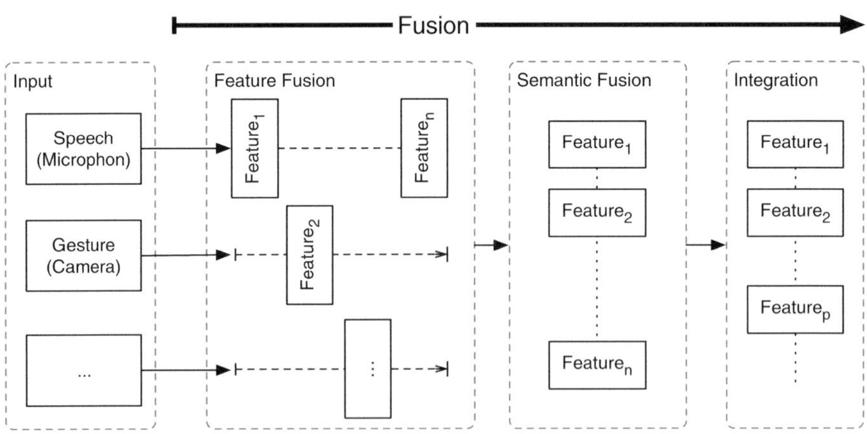

Fig. 19.3 Unification-based multimodal fusion

$$\begin{aligned} C_k &= I_{M_S} \oplus I_{M_G} \\ &= \{S_{\text{action}}(\text{“close”}), S_{\text{object}}()\} \oplus \\ &\quad \{S_{\text{action}}(), S_{\text{object}}(\text{“shutter”})\} \\ &= \{S_{\text{action}}(\text{“close”}), S_{\text{object}}(\text{“shutter”})\} \end{aligned}$$

In case the user also made a move up gesture in addition to pointing to the shutter to indicate opening it, two results would be produced.

$$\begin{aligned} C_k &= I_{M_S} \oplus I_{M_G} \\ &= \{S_{\text{action}}(\text{“close”}), S_{\text{object}}()\} \oplus \\ &\quad \{S_{\text{action}}(\text{“open”}), S_{\text{object}}(\text{“shutter”})\} \\ &= \{(S_{\text{action}}(\text{“close”}), S_{\text{object}}(\text{“shutter”})), S_{\text{action}}(\text{“open”}), S_{\text{object}}(\text{“shutter”})\} \end{aligned}$$

In contrast to frame-based fusion, confidence values are not able to select one meaning over the other. As a result, the dialog manager will have to disambiguate both possible meanings.

19.2.3 Symbolic/statistical Fusion

Symbolic/statistical fusion was introduced by Wu et al. [27] with the metaphor of members to teams to committee. The concept is shown in Fig. 19.4.

The fusion is based on a-posteriors. Let $I = \{I_1, I_2, \ldots, I_n\}$ be a set of input features and $T = \{T_1, T_2, \ldots, T_m\}$ a set of recognition targets. Then, the goal of multimodal fusion, described as the a-posteriori probability per target, i.e. the mapping of input to a defined recognition target with the largest probability, is defined by

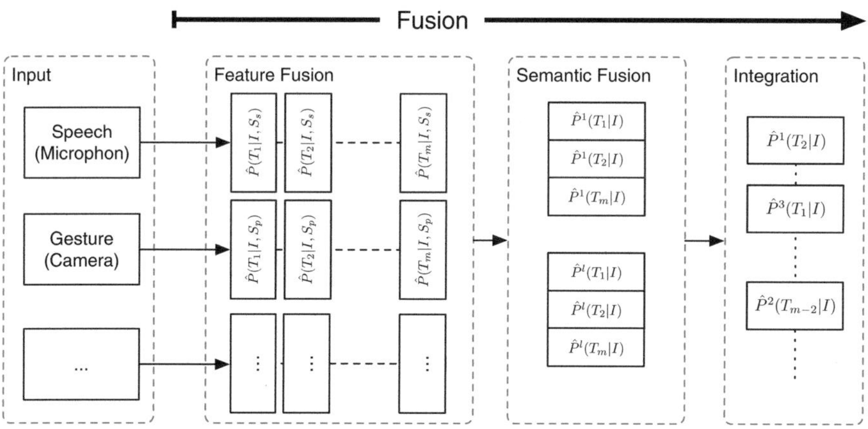

Fig. 19.4 Symbolic/statistical multimodal fusion

$$P(T_k|I), \text{ for } K = 1, \ldots, m \tag{19.8}$$

Wu et al. propose a 3-stage calculation as described below.

Fusion data is calculated in the *feature fusion* stage per modality. This is also called the member stage. Therefore, Wu et al. rely on *modeling specifications*. Such a specification includes (1) the model type, (2) the model complexity, (3) the extraction of input features,(4) the training and validation data, and (5) the learning algorithm.

In this case, an estimate for the a-posteriors for M modalities and S_i being the ith modeling specification is defined by

$$\widehat{P}(T_k|I, S_i), \text{ for } K = 1, \ldots, m \text{ and } i = 1, \ldots, M \tag{19.9}$$

Semantic fusion integrates the a-posteriori probabilities into *teams*. Let $P_k(S - I)$ be the mode probability of the kth target associated by the ith combination of modeling specifications.

$$\widehat{P}(T_k|I) = \sum_{i=1}^{M} \widehat{P}(T_k|I, S_i) P_k(S_i)), \text{ for } k = 1, \ldots, M \tag{19.10}$$

This results in a mode probability matrix with entries if the lth way of determining the mode probability out of L

$$\widehat{P}^{(l)}(T_k|I), \text{ for } K = 1, \ldots, m \text{ and } l = 1, \ldots, L \tag{19.11}$$

This is ranked within the *committee* based on the empirical distribution of their a-posteriori probabilities to an N-best list of multimodal commands. Therefore, a significance matrix H is determined as an $m \times m$ matrix for the recognition targets T_k. The entries h_{ij} are calculated as

$$h_{ij} = \begin{cases} 1, & \widehat{P}^{(l)}(T_i|I) \triangleright \widehat{P}^{(l)}(T_j|I) \\ -1, & \widehat{P}^{(l)}(T_i|I) \triangleleft \widehat{P}^{(l)}(T_j|I) \\ 0, & \text{else} \end{cases} \tag{19.12}$$

where $a \triangleright b$ indicates that a is significantly greater than b.

Now, a significance vector V is computed by summing up the values in the rows to the significance values v_i.

$$v_i = \sum_{k=1}^{m} h_{ik} \tag{19.13}$$

In a final step the significance values are ranked. The best value is $m - 1$, indicating that the input matches the target value. Higher values result in higher rankings. If all values are smaller than $m - 1$, no confidence can be calculated without requiring further external information.

In contrast to the previously described two approaches, this type of multimodal fusion is able to work with the raw data coming from the various modalities. In our illustrative use case, these are, for instance, the room coordinates of the left index finger for the gesture modality and the recorded audio signal for spoken input.

This would require applying machine learning to train the a-posteriors of the team members. A description of how this can be done for 2D gesture strokes to recognize handwriting is given by Wu et al. [27]. Essentially, we are interested in obtaining the a-posteriors per modality.

$$\widehat{P}(T_G^i|I_G), \text{ for } i = 1, \ldots, m_G \tag{19.14}$$

$$\widehat{P}(T_S^j|I_S), \text{ for } j = 1, \ldots, m_S \tag{19.15}$$

This does not necessarily have to be raw data but, e.g., in the case of spoken input, can rely on a word sequence as ASR output.

The team integrates the output of both to

$$\widehat{P}^{(1)}(T_k|I_G, I_S) = \widehat{P}(T_G^i|I_G) + \widehat{P}(T_S^j|I_S) \tag{19.16}$$

for a single team member ($L = 1$) and m commands to recognize ($k = 1, \ldots, m$).

Imagine that there are four commands, i.e., the targets T_k to distinguish (1) turn the light on (T_1), (2) turn the light off (T_2), (3) open the shutter (T_3), and (4) close the shutter (T_4). In case the significance matrix H in the committee could be

$$H = \begin{pmatrix} 0 & -1 & 1 & 0 \\ 1 & 0 & 1 & 1 \\ -1 & -1 & 0 & 0 \\ 0 & 1 & 0 & 0 \end{pmatrix} \tag{19.17}$$

then, the significance vector V and rank are

$$V = \begin{pmatrix} 0 \\ 3 \\ -2 \\ 1 \end{pmatrix}, \text{ Rank} = \begin{pmatrix} 2 \\ 1 \\ 4 \\ 3 \end{pmatrix} \tag{19.18}$$

The highest significance values of 4, results in selection of the third command T_3 as the most probable one and the fourth command T_4 as the best alternative. Consequently, the user most likely wanted to close the shutter but it is possible he/she maybe wanted to open it.

The differentiation of a false recognition of the gesture is already integrated with this approach and needs no further elaboration.

Note that passing values for entities that are hard to express by targets, e.g., the amount to increase and decrease when dimming the light, cannot be handled with this approach. In these cases, a subsequent step employing unification-based multimodal fusion is suggested.

19.3 Multimodal Fusion within the W3C MMI Architecture

While multimodal fusion is not covered in the current W3C MMI architecture, it was part of the earlier multimodal framework [12]. The envisioned concept was pretty close to the state of the art in multimodal architectures of that time. An overview is given in Fig. 19.5.

The authors also thought of dedicated components for session handling and to access the environment. Application specific functions were added to the central interaction manager. It also employed fusion as a three-tiered process (1) recognition, (2) semantic interpretation, and (3) integration. The fusion concepts mentioned above rely on similar concepts. However, this work was discontinued and eventually replaced by the W3C MMI architecture [2]. The concept of a modality component can be more elaborate than a pure recognizer per modality. Hence, it can also be responsible for a preliminary semantic interpretation for the modality to be forwarded to the upper interaction manager. The current architectural pattern also allows for multiple interaction managers, despite the topmost instance. Following the principle of a Russian doll, each modality component can interact as an upper interaction manager to lower modality components. Consequently, multimodal fusion may happen at each node, i.e. interaction manager, in the tree-like structure.

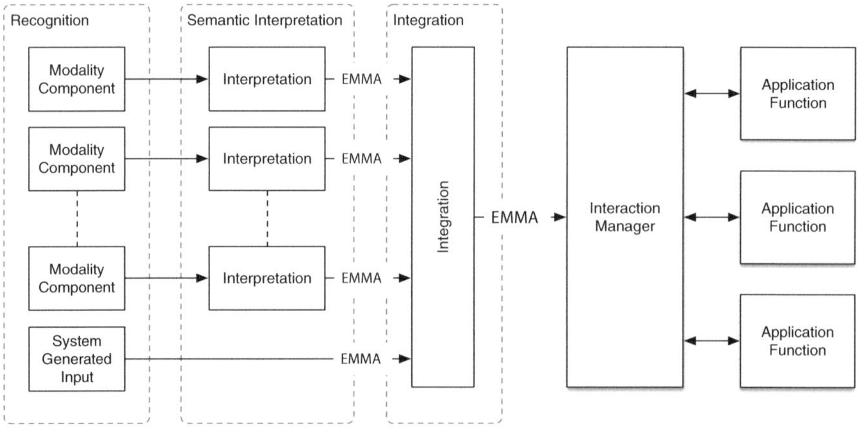

Fig. 19.5 Input components with the W3C multimodal framework (after [12])

Following the architecture described in [23] and [25] we suggest a dedicated modality component MC_{fus} to cope with multimodal fusion as shown in Fig. 19.6.

The internal handling highly depends on the employed type of multimodal fusion. In the following, we will discuss the implementation within the W3C MMI architecture for the three fusion types described above.

19.3.1 Frame-Based Fusion

For frame-based multimodal fusion, we follow the approach of Schnelle-Walka et al. [23]. The output of the various modality components is first transformed into a modality independent format. Following the recommendation of the W3C MMI architectural pattern, this happens in the data attribute of an extension or done notification.

For instance, the notification of the speech modality component in our use case to notify the recognition of the *close* action in MC_{voice} to MC_{fus} is shown in Listing 19.1.

Listing 19.1 EMMA notification for Frame-based multimodal fusion

```
<mmi:mmi xmlns:mmi="http://www.w3.org/2008/04/mmi-arch" version="1.0"
 xmlns:emma="http://www.w3.org/2003/04/emma">
 <mmi:ExtensionNotification mmi:Source="http://localhost/VoiceXML"
 mmi:Target="http://localhost/Fusion" mmi:Context="42" mmi:RequestID="43">
 <mmi:Data>
   <emma:emma version="2.0">
    <emma:interpretation id="asr-1" emma:medium="acoustic"
     emma:confidence=".75" emma:mode="voice"
     emma:tokens="close it">
       <action>close</action>
    </emma:interpretation>
   </emma:emma>
  </mmi:Data>
 </mmi:ExtensionNotification>
</mmi:mmi>
```

The EMMA `interpretation` tag is well suited to capture the semantic interpretation as slots. Ideally, slots appear as tags for easy access. We employed JBoss Drools[1] as a rule engine for the actual fusion of the frame values arriving from the modality components.

Once MC_{fus} receives information to be fused, it stores it in a knowledge base, modeled as a data component, for later retrieval. This is important to wait for

[1]http://www.drools.org/.

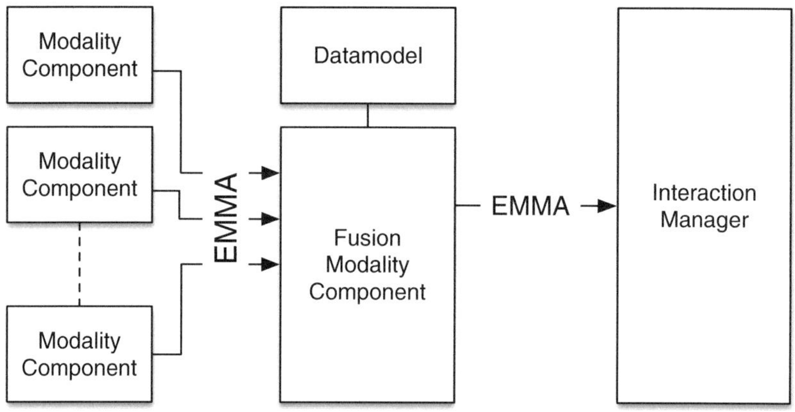

Fig. 19.6 Frame-based-multimodal fusion within the W3C MMI architectural pattern

possible addition input from the other modalities. The timeout is started only after the first EMMA content was received to avoid further delays. When the timeout expires, semantic fusion starts. Therefore, all available information is retrieved from the knowledge base to create the union of values per slot.

The integration rules select those values with the maximum confidence per slot as the hypothesis that should be forwarded to the upper interaction manager *IM*. The forwarded EMMA document can also make use of the `derived-from` annotation element to indicate the sources, as shown in Listing 19.2.

Listing 19.2 EMMA notification of an integrated fusion result

```
<mmi:mmi xmlns:mmi="http://www.w3.org/2008/04/mmi-arch" version="1.0"
 xmlns:emma="http://www.w3.org/2003/04/emma">
 <mmi:ExtensionNotification mmi:Source="http://localhost/VoiceXML"
  mmi:Target="http://localhost/Fusion" mmi:Context="42" mmi:RequestID="43">
  <mmi:Data>
   <emma:emma version="2.0">
   <emma:interpretation id="asr-1" emma:medium="acoustic"
      emma:confidence=".75" emma:mode="voice"
       emma:tokens="close it">
        <action>close</action>
   </emma:interpretation>
   <emma:interpretation id="gesture-1" emma:medium="tactile"
      emma:confidence=".9" emma:mode="gesture">
        <object>shutter</object>
   </emma:interpretation>
   <emma:interpretation id="gesture-1" emma:medium="acoustic tactile"
      emma:confidence=".9" emma:mode="speech gesture">
        <action>close</action>
        <object>shutter</object>
```

```
      <emma:derived-from resource="#asr-1" composite="true"/>
      <emma:derived-from resource="#gesture-1" composite="true"/>
    </emma:interpretation>
   </emma:emma>
  </mmi:Data>
 </mmi:ExtensionNotification>
</mmi:mmi>
```

This may be exploited to also forward, e.g., the recognized utterance to be displayed.

It is important to cleanup the knowledge-base after the message has been sent out to avoid confusion with follow-up input. There are, however, scenarios where it may be desirable to keep the information. If the user did not provide all the required information right away, the application should ask for the missing information. In these cases, Schnelle-Walka et al. suggest to forward the obtained information after a predefined timeout expired [23]. Not unlike grounding in the information state update approach [13], the information will be kept, thus requiring the user to deliver only the missing part of information. This can be fused with the information from the previous dialog turn as described above.

19.3.2 Unification-Based Fusion

The situation to implement unification-based multimodal fusion is comparable to what has been described in the previous section for frame-based fusion.

The biggest difference lies in the way multimodal fusion is actually handled, with the same messages arriving from the modality components shown in Fig. 19.6. Here, unification is needed which is available, e.g., with Prolog. A first approach to integrate Prolog as a datamodel with SCXML was described by Radomski et al. [18]. Since SCXML is a suggested W3C standard for interaction managers, it is a good candidate to be the controlling instance in an MC_{fus} for unification-based multimodal fusion. MC_{fus} can be modeled with three states as shown in Fig. 19.7.

The corresponding SCXML code is shown in the following listing.

Listing 19.3 SCXML for unification-based multimodal fusion

```
<scxml datamodel="prolog" name="mc-fusion">
  <datamodel>
    <data src="fusion.pl"/>
  </datamodel>
  <state id="Start">
    <invoke type="umundo" id="dialogInput">
      <param name="type" expr="ExtensionNotification.proto" />
      <param name="channel" expr="fusioninput" />
      <finalize>
```

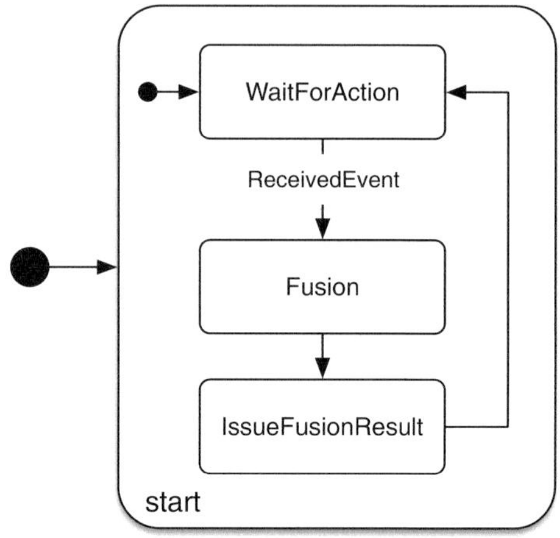

Fig. 19.7 SCXML statechart for unification-based multimodal fusion

```
      <log expr="event(X)" />
    </finalize>
  </invoke>
  <invoke type="umundo" id="dialogOutput">
    <param name="type" expr="ExtensionNotification.proto" />
    <param name="channel" expr="fusionoutput" />
    <finalize>
      <log expr="result(X)" />
    </finalize>
  </invoke>
  <state id="WaitForAction">
    <onentry>
      <log label="Waiting for actions..." />
    </onentry>
    <transition target="Fusion"
       event="umundo.rcvd">
    </transition>
  </state>
  <state id="Fusion">
    <onentry>
      <script type="query">feature_fusion.</script>
      <script type="query">semantic_fusion.</script>
      <script type="query">integration.</script>
    </onentry>
    <transition target="IssueFusionResult" />
  </state>
  <state id="IssueFusionResult">
    <onentry>
      <send target="#_dialogOutput">
```

```
          <param name="type" expr="event(data(Type))"/>
          <content expr="result(X)" />
        </send>
      </onentry>
      <transition target="WaitForAction" />
    </state>
  </state>
  <state id="End" final="true" />
</scxml>
```

We employed uMundo[2] for the event and transport layer to deliver MMI messages encoded as Protobuf[3] messages. Incoming events can be obtained via `event(X)` to resolve *X* to the incoming MMI message. Hence, the EMMA code in the data attribute *Y* is resolved by `event(data(Y))`. Furthermore, the implementation allows to script Prolog expressions, where we provided three predicates (1) `feature_fusion/0`, (2) `semantic_fusion/0`, and (3) `integration/0` for each of the fusion stages. Feature fusion will assert the slot values per modality. In subsequent steps they will be fused and integrated. If *result*(*X*) resolves to a suitable fusion result, a corresponding MMI extension notification will be sent. Note that a `foreach` tag will be needed to cope with multiple solutions. We left it out to reduce complexity.

19.3.3 Symbolic/Statistical Fusion

One of the key concerns for this type of fusion is which data to deliver from the modality components to MC_{fus}. Two options are available.

1. The modality component sends the raw data to the fusion modality component for further processing or
2. the modality component sends semantically derived data.

This includes the decision if, e.g., a gesture modality component sends their raw 3D data to MC_{fus} or the detected gesture. This is in essence the decision, if modality components are eligible for the evaluation values as team members. As a positive aspect, modality recognizers that have already been developed could be used as is. On the other hand, the modality components will have to be aware of fusion targets. In the other case, implementers will need to train their own recognizers or move them from the modality component to the fusion engine. Also, more data will have to be transmitted. As a positive aspect, modality components can be kept dumb without the need for further knowledge. Both options are possible.

[2]https://github.com/tklab-tud/umundo.

[3]https://developers.google.com/protocol-buffers/.

The associated data can easily be transmitted via EMMA. For now we are not able to come up with a recommendation.

19.4 Multimodal Fission

Multimodal fission has also been evolving since the nineties, with the earlier systems combining text and graphics (e.g., COMET [7]). Recent systems consider a larger number of modalities: speech, haptic, graphics, text, 2D and 3D animations, or avatars (e.g., SmartKom [19], MIAMM [20]). Still, most applications continue to use a limited number of output modalities, meaning that the fission techniques applied are straightforward. Those dealing with the above-mentioned combination of outputs need to process complex presentations that are difficult to coordinate and in need of ensuring their coherence. To guarantee these objectives are met, Oviatt formalized the three tasks fission engines should follow [14]:

- **Message construction**—This includes the steps required for selection and structuring of the presentation content. In this task, it is necessary to decompose the semantic information provided by the dialog manager into the elementary data that will be presented to the user. Two main approaches for content selection and structuring have been identified: (1) schema-based [6] and (2) plan-based [3].
- **Modality selection**—After message construction, its presentation must be planned, i.e., each elementary data must be allocated to a multimodal presentation according to the interaction context. The planning process follows a behavioral model specifying the components (mode, modality and medium) that are to be used. The selection and coordination of the available modalities should be informed by the type of information they can handle, the perceptual tasks they support, the characteristics of the information to present, the user's profile, and the resource limitations. For this process, three approaches have been typically considered: (1) rule based [1], (2) composite based [6], and (3) agent based [10].
- **Output coordination**—After presentation planning, the output must be instantiated, i.e., accessing the lexical and syntactic content and the modalities' attributes. The process begins with the selection of the concrete content to render, and is followed by deciding attributes, such as modality attributes and spatial and temporal parameters for coordination purposes.

The above approach as shown in Fig. 19.8 was conceptualized as What-Which-How-Then (WWHT) by Rousseau et al. [21]. According to this model, the fission engine must know what information to present, which modalities to choose to present it, how to present the information with those modalities and to coordinate the flow of the presentation. This conceptual model proved to be general enough to accommodate the requirements of an adaptive multimodal fission engine, as the one used in the GUIDE project [4].

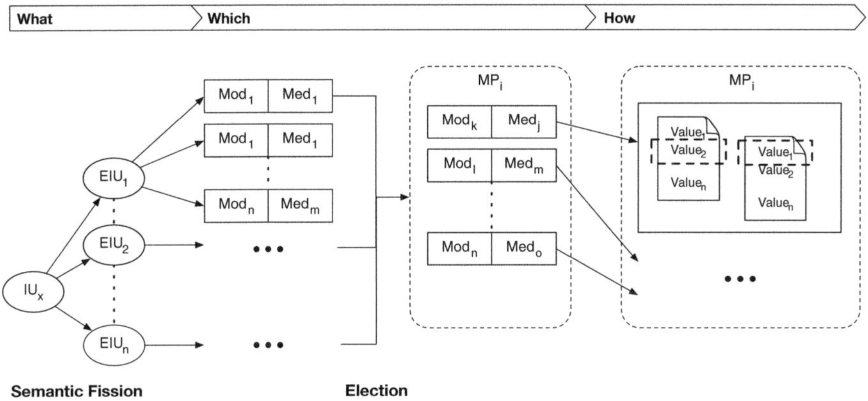

Fig. 19.8 The stages of WWHT (adapted from [21])

As an alternate approach to WWHT, Pitsikalis et al. [15] trained Hidden Markov Models for multimodal fission. HMMs also proved to be useful when fusing input from multiple modalities. Potamianos et al. [17] rely on HMM for audiovisual ASR, i.e. multimodal fusion. In order to actually train the models, sufficient data is required which may be obtained by the rule based approach described later on.

19.5 Multimodal Fission within the W3C MMI Architecture

Similar to the input components for multimodal fusion described in Sect. 19.3 the W3C MMI framework [12] also described output components for the multimodal fission. An overview is shown in Fig. 19.9.

The authors thought of three stages: (1) generation, (2) styling, and (3) rendering. Compared with the WWHT approach, *generation* would comprise the stages *what* and *which*, *styling* would be *how* and *rendering* would be mapped to *then*.

For the W3C MMI architecture, a fission modality component MC_{fis} can be integrated as shown in Fig. 19.10.

As a proof of concept we employed JBoss Drools for the multimodal fission, as described in [23]. Here, we implemented knowledge about the available modalities and user preferences as rules.

Imagine the simple scenario where a user, watching television, shall be informed that someone is at the door. After MC_{fis} received a `StartRequest` with the output information encoded as EMMA in the `Data` attribute, it stores the information in the knowledge base for further processing. Decomposition of this semantic information

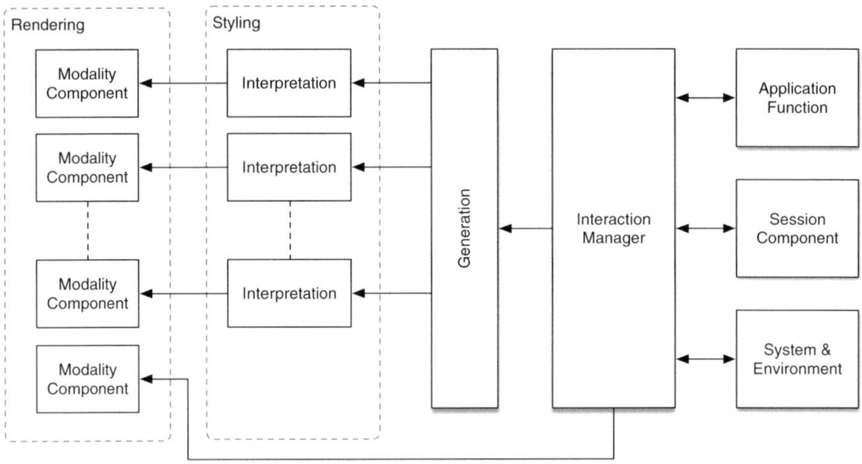

Fig. 19.9 Output components with the W3C multimodal framework (after [12])

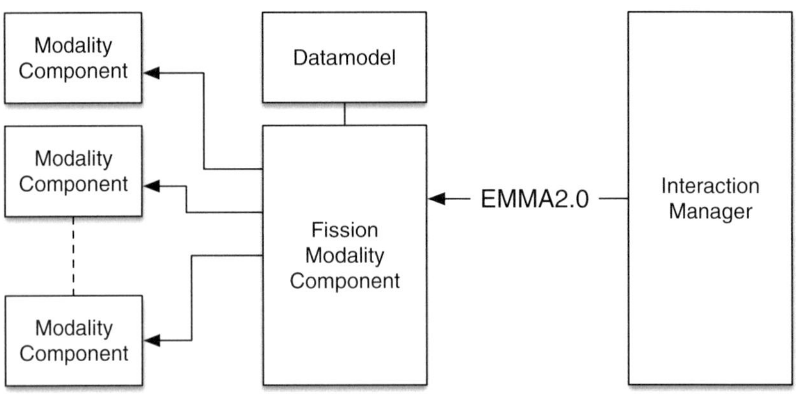

Fig. 19.10 Multimodal fission within the W3C MMI architectural pattern

unit *IU* comprises two elementary information units *EIU*s: (1) presence of a visitor and (2) identity of the visitor if it can be determined, e.g. by a camera and face recognition.

Depending on the abilities of the user (level of visual and hearing impairment, etc.) these *EIU*s will have to be mapped in the *which* stage to the available devices as modality (Mod)–medium (Med) pairs: (1) acoustically via the door bell, (2) textual on the TV screen, or (3) image on the TV screen. In the subsequent *how* stage, the system will select the best values for each medium-modality pair. For instance, the system may decide that it will be best to show a well-known photo of the visitor on the TV screen and display the textual information *Horst is at the door*.

The decision may be based on the fact that it is already late and the user's wife is already sleeping. For the *then* stage, MC_{fis} will issue a `StartRequest` with the information to be rendered to the TV modality component.

19.6 MMI Messaging with Fusion and Fission Components

Multimodal fusion and fission requires a clear separation of input and output modality components. If a modality component is responsible for both input and output, it will have to be addressed by MC_{fus} and MC_{fis}. This contradicts the demand of the MMI architectural pattern for a tree like structure. In this case, a modality component will have more than a single upper interaction manager. This was already criticized by Schnelle-Walka et al. [24]. However, this has only minor consequences for the overall messaging concept at the heart of the standard. Table 19.1 lists all the life-cycle events along with their purpose as they are defined by the standard. It is copied from [24] for your convenience.

Message exchange between the modality components and the upper interaction manager will simply be forwarded by MC_{fus} and MC_{fis}. What is lost is the actual issuer of the message, since the value of the `Source` attribute will be overwritten by the fusion and fission engines. This is in line with the intention of the W3C to be able to respond to the message by using this value as the `Target` of a message. Fusion and fission components will have to take care to forward messages to the intended modality component. This is also something that a modality component with the role of an interaction manager would do. While this can easily be achieved with the help of the `RequestID` for the `NewContextResponse` it may become trickier for the other requests to modality components residing as children of MC_{fus}. Before introducing the fusion modality component, the interaction manager was able to *only* start, e.g., keyboard input since voice input became inappropriate because of a noisy environment. MC_{fus} has no means to decide upon that. While a pragmatic solution may be to employ the `Data` attribute to carry the information, the decision logic may better be handled by MC_{fis}. It already has to consider similar issues for the output.

Consequently, we suggest that all requests from *IM* to any lower *MC* should be sent to MC_{fis}. The decision logic to address specific modalities for input will be shifted from *IM* to MC_{fis} as a specialist.

This creates a messaging circle as shown in Fig. 19.11.

Responses to, e.g., `StartRequests`, that still go back over MC_{fis} from the modality components are not shown.

Another important aspect to consider is that this messaging circle may be present at several nodes in the MMI nested structure. This is in contrast to *traditional* approaches where this happened only before the events were passed to and from the central dialog manager.

Table 19.1 Life-cycle events between IMs and MCs in the W3C MMI architecture (taken from [24])

Event	Origin	Description
IM to MC		
Prepare Request	IM	Initialize and preload data. Can be sent multiple times prior to starting.
Start Request	IM	Initiate processing of the document given as part of the request or per URL.
Pause Request	IM	Suspend processing of the current start request.
Resume Request	IM	Resume processing of the current start request.
Cancel Request	IM	Cancel processing of the current start request.
ClearContext Request	IM	Context no longer needed, free resources and terminate if appropriate.
Status Request	IM	Keep-alive request.
NewContext Response	MC	Acknowledgement of success or failure for a NewContext Request.
MC to IM		
Prepare Response	IM	If successful, the MC must respond with minimal delay to start requests.
Start Response	IM	Acknowledgement of success or failure.
Pause Response	IM	Acknowledge suspension.
Resume Response	IM	Acknowledgement of success or failure.
Cancel Response	IM	Acknowledgement of cancellation.
ClearContext Response	IM	Acknowledge end of context.
Status Response	IM	Keep-alive response if context is known, undefined otherwise.
Done Notification	MC	End of processing reached.
NewContext Request	MC	Request for a new context from the interaction manager.
Any Direction		
Extension Notification	Any	Application specific extensions with arbitrary data.

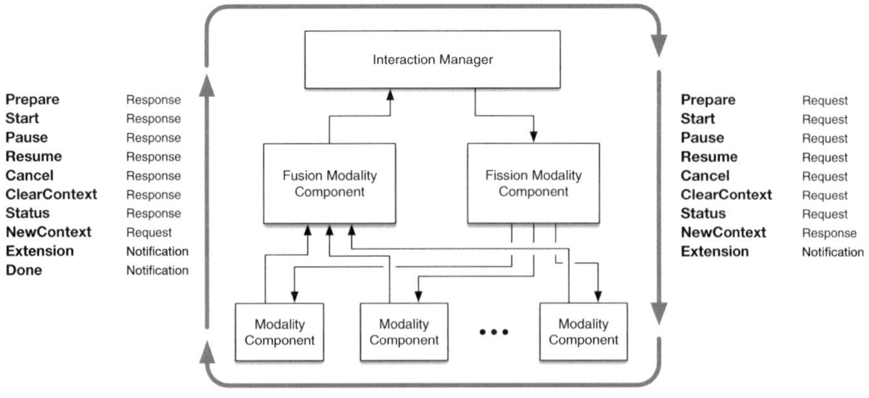

Fig. 19.11 Messaging circle with W3C fusion and fission modality components

19.7 Summary

In this chapter we introduced a way to integrate multimodal fusion and fission with the W3C MMI architectural pattern. As an example we focused on three established multimodal fusion approaches that we consider highly compatible with EMMA as a means of transporting the needed information: 1. frame-based fusion, 2. unification-based fusion, and 3. symbolical/statistical fusion. They range from well-known slot filling, over unification, to more modern statistical approaches. For each of them, we provided detailed descriptions on the internal handling of the usual three fusion stages: (1) modality-specific feature fusion, (2) semantic fusion, and (3) integration.

For multimodal fission we were in favor of the WWHT approach. It features the following stages:

(1) **W**hat is the information to process,
(2) **W**hich modalities should we use to present this information,
(3) **H**ow to present the information using these modalities and
(4) and **T**hen, how to handle the evolution of the resulting presentation.

We also described an integration into the W3C MMI architecture and described the consequences for the original structuring of elements and messaging of the standard. Here, we focused on architectural aspects as some first steps in this direction. In order to arrive at a standard, an evaluation of how they behave in real-world applications is needed. This also includes strategies to deal with conflicting user input.

Another aspect that we did not address in this chapter is the inclusion of contextual knowledge for fusion engines. This is needed to know, e.g. if the window is subject or object of an action and consequently fill in the right slot-value pairs. Contextual knowledge can be shared through attached data models, which we need to investigate in future work.

References

1. Bateman, J., Kleinz, J., Kamps, T., & Reichenberger, K. (2001). Towards constructive text, diagram, and layout generation for information presentation. *Computational Linguistics, 27*(3), 409–449. doi:10.1162/089120101317066131. http://dx.doi.org/10.1162/089120101317066131.
2. Bodell, M., Dahl, D., Kliche, I., Larson, J., Porter, B., Raggett, D., et al. (2012). *Multimodal architecture and interfaces*. W3C Recommendation, W3C. http://www.w3.org/TR/mmi-arch/.
3. Duarte, C. (2008). *Design and Evaluation of Adaptive Multimodal System*. Ph.D. thesis, University of Lisbon.
4. Duarte, C., Costa, D., Feiteira, P., & Costa, D. (2013). Building an adaptive multimodal framework for resource constrained systems. In P. Biswas, C. Duarte, P. Langdon, L. Almeida, & C. Jung (Eds.), *A multimodal end-2-end approach to accessible computing. Human–computer interaction series* (pp. 155–173). London: Springer. doi:10.1007/978-1-4471-5082-4_8. http://dx.doi.org/10.1007/978-1-4471-5082-4_8.

5. Dumas, B., Lalanne, D., & Oviatt, S. (2009). Multimodal interfaces: A survey of principles, models and frameworks. In *Human machine interaction* (pp. 3–26). Berlin, Heidelberg: Springer. doi:10.1007/978-3-642-00437-7_1. http://dx.doi.org/10.1007/978-3-642-00437-7_1
6. Fasciano, M., & Lapalme, G. (2000). Intentions in the coordinated generation of graphics and text from tabular data. *Knowledge and Information Systems, 2*(3), 310–339. doi:10.1007/PL00011645. http://dx.doi.org/10.1007/PL00011645
7. Feiner, S. K., & McKeown, K. R. (1993). Automating the generation of coordinated multimedia explanations. In *Intelligent multimedia interfaces* (pp. 117–138). Menlo Park, CA: American Association for Artificial Intelligence. http://dl.acm.org/citation.cfm?id=162477.162493
8. Goubran, R. A., & Wood, C. (1996). Building an application framework for speech and pen input integration in multimodal learning interfaces. In *Proceedings of the Acoustics, Speech, and Signal Processing, 1996. On Conference Proceedings, 1996 I.E. International Conference - Volume 06, ICASSP '96* (pp. 3545–3548). Washington, DC: IEEE Computer Society. doi:10.1109/ICASSP.1996.550794. http://dx.doi.org/10.1109/ICASSP.1996.550794
9. Hall, D., & Llinas, J. (2001). Multisensor data fusion. In Handbook of multisensor data fusion (pp. 1–10). Boca Raton: CRC Press.
10. Han, Y., & Zukerman, I. (1997). A mechanism for multimodal presentation planning based on agent cooperation and negotiation. *International Journal of Human-Computer Interaction, 12* (1), 187–226. doi:10.1207/s15327051hci1201&2_6. http://dx.doi.org/10.1207/s15327051hci1201&2_6
11. Johnston, M., Cohen, P. R., McGee, D., Oviatt, S. L., Pittman, J. A., & Smith, I. (1997). Unification-based multimodal integration. In *Proceedings of the 35th Annual Meeting of the Association for Computational Linguistics and Eighth Conference of the European Chapter of the Association for Computational Linguistics, ACL '98* (pp. 281–288). Stroudsburg, PA: Association for Computational Linguistics. doi:10.3115/976909.979653. http://dx.doi.org/10.3115/976909.979653
12. Larson, J. A., Raggett, D., & Raman, T. V. (2003). W3C multimodal interaction framework. W3C Note, W3C. http://www.w3.org/TR/2003/NOTE-mmi-framework-20030506/
13. Larsson, S. (2002). *Issue-based Dialogue Management*. Ph.D. thesis, University of Gothenburg.
14. Oviatt, S. (2003). Multimodal information fusion. In *The human–computer interaction handbook: Fundamentals, evolving technologies and emerging application* (pp. 286–304). Hillsdale: L. Erlbaum Associates Inc.
15. Pitsikalis, V., Katsamanis, A., & Papandreou, G. (2009). Adaptive multimodal fusion by uncertainty compensation. In *IEEE Transactions on Audio, Speech, and Language Processing*
16. Poh, N., Bourlai, T., & Kittler, J. (2010). Multimodal information fusion. In *Multimodal signal processing theory and applications for human computer interaction* (p. 153). London: Academic.
17. Potamianos, G., Huang, J., Marcheret, E., Libal, V., Balchandran, R., Epstein, M., et al. (2008). Far-field multimodal speech processing and conversational interaction in smart spaces. In *Hands-Free Speech Communication and Microphone Arrays, 2008. HSCMA 2008* (pp. 119–123). doi:10.1109/HSCMA.2008.4538701
18. Radomski, S., Schnelle-Walka, D., & Radeck-Arneth, S. (2013). A prolog datamodel for state chart XML. In *Proceedings of the SIGDIAL 2013 Conference* (pp. 127–131)
19. Reithinger, N., Alexandersson, J., Becker, T., Blocher, A., Engel, R., Löckelt, M., et al. (2003). Adaptive and flexible multimodal access to multiple applications. In *Proceedings of the 5th International Conference on Multimodal Interfaces, ICMI '03* (pp. 101–108). New York, NY: ACM. doi:10.1145/958432.958454. http://doi.acm.org/10.1145/958432.958454
20. Reithinger, N., Fedeler, D., Kumar, A., Lauer, C., Pecourt, E., & Romary, L. (2005). Miamm - A multimodal dialogue system using haptics. In J. van Kuppevelt, L. Dybkjã, & N. Bernsen (Eds.), *Advances in natural multimodal dialogue systems. Text, speech and language technology* (Vol. 30, pp. 307–332). Netherlands: Springer. doi:10.1007/1-4020-3933-6_14. http://dx.doi.org/10.1007/1-4020-3933-6_14

21. Rousseau, C., Bellik, Y., Vernier, F., & Bazalgette, D. (2006). A framework for the intelligent multimodal presentation of information. *Signal Processing, 86*(12), 3696–3713. doi:10.1016/j.sigpro.2006.02.041. http://dx.doi.org/10.1016/j.sigpro.2006.02.041
22. Sanderson, C., & Paliwal, K. K. (2004). Information fusion and person verification using speech & face information. *Digital Signal Processing, **14***(5), 449–480. doi:10.1016/j.dsp.2004.05.001.
23. Schnelle-Walka, D., Radeck-Arnet, S., & Striebinger, J. (2015). Multimodal dialogmanagement in a smart home context with SCXML. In *Proceedings of the 2nd Workshop on Engineering Interactive Systems with SCXML.*
24. Schnelle-Walka, D., Radomski, S., & Mühlhäuser, M. (2013). JVoiceXML as a modality component in the W3C multimodal architecture. *Journal on Multimodal User Interfaces, 7*(3), 183–194. doi:10.1007/s12193-013-0119-y. http://dx.doi.org/10.1007/s12193-013-0119-y
25. Schnelle-Walka, D., Radomski, S., & Mühlhäuser, M. (2014). Multimodal fusion and fission within W3C standards for nonverbal communication with blind persons. In K. Miesenberger, D. Fels, D. Archambault, P. Peňáz, & W. Zagler (Eds.), *Computers helping people with special needs. Lecture notes in computer science* (Vol. 8547, pp. 209–213). Paris, Cham: Springer International Publishing. doi:10.1007/978-3-319-08596-8_33. http://dx.doi.org/10.1007/978-3-319-08596-8_33
26. Sharma, R., Pavlovic, V., & Huang, T. (1998). Toward multimodal human–computer interface. *Proceedings of the IEEE, 86*(5), 853–869. doi:10.1109/5.664275.
27. Wu, L., Oviatt, S. L., & Cohen, P. R. (2002). From members to teams to committee—A robust approach to gestural and multimodal recognition. *IEEE Transactions on Neural Networks, 13* (4), 972–982. doi:10.1109/TNN.2002.1021897. http://dx.doi.org/10.1109/TNN.2002.1021897

Index

D.A. Dahl (ed.), *Multimodal Interaction with W3C Standards*,
DOI 10.1007/978-3-319-42816-1

MIX
Papier aus verantwortungsvollen Quellen
Paper from responsible sources
FSC® C105338

If you have any concerns about our products,
you can contact us on
ProductSafety@springernature.com

In case Publisher is established outside the EU,
the EU authorized representative is:
Springer Nature Customer Service Center GmbH
Europaplatz 3, 69115 Heidelberg, Germany

Printed by Libri Plureos GmbH
in Hamburg, Germany